HSC Year 12

MATHEMATICS STANDARD 2

TANIA EASTCOTT | RACHEL EASTCOTT

SERIES EDITOR: ROBERT YEN

A+

2020 UPDATED SYLLABUS • 2020 UPDATED SYLLABUS •

- topic summaries
- graded practice questions with worked solutions
- HSC exam topic grids (2011–2020)

STUDY NOTES

A+ HSC Mathematics Standard 2 Study Notes
1st Edition
Tania Eastcott
Rachel Eastcott
ISBN 9780170459204

Publishers: Robert Yen, Kirstie Irwin
Project editor: Tanya Smith
Cover design: Nikita Bansal
Text design: Alba Design
Project designer: Nikita Bansal
Permissions researcher: Corrina Gilbert
Production controller: Karen Young
Typeset by: Nikki M Group Pty Ltd

Any URLs contained in this publication were checked for currency during the production process. Note, however, that the publisher cannot vouch for the ongoing currency of URLs.

For product information and technology assistance,
in Australia call **1300 790 853**;
in New Zealand call **0800 449 725**

For permission to use material from this text or product, please email
aust.permissions@cengage.com

ISBN 978 0 17 045920 4

Cengage Learning Australia
Level 7, 80 Dorcas Street
South Melbourne, Victoria Australia 3205

Cengage Learning New Zealand
Unit 4B Rosedale Office Park
331 Rosedale Road, Albany, North Shore 0632, NZ

For learning solutions, visit **cengage.com.au**

Printed in China by 1010 Printing International Limited.
1 2 3 4 5 6 7 25 24 23 22 21

ABOUT THIS BOOK

Introducing *A+ HSC Year 12 Mathematics*, a new series of study guides designed to help students revise the topics of the new HSC maths courses and achieve success in their exams. *A+* is published by Cengage, the educational publisher of *Maths in Focus* and *New Century Maths*.

For each HSC maths course, Cengage has developed a STUDY NOTES book and a PRACTICE EXAMS book. These study guides have been written by experienced teachers who have taught the new courses, some of whom are involved in HSC exam marking and writing. This is the first study guide series to be published after the first HSC exams of the new courses in 2019–2020, so it incorporates the latest changes to the syllabus and exam format.

This book, *A+ HSC Year 12 Mathematics Standard 2 Study Notes*, contains topic summaries and graded practice questions, grouped into 6 broad topics, addressing the outcomes in the Mathematics Standard 2 Syllabus. The topic-based structure means that this book can be used for revision after a topic has been covered in the classroom, as well as for course review and preparation for the trial and HSC exams. Each topic chapter includes a review of the main mathematical concepts, and multiple-choice and short-answer questions with worked solutions. Past HSC exam questions have been included to provide students with the opportunity to see how they will be expected to show their mathematical understanding in the exams. An HSC exam topic grid (2011–2020) guides students to where and how each topic has been tested in past HSC exams.

Mathematics Standard 2 Year 12 topics

1. Linear and non-linear relationships
2. Trigonometry
3. Rates and ratios
4. Investments, loans and annuities
5. Bivariate data and the normal distribution
6. Networks

This book contains:

- Concept map (see p. 2 for an example)
- Glossary and digital flashcards (see p. 3 for an example)
- Topic summary, addressing key outcomes of the syllabus (see p. 5 for an example)
- Practice set 1: 20 multiple-choice questions (see p. 20 for an example)
- Practice set 2: 20 short-answer questions (see p. 25 for an example)
- Questions graded by level of difficulty: foundation ●○○, moderate ●●○, complex ●●●
- Worked solutions to both practice sets
- HSC exam topic grid (2011–2020) (see p. 38 for an example)

The companion A+ PRACTICE EXAMS book contains topic exams and practice HSC exam papers both of which are written and formatted in the style of the HSC exam, with spaces for students to write answers. Worked solutions are provided, along with the author's expert comments and advice, including how each exam question is marked. As a special bonus, the worked solutions to the 2020 HSC exam paper have been included.

This A+ STUDY NOTES book will become a staple resource in your study in the lead-up to your final HSC exam. Revisit it throughout Year 12 to ensure that you do not forget key concepts and skills. Good luck!

CONTENTS

ABOUT THIS BOOK III
YEAR 12 COURSE OVERVIEW VI
SYLLABUS REFERENCE GRID VIII
ABOUT THE AUTHORS VIII
A+ DIGITAL FLASHCARDS VIII
HSC EXAM FORMAT IX
STUDY AND EXAM ADVICE X
MATHEMATICAL VERBS XIV
SYMBOLS AND ABBREVIATIONS XV

CHAPTER 1

LINEAR AND NON-LINEAR RELATIONSHIPS

Concept map 2
Glossary 3
Topic summary 5
Practice set 1: Multiple-choice questions 20
Practice set 2: Short-answer questions 25
Practice set 1: Worked solutions 32
Practice set 2: Worked solutions 34
HSC exam topic grid (2011–2020) 38

TRIGONOMETRY

Concept map 40
Glossary 41
Topic summary 42
Practice set 1: Multiple-choice questions 51
Practice set 2: Short-answer questions 56
Practice set 1: Worked solutions 62
Practice set 2: Worked solutions 65
HSC exam topic grid (2011–2020) 70

RATES AND RATIOS

Concept map 72
Glossary 73
Topic summary 74
Practice set 1: Multiple-choice questions 88
Practice set 2: Short-answer questions 91
Practice set 1: Worked solutions 95
Practice set 2: Worked solutions 97
HSC exam topic grid (2011–2020) 100

9780170459204

CHAPTER 4

INVESTMENTS, LOANS AND ANNUITIES

Concept map 102

Glossary 103

Topic summary 104

Practice set 1: Multiple-choice questions 115

Practice set 2: Short-answer questions 120

Practice set 1: Worked solutions 125

Practice set 2: Worked solutions 127

HSC exam topic grid (2011–2020) 130

CHAPTER 5

BIVARIATE DATA AND THE NORMAL DISTRIBUTION

Concept map 132

Glossary 133

Topic summary 134

Practice set 1: Multiple-choice questions 144

Practice set 2: Short-answer questions 149

Practice set 1: Worked solutions 155

Practice set 2: Worked solutions 157

HSC exam topic grid (2011–2020) 162

CHAPTER 6

NETWORKS

Concept map 164

Glossary 165

Topic summary 167

Practice set 1: Multiple-choice questions 176

Practice set 2: Short-answer questions 182

Practice set 1: Worked solutions 190

Practice set 2: Worked solutions 191

HSC exam topic grid (2019–2020) 195

HSC EXAM REFERENCE SHEET 196

INDEX 197

YEAR 12 COURSE OVERVIEW

See each concept map printed in full size at the beginning of each chapter.

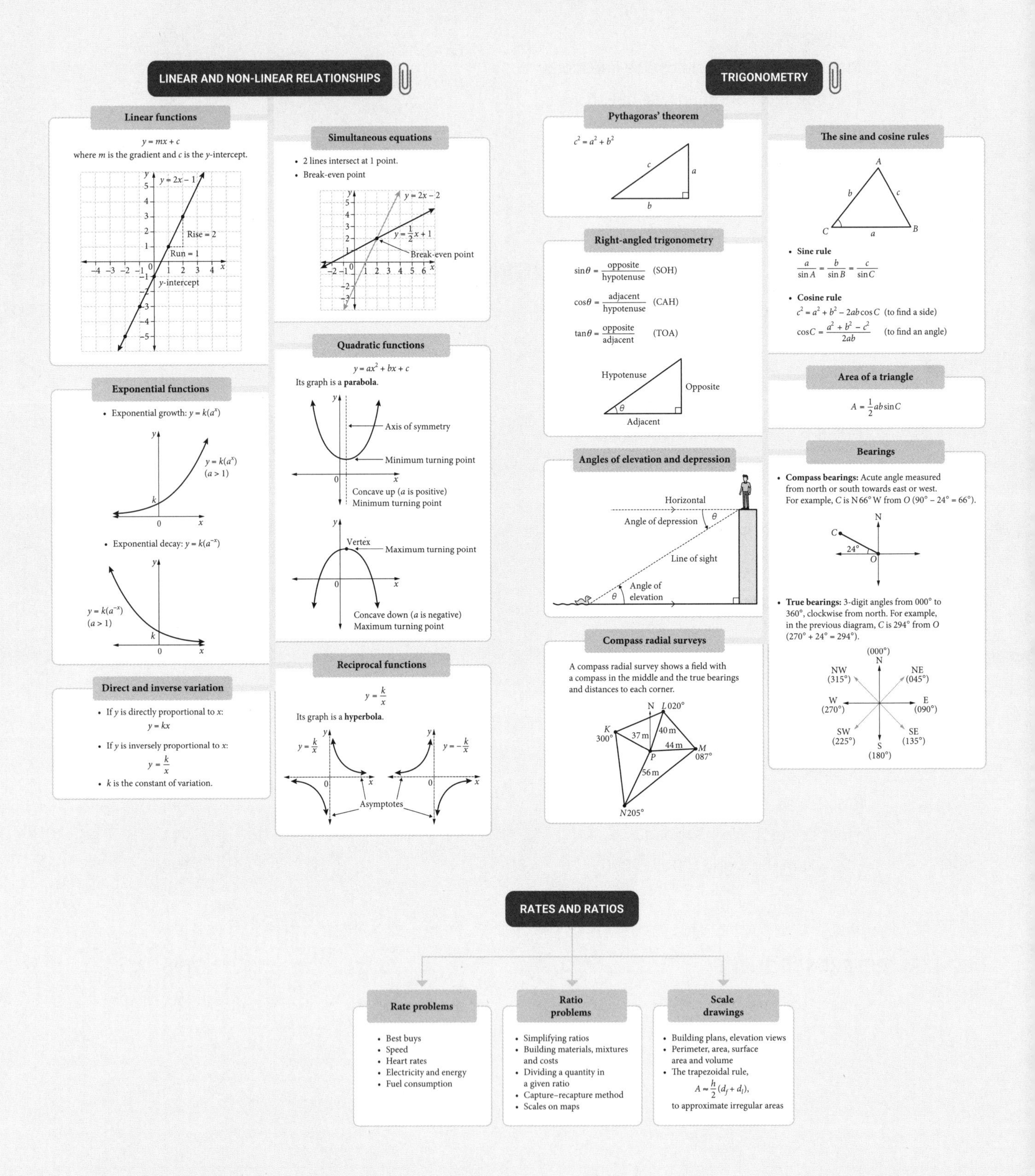

INVESTMENTS, LOANS AND ANNUITIES

Investments

- Simple interest:
  $I = Prn$
- Compound interest:
  future value (FV), present value (PV)
  $FV = PV(1 + r)^n$
  $I = FV - PV$
- Comparing simple interest and compound interest
- Inflation and appreciation
- Shares: dividend and dividend yield

  Dividend yield = $\frac{\text{dividend per share}}{\text{market value per share}} \times 100\%$

Depreciation

- Straight-line method of depreciation:
  $S = V_0 - Dn$
- Declining-balance method of depreciation:
  $S = V_0(1 - r)^n$

Annuities

- Annuities: modelling as a recurrence relation
- Present and future value
- Present and future value tables

Loans and credit cards

- Reducing balance loans
- Credit cards: compound interest

BIVARIATE DATA AND THE NORMAL DISTRIBUTION

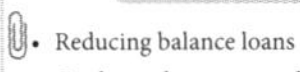

Scatterplots

No pattern | Linear pattern | Clustering | Linear pattern | Outliers | Non-linear pattern

Relationship/correlation

Perfect positive | Moderate negative | Strong negative | No relationship

Line of best fit

- Line of best fit by eye
- Least-squares regression line by calculator
- Interpolation: data prediction within the set
- Extrapolation: data prediction outside the set

The normal distribution

Mean
Median
Mode

Symmetry

50% 50%

−3 −2 −1 0 1 2 3 z

- Approximately 68% of scores have z-scores between −1 and 1.
- Approximately 95% of scores have z-scores between −2 and 2.
- Approximately 99.7% of scores have z-scores between −3 and 3.

z-scores

$z\text{-score} = \frac{\text{score} - \text{mean}}{\text{standard deviation}}$

$z = \frac{x - \mu}{\sigma}$

Pearson's correlation coefficient (r)

−1 −0.75 −0.5 −0.25 0 0.25 0.5 0.75 1

No correlation

Strong, negative correlation | Moderate, negative correlation | Weak, negative correlation | Weak, positive correlation | Moderate, positive correlation | Strong, positive correlation

NETWORKS

Networks terminology

Loop, Vertex, Edge

Trail | Path | Circuit | Cycle

- Directed networks
- Weighted networks
- Connected networks

Minimum spanning trees

- Kruskal's algorithm
- Prim's algorithm

Shortest path problems

Flow networks

- 'Maximum-flow, minimum-cut' theorem
- Flow capacity

Critical path analysis

- Activity charts and network diagrams

Activity	Predecessor(s)
A	–
B	–
C	A
D	A, B
E	C

Start, Finish

- Forward and backward scanning, float times, critical paths

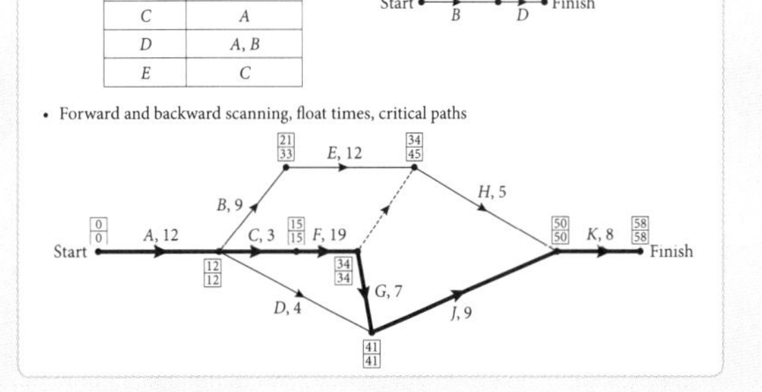

SYLLABUS REFERENCE GRID

Topic and subtopics	*A+ HSC Year 12 Mathematics Standard 2 Study Notes* chapter
ALGEBRA	
MS-A4 Types of relationships A4.1 Simultaneous linear equations A4.2 Non-linear relationships	1 Linear and non-linear relationships
MEASUREMENT	
MS-M6 Non-right-angled trigonometry	2 Trigonometry
MS-M7 Rates and ratios	3 Rates and ratios
FINANCIAL MATHEMATICS	
MS-F4 Investments and loans F4.1 Investments F4.2 Depreciation and loans	4 Investments, loans and annuities
MS-F5 Annuities	4 Investments, loans and annuities
STATISTICAL ANALYSIS	
MS-S4 Bivariate data analysis	5 Bivariate data and the normal distribution
MS-S5 The normal distribution	5 Bivariate data and the normal distribution
NETWORKS	
MS-N2 Network concepts N2.1 Networks N2.2 Shortest paths	6 Networks
MS-N3 Critical path analysis	6 Networks

ABOUT THE AUTHORS

Tania Eastcott teaches at St Mary's Catholic College, Gateshead, Newcastle and has taught at Barker College, Hornsby, Morriset High School and St Charbel's College, Punchbowl. She has been involved in HSC marking, judging and curriculum development, and run HSC workshops for students. Tania co-wrote the *New Century Maths 11–12 General Mathematics* workbooks and topic tests for the *Maths in Focus 11–12 Mathematics Advanced* and *Extension 1*.

Rachel Eastcott teaches at Lisarow High School on the Central Coast. She was awarded the first Great Teaching, Inspired Learning Scholarship and Cadetship from the NSW Department for Education while studying at the University of Newcastle, NSW and the University of Arizona, USA.

A+ DIGITAL FLASHCARDS

Revise key terms and concepts online with the A+ Flashcards. Each topic glossary in this book has a corresponding deck of digital flashcards you can use to test your understanding and recall. Just scan the QR code or type the URL into your browser to access them.

Note: You will need to create a free NelsonNet account.

HSC EXAM FORMAT

Mathematics Standard 2 HSC exam

The information below about the Mathematics Standard 2 HSC exam was correct at the time of printing in 2021. Please check the NESA website in case it has changed. Visit www.educationstandards.nsw.edu.au, select 'Year 11 – Year 12', 'Syllabuses A–Z', 'Mathematics Standard', then 'Assessment and Reporting in Mathematics Standard Stage 6'. Scroll down to the heading 'HSC examination specifications for Mathematics Standard 2'.

	Questions	Marks	Recommended time
Section I	15 multiple-choice questions Mark answers on the multiple-choice answer sheet.	15	25 min
Section II	Approx. 25 short-answer questions, including 2 or more questions worth 4 or 5 marks. Write answers on the lines provided on the paper.	85	2 h 5 min
Total		100	2 h 30 min

Exam information and tips

- Reading time: 10 minutes; use this time to preview the whole exam.
- Working time: 2 hours and 30 minutes
- Questions focus on Year 12 outcomes but Year 11 knowledge may be examined.
- Answers are to be written on the question paper.
- A reference sheet is provided at the back of the exam paper, containing formulas.
- Common questions with the Mathematics Advanced HSC exam: 20–25 marks
- Common questions with the Mathematics Standard 1 HSC exam: 20–25 marks
- The 4- or 5-mark questions are usually complex problems that require many steps of working and careful planning.
- To help you plan your time, the halfway point of Section II is marked by a notice at the bottom of the relevant page; for example, 'Questions 16–27 are worth 44 marks in total'.
- Having 2 hours and 30 minutes for a total of 100 marks means that you have an average of 1.5 minutes per mark (or 3 minutes for 2 marks).
- If you budget 20 minutes for Section I and 55 minutes for each half of Section II, then you will have 20 minutes at the end of the exam to check over your work and complete questions you missed.

STUDY AND EXAM ADVICE

A journey of a thousand miles begins with a single step.

Lao Tzu (c. 570–490 BCE), Chinese philosopher

I've always believed that if you put in the work, the results will come.

Michael Jordan (1963–), American basketball player

Four PRACtical steps for maths study

1. Practise your maths

- Do your homework.
- Learning maths is about mastering a collection of skills.
- You become successful at maths by doing it more, through regular practice and learning.
- Aim to achieve a high level of understanding.

2. Rewrite your maths

- Homework and study are not the same thing. Study is your private 'revision' work for strengthening your understanding of a subject.
- Before you begin any questions, make sure you have a thorough understanding of the topic.
- Take ownership of your maths. Rewrite the theory and examples in your own words.
- Summarise each topic to see the 'whole picture' and know it all.

3. Attack your maths

- All maths knowledge is interconnected. If you don't understand one topic fully, then you may have trouble learning another topic.
- Mathematics is not an HSC course you can learn 'by halves' – you have to know it all!
- Fill in any gaps in your mathematical knowledge to see the 'whole picture'.
- Identify your areas of weakness and work on them.
- Spend most of your study time on the topics you find difficult.

4. Check your maths

- After you have mastered a maths skill, such as graphing a quadratic equation, no further learning or reading is needed, just more practice.
- Compared to other subjects, the types of questions asked in maths exams are conventional and predictable.
- Test your understanding with revision exercises, practice papers and past exam papers.
- Develop your exam technique and problem-solving skills.
- Go back to steps 1–3 to improve your study habits.

Topic summaries and concept maps

Summarise each topic when you have completed it, to create useful study notes for revising the course, especially before exams. Use a notebook or folder to list the important ideas, formulas, terminology and skills for each topic. This book is a good study guide, but educational research shows that effective learning takes place when you rewrite learned knowledge in your own words.

A good topic summary runs for 2 to 4 pages. It is a condensed, personalised version of your course notes. This is your interpretation of a topic, so include your own comments, symbols, diagrams, observations and reminders. Highlight important facts using boxes and include a glossary of key words and phrases.

A concept map or mind map is a topic summary in graphic form, with boxes, branches and arrows showing the connections between the main ideas of the topic. This book contains examples of concept maps. The topic name is central to the map, with key concepts or subheadings listing important details and formulas. Concept maps are powerful because they present an overview of a topic on one large sheet of paper. Visual learners absorb and recall information better using concept maps.

When compiling a topic summary, use your class notes, your textbook and this study guide. Ask your teacher for a copy of the course syllabus or the school's teaching program, which includes the objectives and outcomes of every topic in dot point form.

Attacking your weak areas

Most of your study time should be spent on attacking your weak areas to fill in any gaps in your maths knowledge. Don't spend too much time on work you already know well, unless you need a confidence boost! Ask your teacher, use this book or your textbook to improve the understanding of your weak areas and to practise maths skills. Use your topic summaries for general revision, but spend longer study periods on overcoming any difficulties in your mastery of the course.

Practising with past exam papers

Why is practising with past exam papers such an effective study technique? It allows you to become familiar with the format, style and level of difficulty expected in an HSC exam, as well as the common topic areas tested. Knowing what to expect helps alleviate exam anxiety. Remember, mathematics is a subject in which the exam questions are fairly predictable. The exam writers are not going to ask too many unusual questions. By the time you have worked through many past exam papers, this year's HSC paper won't seem that much different.

Don't throw your old exam papers away. Use them to identify your mistakes and weak areas for further study. Revising topics and then working on mixed questions is a great way to study maths. You might like to complete a past HSC exam paper under timed conditions to improve your exam technique.

Past HSC exam papers are available at the NESA website: visit www.educationstandards.nsw.edu.au and select 'Year 11 – Year 12', 'HSC exam papers'. NESA marking feedback and guidelines can also be viewed there. Cengage has also published *A+ HSC Year 12 Mathematics Standard 2 Practice Exams*, containing topic exams and practice HSC exam papers. You can find past HSC exam papers with solutions online, in bookstores, at the Mathematical Association of NSW (www.mansw.nsw.edu.au) and at your school (ask your teacher) or library.

Preparing for an exam

- Make a study plan early; don't leave it until the last minute.
- Read and revise your topic summaries.
- Work on your weak areas and learn from your mistakes.
- Don't spend too much time studying work you know already.
- Revise by completing revision exercises and past exam papers or assignments.
- Vary the way you study so that you don't become bored: ask someone to quiz you, voice-record your summary, design a poster or concept map, or explain the work to someone.
- Anticipate the exam:
  - How many questions will there be?
  - What are the types of questions: multiple-choice, short-answer, long-answer, problem-solving?
  - Which topics will be tested?
  - How many marks are there in each section?
  - How long is the exam?
  - How much time should I spend on each question/section?
  - Which formulas are on the reference sheet and how do I use them in the exam?

During an exam

1. Bring all of your equipment, including a ruler and calculator (check that your calculator works and is in DEGREES mode for trigonometry). A highlighter pen may help for tables, graphs and diagrams.
2. Don't worry if you feel nervous before an exam – this is normal and helps you to perform better; however, being too casual or too anxious can harm to your performance. Just before the exam begins, take deep, slow breaths to reduce any stress.
3. Write clearly and neatly in black or blue pen, not red. Use a pencil only for diagrams and constructions.
4. Use the **reading time** to browse through the exam to see the work that is ahead of you and the marks allocated to each question. Doing this will ensure you won't miss any questions or pages. Note the harder questions and allow more time for working on them. Leave them if you get stuck, and come back to them later.
5. Attempt every question. It is better to do most of every question and score some marks, rather than ignore questions completely and score 0 for them. Don't leave multiple-choice questions unanswered! Even if you guess, you have a chance of being correct.
6. Easier questions are usually at the beginning, with harder ones at the end. Do an easy question first to boost your confidence. Some students like to leave multiple-choice questions until last so that, if they run out of time, they can make quick guesses. However, some multiple-choice questions can be quite difficult.
7. Read each question and identify what needs to be found and what topic/skill it is testing. The number of marks indicates how much time and working out is required. Highlight any important keywords or clues. Do you need to use the answer to the previous part of the question?
8. After reading each question, and before you start writing, spend a few moments planning and thinking.
9. You don't need to be writing all of the time. What you are writing may be wrong and a waste of time. Spend some time considering the best approach.
10. Make sure each answer seems reasonable and realistic, especially if it involves money or measurement.

11. Show all necessary working, write clearly, draw big diagrams, and set your working out neatly. Write solutions to each part underneath the previous step so that your working out goes down the page, not across.
12. Use a ruler to draw (or read) half-page graphs with labels and axes marked, or to measure scale diagrams.
13. Don't spend too much time on one question. Keep an eye on the time.
14. Make sure you have answered the question. Did you remember to round the answer and/or include units? Did you use all of the relevant information given?
15. If a hard question is taking too long, don't get bogged down. If you're getting nowhere, retrace your steps, start again, or skip the question (circle it) and return to it later with a clearer mind.
16. If you make a mistake, cross it out with a neat line. Don't scribble over it completely or use correction fluid or tape (which is time-consuming and messy). You may still score marks for crossed-out work if it is correct, but don't leave multiple answers! Keep track of your answer booklets and ask for more writing paper if needed.
17. Don't cross out or change an answer too quickly. Research shows that often your first answer is the correct one.
18. Don't round your answer in the middle of a calculation. Round at the end only.
19. Be prepared to write words and sentences in your answers, but don't use abbreviations that you've just made up. Use correct terminology and write 1 or 2 sentences for 2 or 3 marks, not mini-essays.
20. If you have time at the end of the exam, double-check your answers, especially for the more difficult or uncertain questions.

Ten exam habits of the best HSC students

1. Has clear and careful working and checks their answers
2. Has a strong understanding of basic algebra and calculation
3. Reads (and answers) the whole question
4. Chooses the simplest and quickest method
5. Checks that their answer makes sense or sounds reasonable
6. Draws big, clear diagrams with details and labels
7. Uses a ruler for drawing, measuring and reading graphs
8. Can explain answers in words when needed, in 1–2 clear sentences
9. Uses the previous parts of a question to solve the next part of the question
10. Rounds answers at the end, not before

Further resources

Visit the NESA website, www.educationstandards.nsw.edu.au, for the following resources. Select 'Year 11 – Year 12' and then 'Syllabuses A–Z' or 'HSC exam papers'.

- Mathematics Standard 2 Syllabus
- Past HSC exam papers, including marking feedback and guidelines
- Sample HSC questions/exam papers and marking guidelines

Before 2019, 'Mathematics Standard 2' was called 'Mathematics General 2' and, before 2014, 'General Mathematics'. For these exam papers, select 'Year 11 – Year 12', 'Resources archive', 'HSC exam papers archive'.

MATHEMATICAL VERBS

A glossary of 'doing words' common in maths problems and HSC exams

analyse
study in detail the parts of a situation

apply
use knowledge or a procedure in a given situation

calculate
See **evaluate**

classify/identify
state the type, name or feature of an item or situation

comment
express an observation or opinion about a result

compare
show how two or more things are similar or different

complete
fill in detail to make a statement, diagram or table correct or finished

construct
draw an accurate diagram

convert
change from one form to another, for example, from a fraction to a decimal, or from kilograms to grams

decrease
make smaller

describe
state the features of a situation

estimate
make an educated guess for a number, measurement or solution, to find roughly or approximately

evaluate/calculate
find the value of a numerical expression, for example, 3×8^2 or $4x + 1$ when $x = 5$

expand
remove brackets in an algebraic expression, for example, expanding $3(2y + 1)$ gives $6y + 3$

explain
describe why or how

give reasons
show the rules or thinking used when solving a problem. *See also* **justify**

graph
display on a number line, number plane or statistical graph

hence find/prove
calculate an answer or prove a result using previous answers or information supplied

identify
See **classify**

increase
make larger

interpret
find meaning in a mathematical result

justify
give reasons or evidence to support your argument or conclusion. *See also* **give reasons**

measure
determine the size of something, for example, using a ruler to determine the length of a pen

prove
See **show/prove that**

recall
remember and state

show/prove that
(in questions where the answer is given) use calculation, procedure or reasoning to prove that an answer or result is true

simplify
express a result such as a ratio or algebraic expression in its most basic, shortest, neatest form

sketch
draw a rough diagram that shows the general shape or ideas (less accurate than **construct**)

solve
calculate the value(s) of an unknown pronumeral in an equation or inequality

state
See **write**

substitute
replace part of an expression with another, equivalent expression.

verify
check that a solution or result is correct, usually by substituting back into an equation or referring back to the problem

write/state
give an answer, formula or result without showing any working or explanation (This usually means that the answer can be found mentally, or in one step)

SYMBOLS AND ABBREVIATIONS

$=$	is equal to
$\neq$	is not equal to
$\approx$	is approximately equal to
$<$	is less than
$>$	is greater than
$\leq$	is less than or equal to
$\geq$	is greater than or equal to
( )	parentheses, round brackets
[ ]	(square) brackets
{ }	braces
$\pm$	plus or minus
π	pi = 3.141 59...
$0.\dot{1}5\dot{2}$	the recurring decimal 0.152 152...
$^\circ$	degree
$\angle$	angle
Δ	triangle
$\therefore$	therefore
x^2	x squared, $x \times x$
x^3	x cubed, $x \times x \times x$
$\sqrt{\ }$	square root
$\sqrt[3]{\ }$	cube root
S 37° W	a compass bearing
217°	a true bearing
$P(E)$	the probability of event E occurring
$P(\tilde{E})$	the probability of event E not occurring
LHS	left-hand side
RHS	right-hand side
%	percentage
p.a.	per annum (per year)
cos	cosine ratio
sin	sine ratio
tan	tangent ratio
$\overline{x}$	the mean
σ_n	the standard deviation
Σ	the sum of, sigma
Q_1	first quartile or lower quartile
Q_2	median (second quartile)
Q_3	third quartile or upper quartile
IQR	interquartile range
α	alpha
θ	theta
μ	micro-, mu, population mean
m	gradient

A+ HSC YEAR 12 MATHEMATICS

STUDY NOTES

Authors:
Tania Eastcott
Rachel Eastcott

Sarah Hamper

Karen Man
Ashleigh Della Marta

Jim Green
Janet Hunter

PRACTICE EXAMS

Authors:
Adrian Kruse

Simon Meli

John Drake

Jim Green
Janet Hunter

CHAPTER 1
LINEAR AND NON-LINEAR RELATIONSHIPS

MS-A4 Types of relationships 10

A4.1 Simultaneous linear equations 10

A4.2 Non-linear relationships 12

1

LINEAR AND NON-LINEAR RELATIONSHIPS

All content for this topic is common with the Mathematics Advanced course.

Linear functions

$$y = mx + c$$

where m is the gradient and c is the y-intercept.

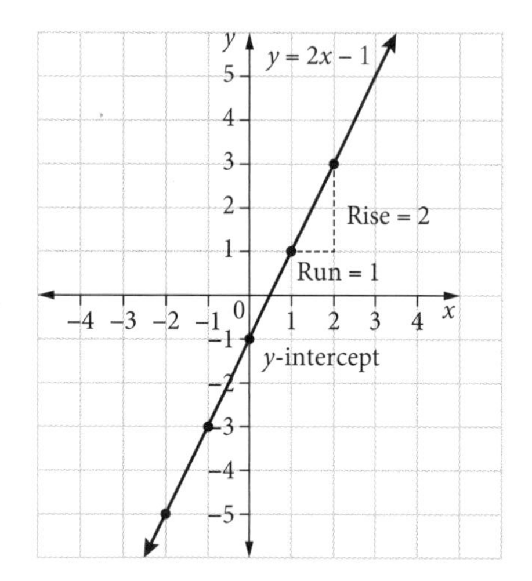

Simultaneous equations

- 2 lines intersect at 1 point.
- Break-even point

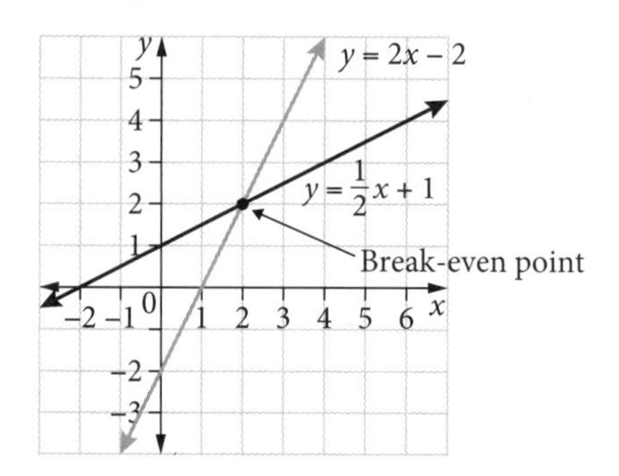

Exponential functions

- Exponential growth: $y = k(a^x)$

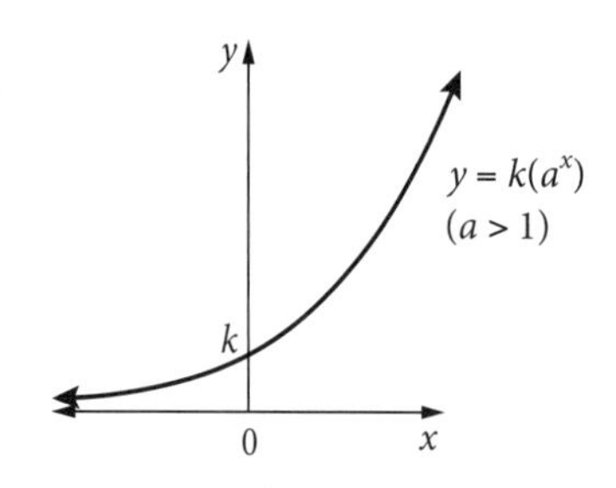

- Exponential decay: $y = k(a^{-x})$

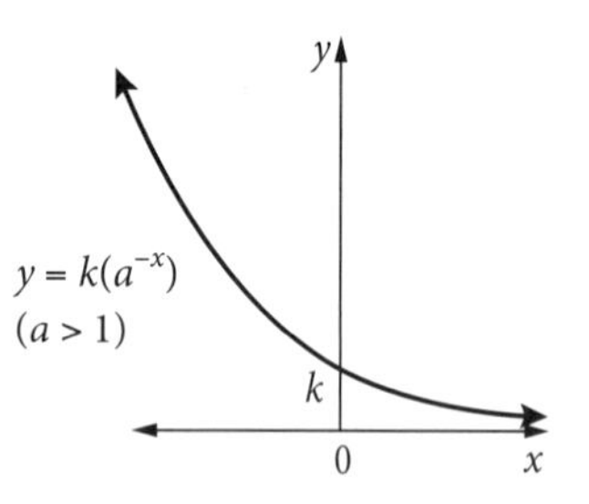

Quadratic functions

$$y = ax^2 + bx + c$$

Its graph is a **parabola**.

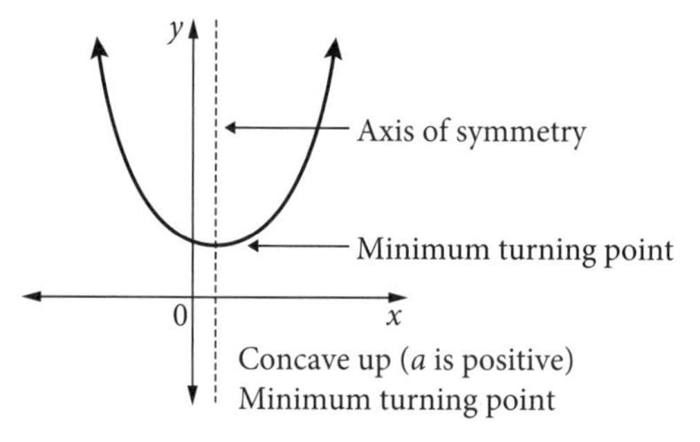

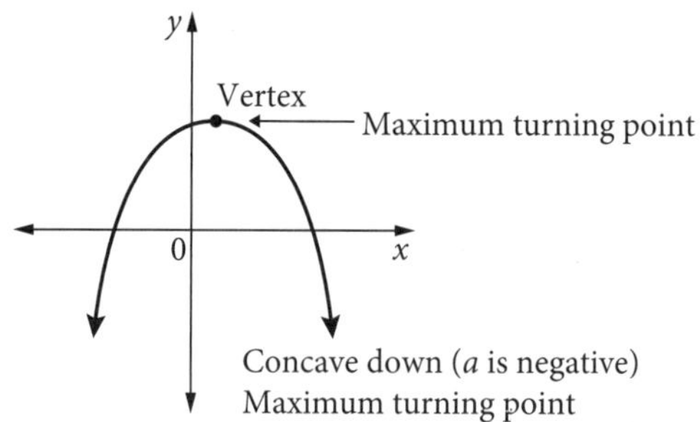

Direct and inverse variation

- If y is directly proportional to x:
  $y = kx$
- If y is inversely proportional to x:
  $y = \frac{k}{x}$
- k is the constant of variation.

Reciprocal functions

$$y = \frac{k}{x}$$

Its graph is a **hyperbola**.

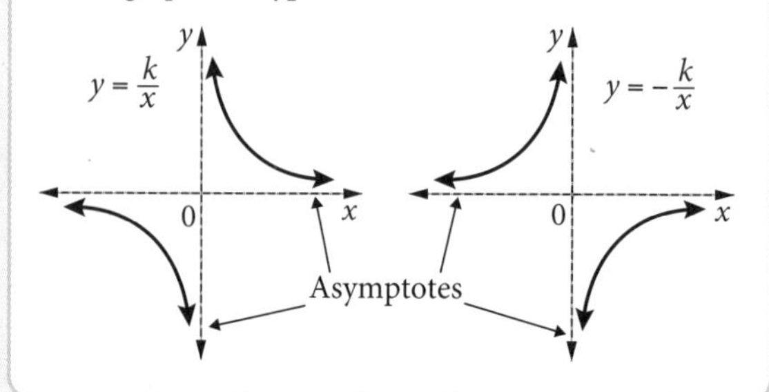

Glossary

algebraic modelling
Using algebraic functions to represent real-life situations.

asymptote
A line that a curve approaches but never touches. For this exponential curve, the x-axis is an asymptote.

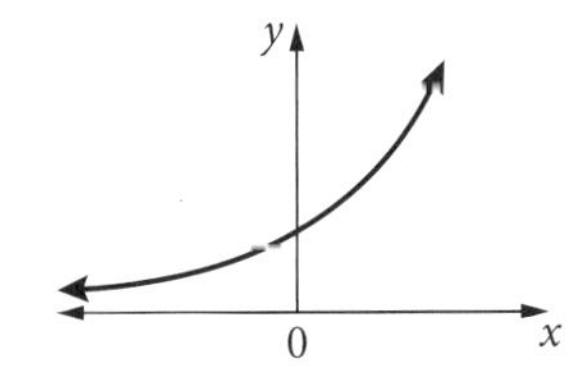

break-even point
The point where the value of sales for a business equals its costs.

concave up
Describes a parabola that points upwards; a 'smiling' parabola.

concave down
Describes a parabola that points downwards; a 'frowning' parabola.

constant term
A number by itself (that is not a coefficient of a variable) in an equation or function. For example, for $y = 7x - 2$, the constant term is -2.

direct variation/proportion
A relationship between 2 variables that has the general form $y = \frac{k}{x}$, where x and y are variables and k is a constant (of variation). It is a straight line that passes through the origin $(0, 0)$ of a graph. Differs from **inverse variation**.

equation
A mathematical statement that contains an equals symbol ('=') and the values on both sides are equal.

exponential curve
The graph of an exponential function.

exponential decay
The decrease of a quantity that is rapid at first, but then slows over time. It is described by an exponential function. Differs from **exponential growth**.

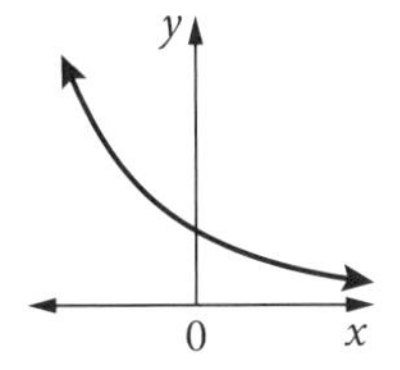

A+ DIGITAL FLASHCARDS
Revise this topic's key terms and concepts by scanning the QR code or typing the URL into your browser.

https://get.ga/a-hsc-maths-standard-2

exponential function
A function with an equation that has x as the power, for example, $y = b(a^x)$. Its graph is an exponential curve.

exponential growth
The increase of a quantity that happens slowly at first but then speeds up over time. It is described by an exponential function. Differs from **exponential decay**.

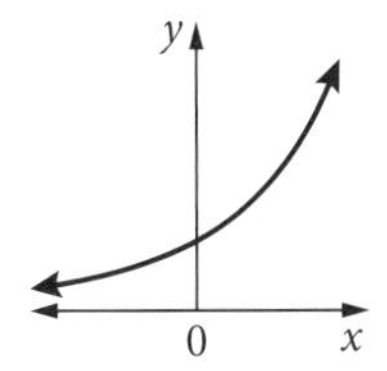

expression
A group of algebraic terms.

function
The relationship between 2 variables, for example, $y = 3x^2 - x + 10$.

gradient
The slope or steepness of a line. Can be found using

$$m = \frac{\text{rise}}{\text{run}} \text{ or } \frac{y_2 - y_1}{x_2 - x_1}.$$

hyperbola
The graph of a reciprocal function, for example, $y = \frac{2}{x}$.

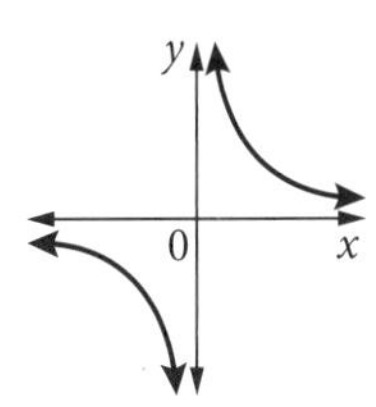

initial value
The 'starting' value in a function or model, which is the value of y found by substituting $x = 0$. On a graph of a function, it is the y-value of the y-intercept. In a linear or quadratic function, it is the constant term, c. In an exponential function, $y = ka^x$, it is the value of k.

inverse variation/proportion
A relationship between 2 variables that has the general form $y = \frac{k}{x}$, where x and y are variables and k is a constant (of variation). Its graph is a hyperbola. Differs from **direct variation**.

GLOSSARY

linear equation/function
An equation that has the form $y = mx + c$, which is a straight line on a graph.

linear model
A model in the form $y = mx + c$.

maximum turning point
The highest point of a concave down parabola, with the vertex as the peak.

maximum/minimum value
The highest/lowest value of a quadratic function, represented by the y-value of the vertex of its graph (parabola).

minimum turning point
The lowest point of a concave up parabola, with the vertex as the trough.

non-linear function
A function whose graph is a curve, not a line, such as a quadratic, exponential or reciprocal function.

parabola
The curve made by a **quadratic function**, such as $y = ax^2 + bx + c$.

point of intersection
The point where 2 or more lines cross each other.

pronumeral
A letter used in algebra to represent an unknown value.

quadratic function
A non-linear function where the highest power of x is 2, for example, $y = ax^2 + bx + c$. Its graph is a parabola.

reciprocal function
A non-linear function where the denominator is x: $y = \frac{k}{x}$. On a graph it is a hyperbola.

revenue
The amount of money made by selling a product or service.

rise
The vertical difference between 2 points on a graph.

run
The horizontal difference between 2 points on a graph.

simultaneous equations
Two or more equations that must be solved together so that the solution satisfies both (or all) equations. When the equations are graphed on a number plane, the solution is the coordinates of the point where the lines intersect.

subject (of a formula)
The variable on the left-hand side (LHS) of a formula. For example, the subject of the formula $A = \frac{1}{2}bh$ is A, but the subject can be changed to h by rewriting as $h = \frac{2A}{b}$.

variable
A pronumeral in an equation or function, such as x or y, that can have different values.

vertex
The turning point of a parabola.

x-intercept
Also known as the horizontal intercept. The value at which a line cuts the x-axis (horizontal axis) on a number plane.

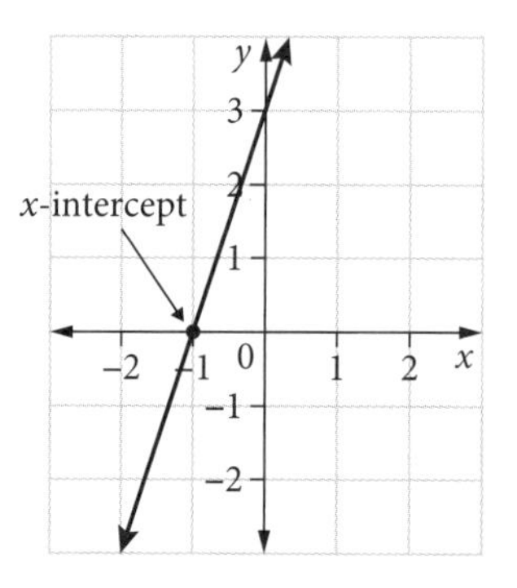

y-intercept
Also known as the vertical intercept. The value at which a line cuts the y-axis (or vertical axis) on a number plane. For example, the vertical intercept of this line is 3.

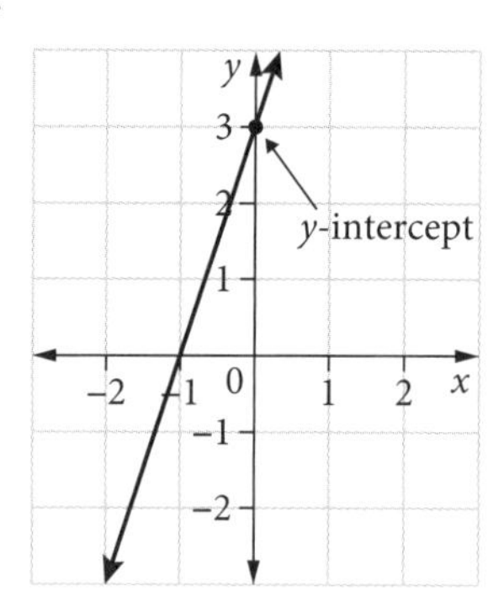

Topic summary

Review of algebra (Year 11)

Equations and expressions

An **equation** is a relation between **variables** and values that contains an equals symbol ('='), for example, $\frac{x-3}{2} = 5$.

An **expression** is a group of algebraic terms. It does not have an equals symbol and it cannot be solved, for example, $7y^2 + 5y - 2$.

A **linear equation** is an equation where the power of the variable is 1, and it contains an equals symbol ('='), for example, $7x + 1 = 15$ and $8 = 4 - \frac{2a}{3}$. The equation $x^2 + 4x = 9$ is not linear.

Solving equations

To solve an equation, use inverse (opposite) operations to 'undo' the equation. The aim is to get the variable by itself on one side of the equals symbol and its value on the other side.

Example 1

Solve each equation.

a $3x - 4 = 23$

b $\frac{2x+4}{5} = x - 1$

Solution

a

$$\begin{aligned} 3x - 4 &= 23 \\ 3x - 4 + 4 &= 23 + 4 && \text{Add 4 to both sides} \\ 3x &= 27 \\ x &= \frac{27}{3} && \text{Divide both sides by 3} \\ &= 9 \end{aligned}$$

b

$$\begin{aligned} \frac{2x+4}{5} &= x - 1 \\ 2x + 4 &= 5(x - 1) && \text{Multiply both sides by 5} \\ 2x + 4 &= 5x - 5 && \text{Expand brackets} \\ 2x &= 5x - 9 && \text{Subtract 4 from both sides} \\ 2x - 5x &= -9 && \text{Subtract } 5x \text{ from both sides} \\ -3x &= -9 \\ x &= \frac{-9}{-3} && \text{Divide both sides by } -3 \\ &= 3 \end{aligned}$$

Example 2

The blood alcohol content (BAC) formula for females is $BAC_{female} = \frac{10N - 7.5H}{5.5M}$, where N is the number of standard drinks consumed, H is the number of hours spent drinking and M is the mass of the female in kilograms.

Karen weighs 66 kg and has consumed 5 standard drinks in the past 3 hours. Calculate her BAC, correct to three decimal places.

Solution

Substitute $N = 5$ (drinks), $H = 3$ (hours) and $M = 66$ (kg) into the formula.

$$\begin{aligned} BAC &= \frac{10 \times 5 - 7.5 \times 3}{5.5 \times 66} \\ &= 0.075\,75\ldots \\ &\approx 0.076 \end{aligned}$$

Example 3

The formula for the volume of a cone is $V = \frac{1}{3}\pi r^2 h$, where r is the radius of its base and h is its height.

Find, correct to two decimal places, the radius of the base of a cone if the cone's volume is 360 cm^3 and its height is 12 cm.

Solution

Substitute $V = 360$ and $h = 12$ into $V = \frac{1}{3}\pi r^2 h$ and solve for r.

$$360 = \frac{1}{3}\pi r^2 \times 12$$

$$360 = 4\pi r^2 \quad \text{Simplify RHS}$$

$$\frac{360}{4\pi} = r^2 \quad \text{Divide by } 4\pi$$

$$\frac{90}{\pi} = r^2 \quad \text{Simplify LHS}$$

$$\sqrt{\frac{90}{\pi}} = r \quad \text{Find the square root}$$

$$r = 5.3523\ldots$$

$$\approx 5.35\text{ cm}$$

Changing the subject of a formula

In the formula $v = u + at$, v is called the **subject** of the formula because it is on the left-hand side.

To change the subject of a formula to another variable, use the same rules as for solving an equation. The answer is not a number but an algebraic equation (another formula).

Example 4

Rearrange the equation $3x + 2y - 5 = 0$ to make y the subject.

Solution

Isolate y on the left-hand side of the equation.

$$3x + 2y - 5 = 0$$

$$2y = -3x + 5 \quad \text{Subtract } 3x \text{ and add 5 to both sides}$$

$$y = \frac{-3x + 5}{2} \quad \text{Divide both sides by 2}$$

Review of linear functions (Year 11)

All content for this topic is common with the Mathematics Advanced course.

A **linear function** has the gradient–intercept form $y = mx + c$ and on a graph it is a straight line.

The **gradient**, m, is the rate of change of y relative to x.

$$m = \frac{\text{rise } (\uparrow)}{\text{run } (\rightarrow)} = \frac{\text{change in } y}{\text{change in } x}$$

Vertical **rise** or change in y

Horizontal **run** or change in x

The **y-intercept** or vertical intercept, c, is the value of y when $x = 0$.

9780170459204

A graph with a line sloping upwards has a positive gradient. If the line slopes downwards, it has a negative gradient.

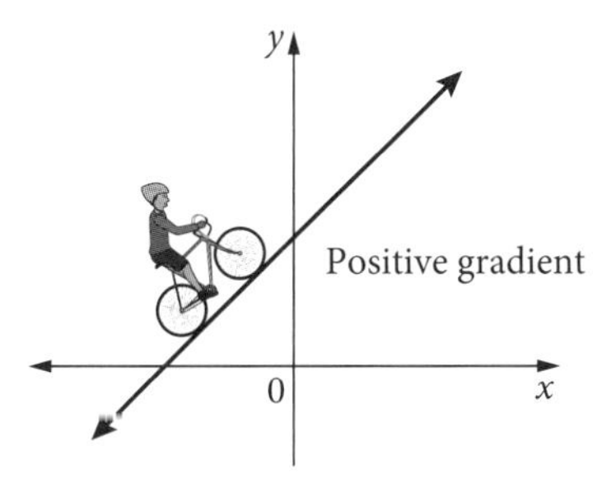

Graph sloping upwards

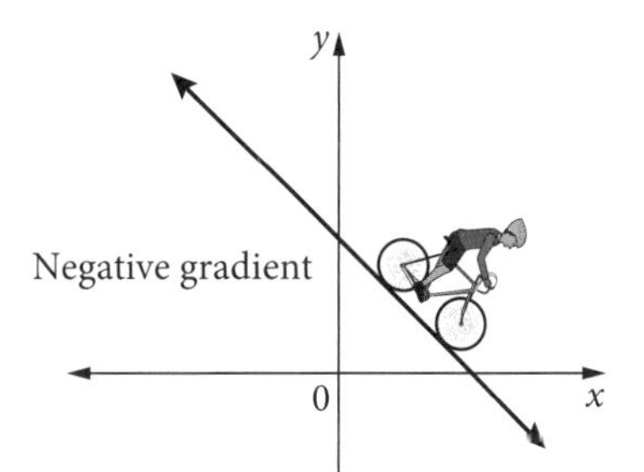

Graph sloping downwards

TOPIC SUMMARY

Finding the equation of a line

1. Find the gradient $m = \frac{\text{rise}}{\text{run}}$ using 2 points, where $m = \frac{y_2 - y_1}{x_2 - x_1}$.
2. Find the y-intercept, c.
3. Rewrite the equation $y = mx + c$ with the values of m and c.

Example 5

Find the equation of this line.

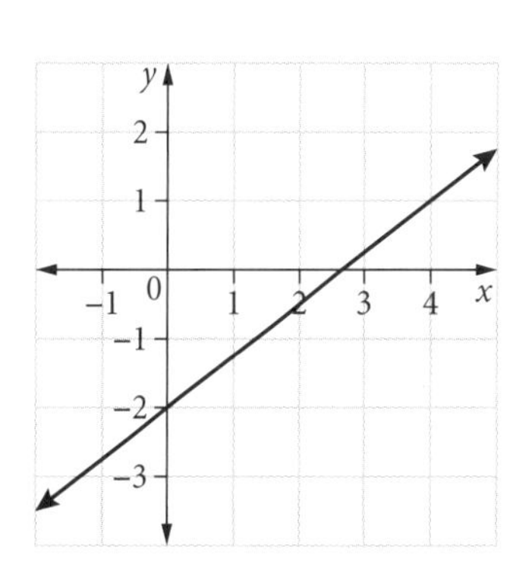

Solution

1. Choose 2 points $(0, -2)$ and $(4, 1)$ on the graph to find the gradient.

$$m = \frac{\text{rise } (\uparrow)}{\text{run } (\rightarrow)} = \frac{\text{change in } y}{\text{change in } x}$$
$$= \frac{1 - (-2)}{4 - 0}$$
$$= \frac{3}{4}$$

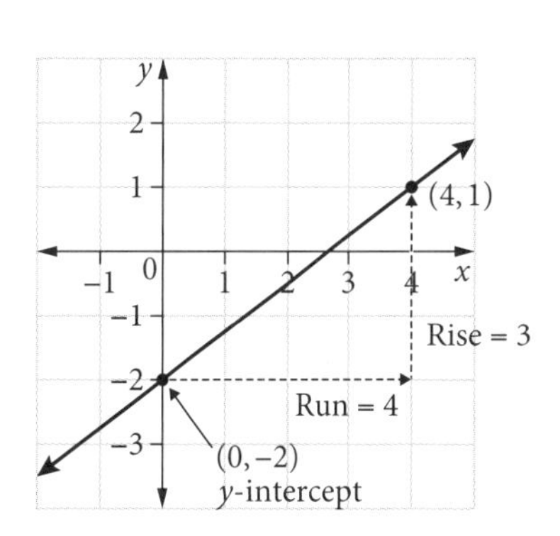

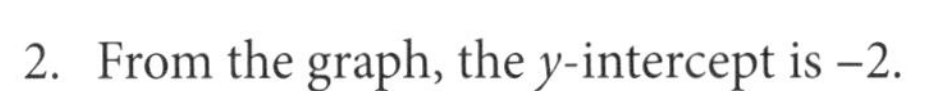

2. From the graph, the y-intercept is -2.
3. The equation of the line is $y = \frac{3}{4}x - 2$.

Example 6

Find the equation of the linear function represented by the table of values.

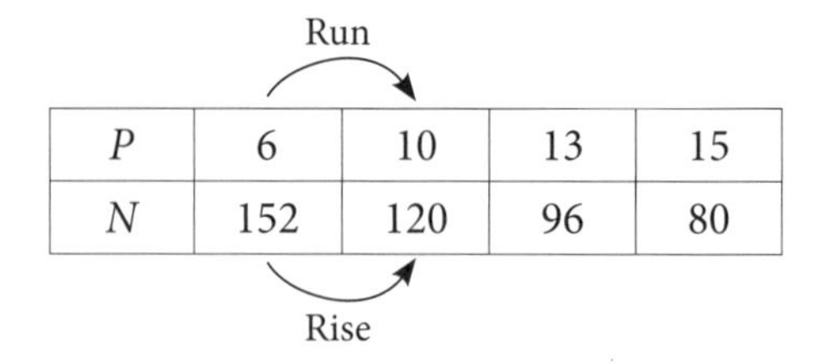

P	6	10	13	15
N	152	120	96	80

Solution

1. Choose 2 points from the table, say (6, 152) and (10, 120), to find the gradient.

$$m = \frac{\text{rise}}{\text{run}} = \frac{120 - 152}{10 - 6} = \frac{-32}{4} = -8$$

   So, $y = mx + c$ becomes $N = mP + c$.

   Because $m = -8$, the equation of the line is $N = -8P + c$.

2. The vertical intercept, c, is the value of N when $P = 0$. This is not shown in the table so, to find c, choose a point to substitute into $N = -8P + c$, such as (10, 120), and solve the equation.

$$120 = -8 \times 10 + c$$
$$120 = -80 + c$$
$$120 + 80 = c$$
$$c = 200$$

3. The equation of the linear function is $N = -8P + 200$.

Graphing linear functions using a table of values

1. Substitute each x-value into the equation to calculate the y-value.
2. Plot points on the number plane.
3. Draw a straight line passing through these points.

Example 7

Graph the line for $y = -2x + 4$ on a number plane.

Solution

Substitute each x-value from the table into the equation to find y and then plot the points.

For:

$x = -1$, $y = -2 \times (-1) + 4 = 6$
$x = 0$, $y = -2 \times 0 + 4 = 4$
$x = 1$, $y = -2 \times 1 + 4 = 2$
$x = 2$, $y = -2 \times 2 + 4 = 0$

x	−1	0	1	2
y	6	4	2	0

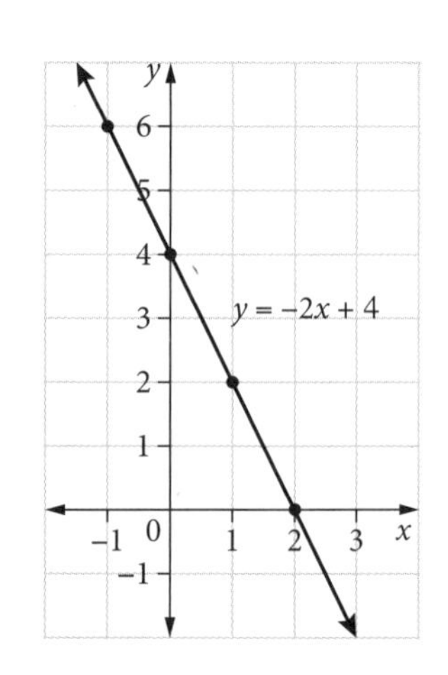

Graphing linear functions using the gradient and y-intercept

1. Plot the y-intercept point $(0, c)$.
2. Use the gradient's rise and run to plot another point.
3. Draw the line passing through these 2 points.

Example 8

Graph the line for $y = 3x - 1$ on a number plane.

Solution

1. The y-intercept is -1, so plot the point $(0, -1)$.
2. Use the gradient $m = \frac{3}{1}$, which means rise = 3 and run = 1, to locate a second point at $(1, 2)$.
3. Draw the line passing through these 2 points.

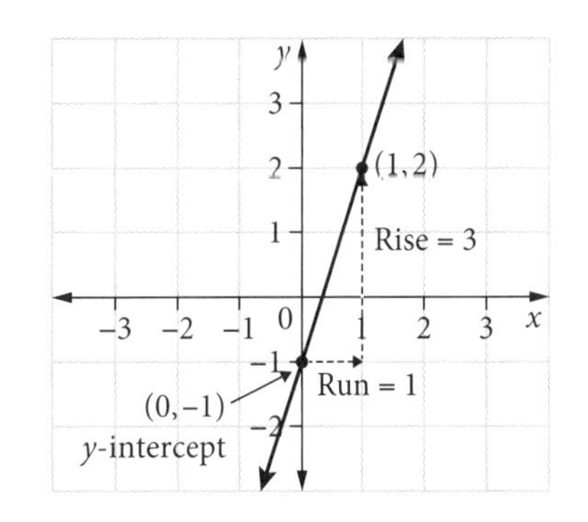

Linear modelling

Researchers observing patterns in science and society often use a mathematical formula to represent a real-life situation. This is called **algebraic modelling**. A model in the form $y = mx + c$ is a **linear model**, where the gradient represents a rate of change and the y-intercept is the **initial value**.

Example 9

A pizza shop sells pizzas for \$12 each. The delivery fee for an order is \$5.

Let n be the number of pizzas bought from the shop and C be the total cost of having these pizzas delivered.

The equation for this can be written in the form $y = mx + c$ as $C = 12n + 5$.

This **linear function** can be represented by a table of values or as a graph.

The table of values, and the points plotted on a number plane, are shown.

n	1	2	3	4	5
C	17	29	41	53	65

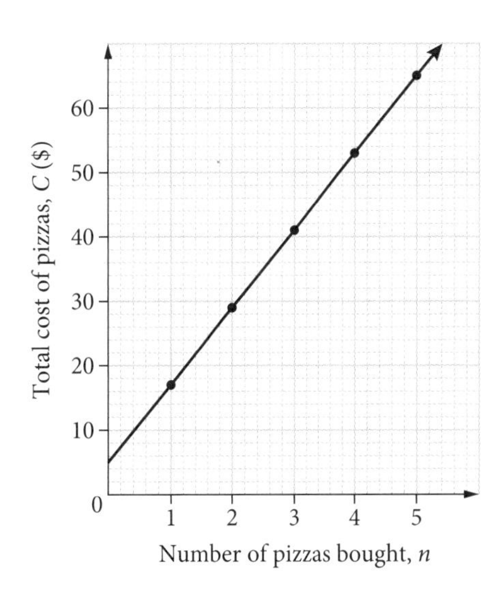

What is the cost to have 10 pizzas delivered?

Solution

Substitute the value of n into the equation.

$$\begin{aligned} C &= 12n + 5 \\ &= 12 \times 10 + 5 \\ &= 125 \end{aligned}$$

The cost is \$125.

Hint
In linear modelling, the gradient is the rate of change.

Note: The gradient, $m = 12$, represents the cost of \$12 per pizza. The y-intercept, $c = 5$, represents the initial delivery fee of \$5.

Types of relationships (MS-A4)

A4.1 Simultaneous linear equations

Solving simultaneous equations graphically

Simultaneous equations are 2 or more equations solved together so that the solution satisfies both (or all) equations. Simultaneous means 'at the same time'.

A pair of simultaneous linear equations can be solved graphically by graphing lines for both equations and finding their **point of intersection**.

Example 10

Solve the simultaneous equations $y = 2x + 4$ and $y = -3x - 1$ graphically.

Solution

Find the values of x and y that solve both equations together.

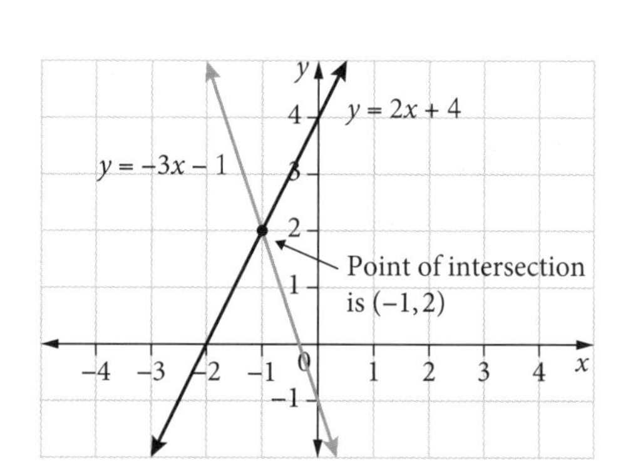

1. Draw the lines for both equations on the same number plane.
2. The 2 lines cross at the point $(-1, 2)$. This point of intersection means that the solution to the simultaneous equations is:

   $x = -1, y = 2$
3. Check that this solution solves both equations.

   For $y = 2x + 4$, substituting $x = -1$ gives:

   $y = 2 \times (-1) + 4$
   $= 2$ ✓

   For $y = -3x - 1$, substituting $x = -1$ gives:

   $y = -3 \times (-1) - 1$
   $= 2$ ✓

Simultaneous equations in business

Simultaneous equations are often used in business to see the effect of cost and **revenue** on profit and the **break-even point**.

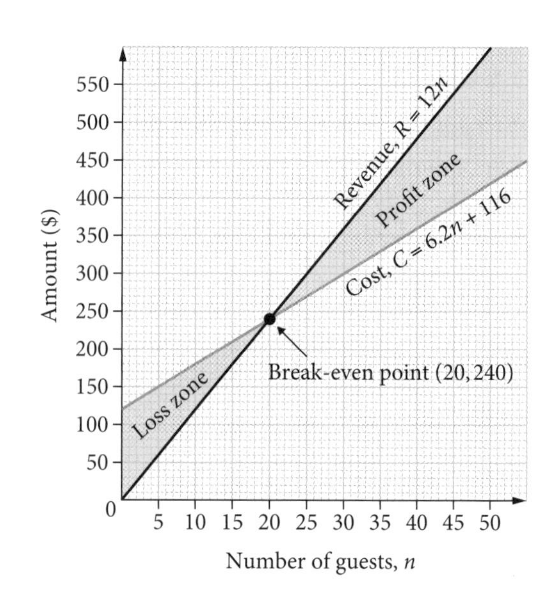

- Cost: the amount of money required for payment of a product or service and of running a business
- Revenue: the amount of money made by selling a product or service
- Profit: how much a business makes after costs have been deducted (subtracted); profit = revenue – cost
- Break-even point: how much of a product or service must be made and sold to 'break even' (not gain or lose money)

To apply simultaneous equations to a practical problem:

1. Define the variables.
2. Write 2 equations using the variables and information in the problem.
3. Graph the equations to find the values of the variables where the lines intersect.
4. Answer the question in the context of the problem.

Example 11

Two hire car companies charge a usage amount per day and a fixed insurance amount for the hire.

Company A charges \$30 per day and the insurance cost is \$160.
Company B charges \$50 per day and the insurance cost is \$60.

a Write a linear equation for each company in the form $y = mx + c$.

b Draw a line for both of these equations on a number plane.

c For which number of days' hire is Company A cheaper than Company B?

Solution

a The gradient of each line is the cost of the car per day, and the vertical intercept is the insurance cost.

$$\text{Company A:}\quad C = 30n + 160$$
$$\text{Company B:}\quad C = 50n + 60$$

where C = cost, in dollars, and n = length of hire, in days.

Hint
The hardest part to linear modelling questions is creating the linear equations. Read the question a few times before answering.

b To graph both equations, complete a table of values and plot at least 3 points for each line.

Company A:

n	0	1	2	3	4	5	6	7
C	160	190	220	250	280	310	340	370

Company B:

n	0	1	2	3	4	5	6	7
C	60	110	160	210	260	310	360	410

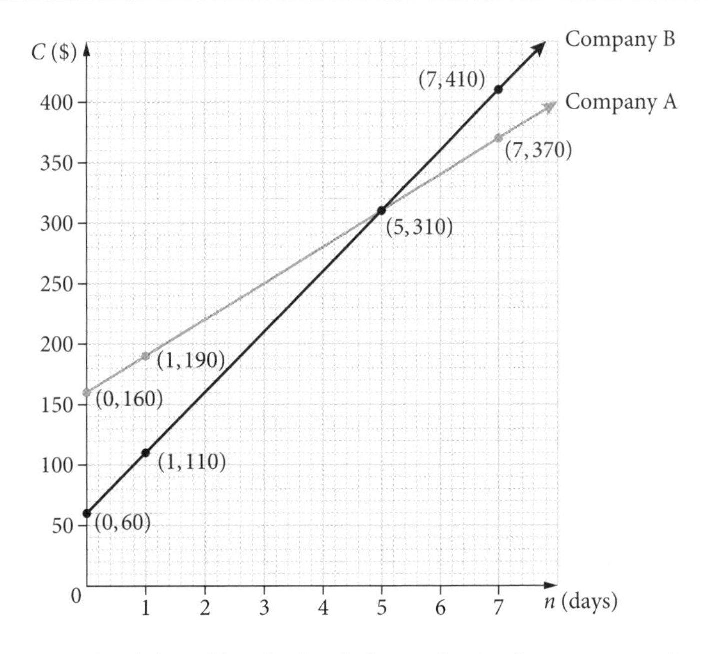

c The point of intersection (and the table of values) shows that both companies charge the same for 5 days' car hire (\$310). Company A is cheaper for more than 5 days' hire.

A4.2 Non-linear relationships

Quadratic functions

A **non-linear function** is a function that is a curve, not a line, when graphed. The 3 non-linear functions studied in this topic are the quadratic function, the exponential function and the reciprocal function.

The **quadratic function** has an equation of the form $y = ax^2 + bx + c$, where a, b and c are **constant terms** and $a \neq 0$. The highest power of x is 2. The graph of a quadratic function is a **parabola**, which is a U-shaped curve with:

- a **vertex** where the graph changes direction, also known as the **maximum/minimum turning point**
- a vertical axis of symmetry passing through the vertex
- a y-intercept, c.

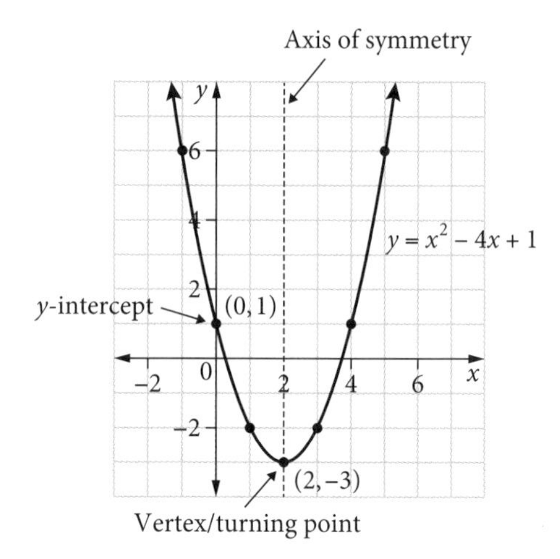

A parabola is **concave up** ('happy face') and points upwards if a is a positive number, for example, $y = 4x^2$.

Vertex

Concave up (a is positive)

A parabola is **concave down** ('sad face') and points downwards if a is a negative number, for example, $y = -2x^2$.

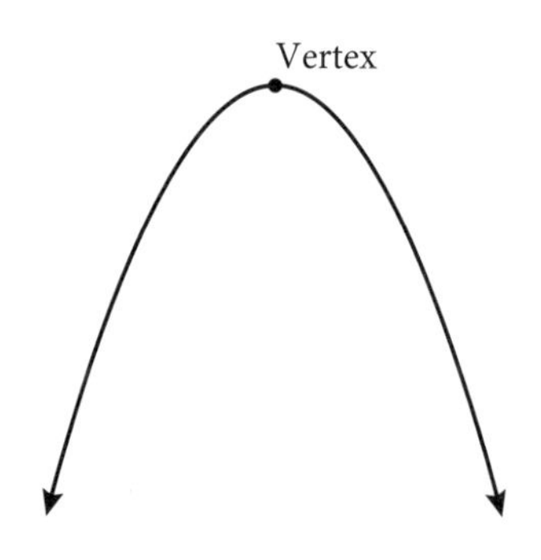

Concave down (a is negative)

Positive = 'smiling' parabola
Negative = 'frowning' parabola

To graph a parabola:

1. Create a table of values.
2. Find y-values by substituting the x-values one by one into the quadratic equation.
3. Draw a suitable number plane.
4. Plot points and join them to sketch a curve.

x					
y					

Hint

It is important that your parabola makes sense. Check that your curve:

- has a y-intercept that is the same as the value of c in the equation
- is 'happy' if a is positive or 'sad' if a is negative
- has a parabolic shape.

If the curve does not have these features, then check your working to see what went wrong.

Maximum and minimum values of a quadratic function

Problems often require you to find the **maximum value** or the **minimum value** of a quadratic function or parabola. The maximum or minimum value will always be the y-value of the vertex.

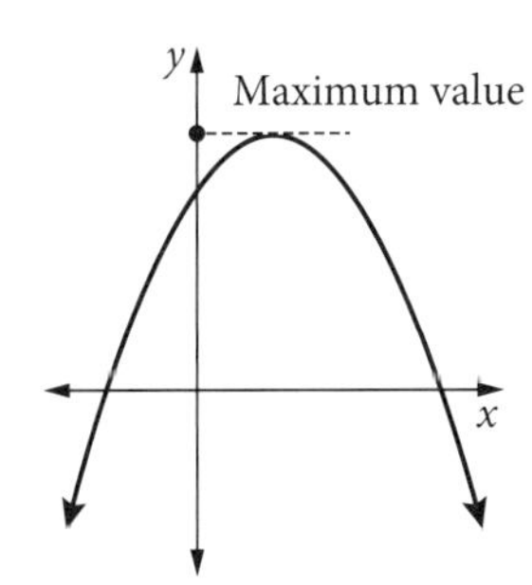

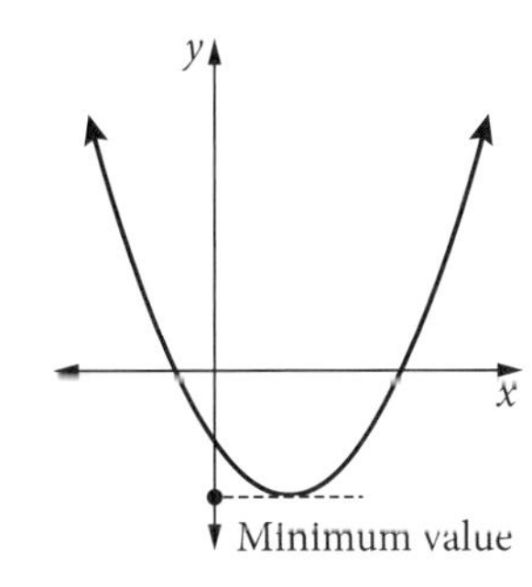

Example 12

A railway bridge underpass has an arch in the shape of a parabola. The quadratic function for the arch is $y = 18 - 2x^2$, where x and y are in metres.

x	−3	−2	−1	0	1	2	3
y							

a Complete the table of values, plot the points on a number plane and sketch the parabola.

b What is the maximum height of the arch of this underpass?

c What are the **x-intercepts** and how wide is the underpass at ground level?

Solution

a Substitute each x-value into the equation to find y, and then plot the points.

For $x = -3$, $y = 18 - 2(-3)^2 = 0$

For $x = -2$, $y = 18 - 2(-2)^2 = 10$, and so on.

x	−3	−2	−1	0	1	2	3
y	0	10	16	18	16	10	0

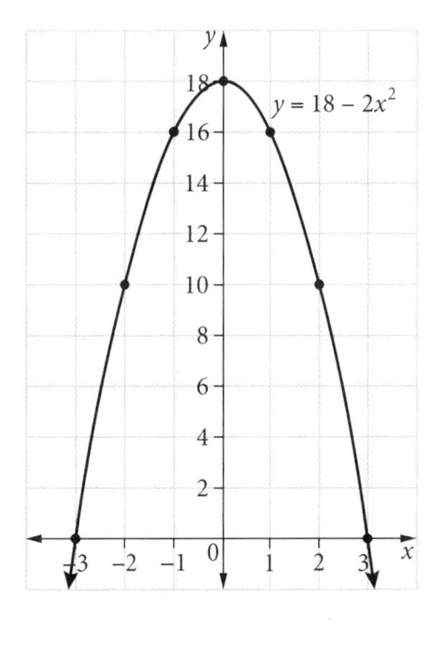

b From the graph and table, the maximum height of the arch is 18 m.

c The x-intercepts are at −3 and 3, so the width of the underpass at ground level is 6 m.

Exponential functions

An **exponential function** has an equation of the form $y = ka^x$ or $y = ka^{-x}$, where x is the power and a is positive ($a > 0$). On a graph, the exponential function is an **exponential curve**. The curve is always above the x-axis – its y-values are always positive.

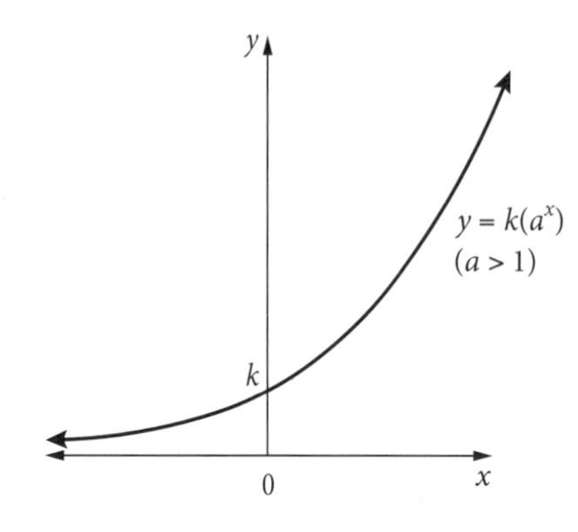

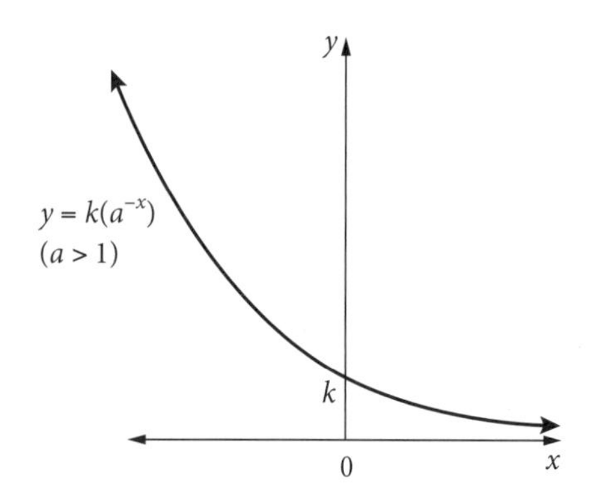

Exponential curves, such as those shown on the graphs, have these features:

- a and k are constants and x is the exponent (index or power of the equation).
- The value of the y-intercept (that is when $x = 0$) is k.
- The x-axis is an **asymptote**: the curve approaches but never touches or crosses the x-axis, so it is always above the x-axis.
- For $y = k(a^x)$, the curve is increasing or 'uphill' (from left to right) if the value of a is greater than 1 and is decreasing or 'downhill' if the value of a is between 0 and 1.
- For $y = k(a^{-x})$, the curve is decreasing if the value of a is greater than 1 and is increasing if the value of a is between 0 and 1.

To graph exponential functions:

1. Create a table of values.
2. Find y-values by substituting the x-values one by one into the exponential equation.
3. Draw a suitable number plane.
4. Plot points and join them to sketch a curve.

Hint

It is important that your exponential curve makes sense. Check that your curve:

- rises or falls correctly compared to the value of a in the equation
- has an exponential shape
- has an asymptote.

If the curve does not have these features, then check your working to see what went wrong.

Example 13

a Graph the exponential function $y = 2(3^x)$.

b State the y-intercept and asymptote of the graph.

Solution

Hint

You do not have to plot all points – some will not fit on the axes.

a

x	−3	−2	−1	0	1	2	3
y	0.07	0.22	0.67	2	6	18	54

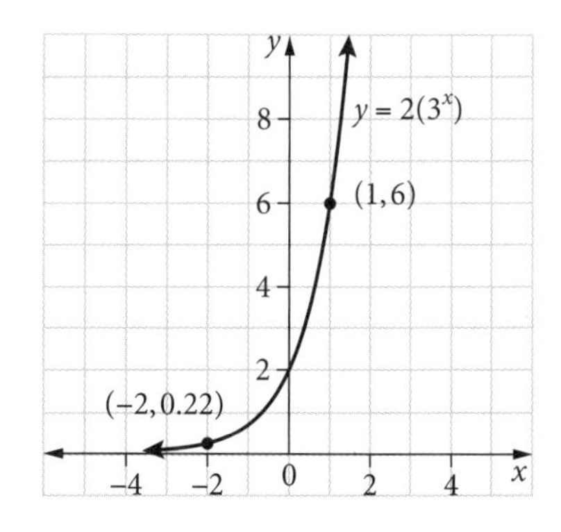

b The y-intercept is 2 and the asymptote is the x-axis.

Exponential growth (increase)

Exponential growth of a quantity happens slowly at first, then increases more rapidly. It is expressed in this form:

$$y = k(a^x), \text{ where } a > 1$$

Some examples are:

- compound interest on investments
- changes in populations of humans and animals
- the spread of viruses, such as COVID-19.

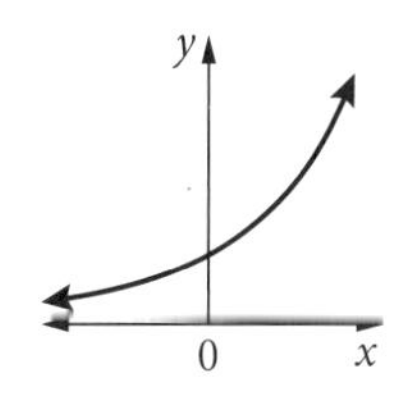

Exponential decay (decrease)

Exponential decay of a quantity happens rapidly at first, but then slows over time. It is expressed in this form:

$$y = k(a^{-x}), \text{ where } a > 1$$
$$\text{or } y = k(a^x), \text{ where } 0 < a < 1.$$

Some examples are:

- depreciation of equipment
- cooling of hot liquids
- decay of radioactive materials.

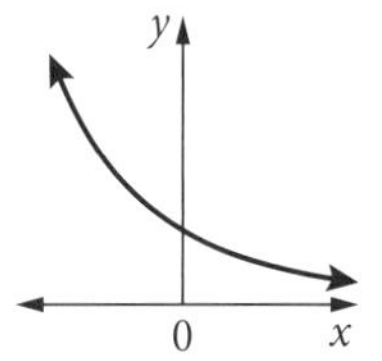

TOPIC SUMMARY

Example 14

Bacteria are being cultivated in a dish in a laboratory. The bacteria are growing exponentially, at a rate of 60% each hour. Initially there were 50 bacteria.

a Write a formula in the form $y = k(a^x)$ using N for number of bacteria and t for time, in hours.

b How many bacteria are cultivated during the first 24 hours? Answer to the nearest thousand.

Solution

a Exponential growth is 100% + 60% = 160% = 1.6

Formula is $N = 50(1.6^t)$

b Substitute $t = 24$.

$$N = 50(1.6^{24})$$
$$= 3\,961\,408.126\ldots$$
$$\approx 3\,961\,000$$

Hint

If the equation you found in part **a** is incorrect but you correctly substitute 24, you will get full marks for part **b**!

Example 15

This graph shows the declining-balance depreciation of a car based on the exponential function $S = 20\,000(0.9)^n$, where S is salvage value, in dollars, and n is age of the car, in years.

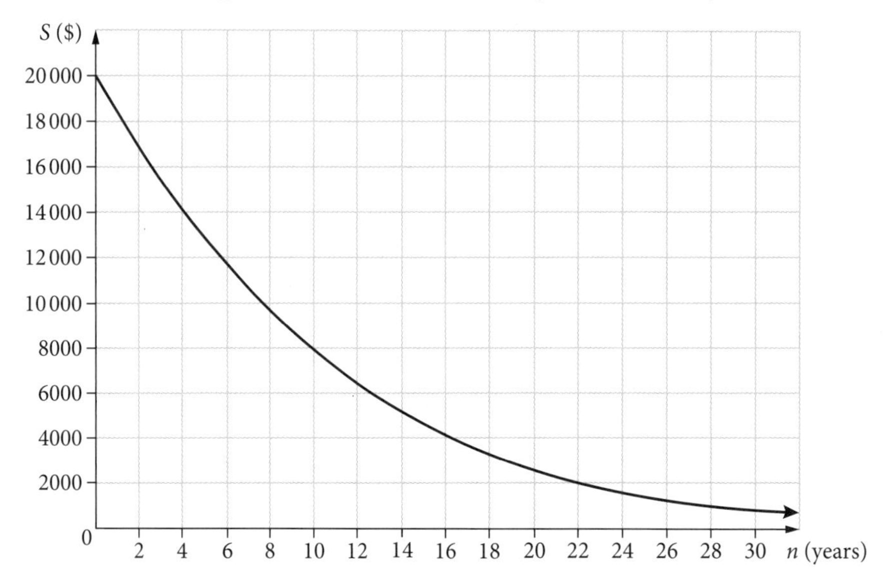

a How much was the car worth initially?

b What is the depreciation rate per annum?

Solution

a The initial value of the car is the value of the y-intercept, \$20 000.

b The salvage value is changing at 0.9 or 90%. This represents a decrease of 100% – 90% = 10%.

So the rate of depreciation is 10% p.a.

This is exponential decay because the function has the form $y = ka^x$, where the value of a is between 0 and 1.

Compound interest and depreciation (see Chapter 4)

- The future value formula for compound interest, $FV = PV(1 + r)^n$, is an example of exponential growth: $y = ka^x$, where the value of a is greater than 1.
- The declining-balance formula of depreciation, $S = V_0(1 - r)^n$, is an example of exponential decay: $y = ka^x$, where the value of a is between 0 and 1.
- The percentage rate, r, is expressed as a decimal in these formulas.

Reciprocal functions and the hyperbola

The **reciprocal function** has an equation of the form $y = \frac{k}{x}$, where x is in the denominator and k is a constant. It represents **inverse variation** and on a graph its curve is a **hyperbola**. There are 2 separate curves in opposite quadrants.

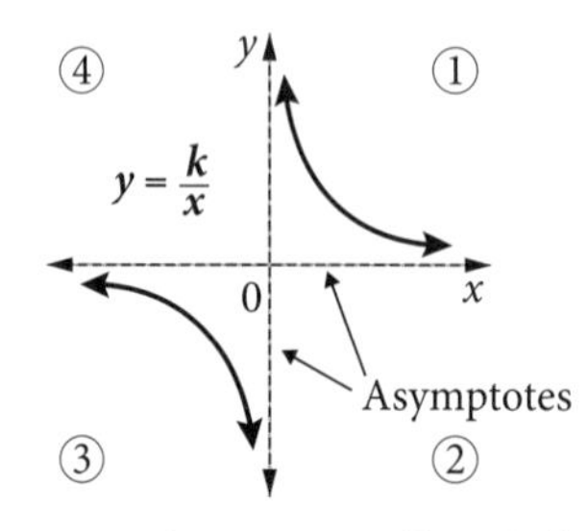

Curves are decreasing or 'downhill' if k is positive (quadrants 1 and 3).

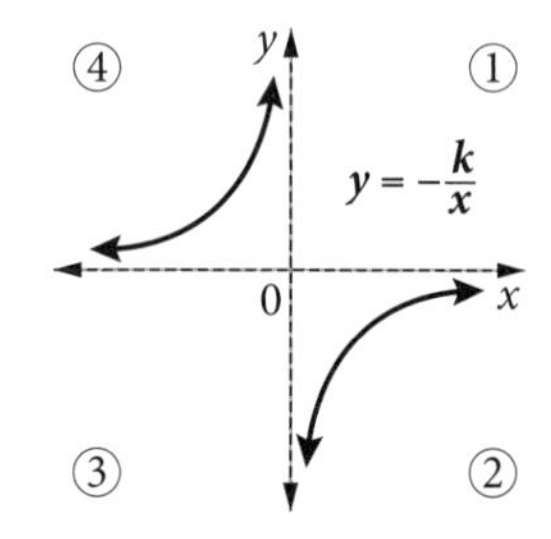

Curves are increasing or 'uphill' if k is negative (quadrants 2 and 4).

Graphs of reciprocal functions have several things in common:

- There is point (rotational) symmetry about the origin.
- Asymptotes are at the x- and y-axes. (Curves in the form of $y = \frac{k}{x}$ never cross the axes.)
- The higher the value of k, the further away the curve is from the origin and the x- and y-axes.

In modelling questions, the curve may only be drawn in the first quadrant for positive x- and y-values if the application involves positive values only, such as in Example 16.

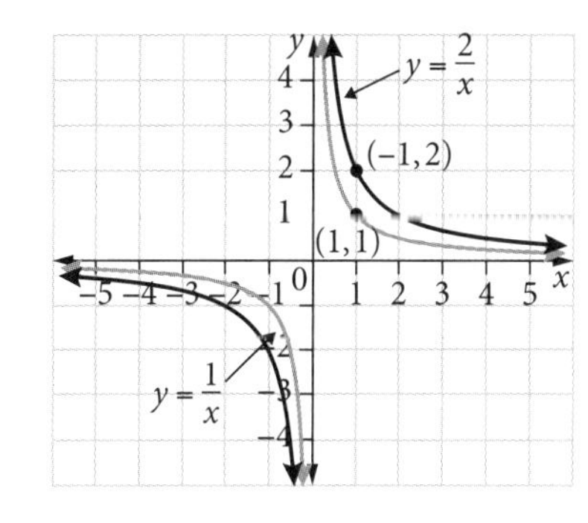

TOPIC SUMMARY

Example 16

The time, T, in months, taken to pay off an interest-free loan is related to the size of each monthly repayment, r, in dollars, by the reciprocal function $T = \frac{1800}{r}$.

a Graph the function for values of r from 100 to 900.

b Use the graph to estimate the time to pay off the loan if the monthly repayment is \$250.

c Use the formula to find the monthly repayment required to pay off the loan in 5 months.

d Why is this model not useful for values of r greater than \$1800?

Solution

a Substitute r-values into $T = \frac{1800}{r}$ and find values for T, correct to two decimal places where necessary.

r (\$)	100	200	300	400	500	600	700	800	900
T (months)	18	9	6	4.5	3.6	3	2.57	2.25	2

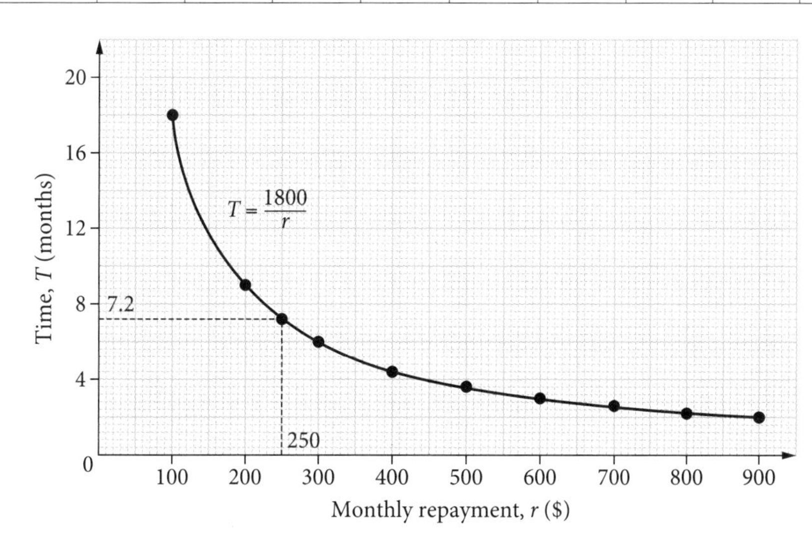

b From the graph, when $r = 250$, $T = 7.2$.

Time taken to pay off the loan is 8 months (rounded up to the nearest month to cover all required payments).

c From the formula, when $T = 5$:

$$5 = \frac{1800}{r}$$
$$5r = 1800$$
$$r = \frac{1800}{5}$$
$$= 360$$

The monthly repayment required is \$360.

d If $r > 1800$, $T < 1$. It is not possible to pay off a loan in less than 1 month using monthly repayments.

Review of direct linear variation (Year 11)

If y is directly proportional to x, then $y = kx$, where k is a constant.

- As x increases, y increases.
- As x decreases, y decreases.
- The graph of $y = kx$ is a straight line that passes through the origin.

For example, the money Jesinta earns, M, varies directly with the number of hours, h, she works. The linear equation is $M = kh$, where k is the constant rate of pay per hour.

In this table, $M = 10h$, where the gradient is the constant rate of pay of \$10 per hour.

The more hours Jesinta works, the more money she earns. $\uparrow h \quad \uparrow M$

h	0	1	2	3	4	5
M	0	10	20	30	40	50

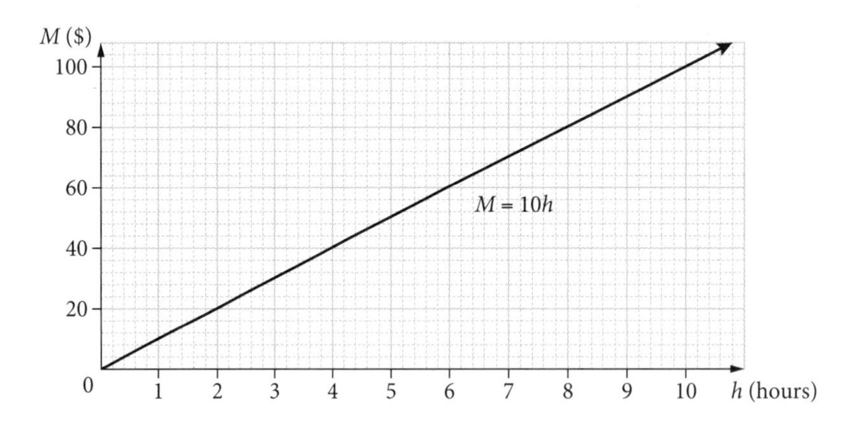

Inverse variation

If y is inversely proportional to x, then $y = \frac{k}{x}$, where k is a constant.

- As x increases, y decreases.
- As x decreases, y increases.
- The graph is a hyperbola.

For example, the higher the speed of a vehicle, then the less time it takes to travel a certain distance.

The equation is $T = \frac{D}{S}$, where S is speed, D is distance and T is time. $\uparrow S \quad \downarrow T$

For a distance of 240 km, $T = \frac{240}{S}$.

S (km/h)	20	40	60	80	100
T (h)	12	6	4	3	2.4

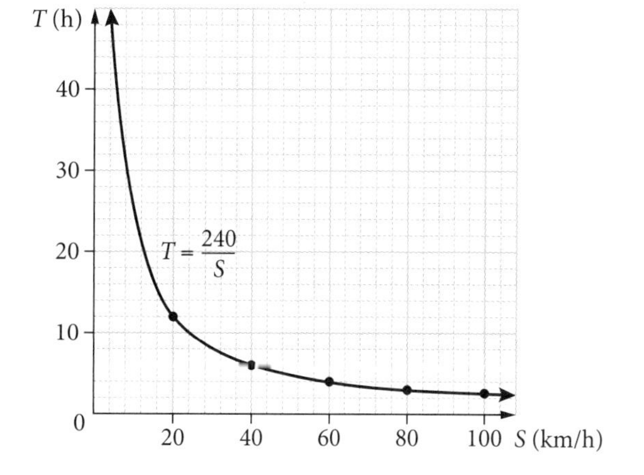

Hint

'Inverse' or 'invert' means 'flip'. Inverse also means 'opposite', which can remind you of x- and y-values increasing or decreasing.

TOPIC SUMMARY

To solve a variation problem:

1. Write a variation equation: $y = kx$ or $y = \frac{k}{x}$.
2. Substitute x- and y-values from the question and solve the equation to find k.
3. Rewrite the variation equation with the value of k substituted in.
4. Solve the problem using the new equation.

Example 17

The number of days (N) taken to paint the walls in the rooms of a school varies inversely with the number of painters (P) employed. The painting takes 5 days if 12 painters are working. How many days will it take with 4 painters?

Solution

Following the steps for a variation problem:

1. $y = \frac{k}{x}$

   $N = \frac{k}{P}$

2. 5 days, 12 painters

   $5 = \frac{k}{12}$

   $k = 5 \times 12$

   $= 60$

3. $N = \frac{60}{P}$

4. Substitute $N = 4$ (painters)

   $N = \frac{60}{4}$

   $= 15$ days

Practice set 1

Multiple-choice questions

Solutions start on page 32.

Question 1

The cost (C dollars) of a taxi trip over a distance of d kilometres is given by $C = 1.6d + 5$.

What is the cost of a 25-kilometre taxi trip?

A \$31.60 **B** \$40 **C** \$45 **D** \$200

Question 2

Solve $\dfrac{x-6}{7} = 2$.

A $x = -20$ **B** $x = -8$ **C** $x = 8$ **D** $x = 20$

Question 3

Find the equation of this line.

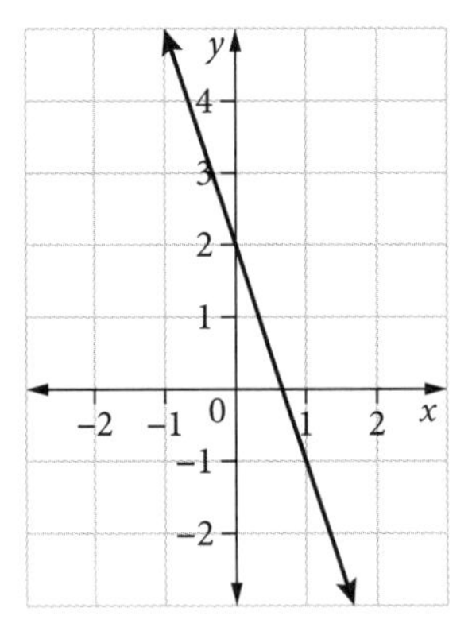

A $y = 2 + 3x$ **B** $y = 2 - 3x$ **C** $y = 2 + \frac{1}{3}x$ **D** $y = 2 - \frac{1}{3}x$

Question 4

What is the point of intersection of these 2 lines?

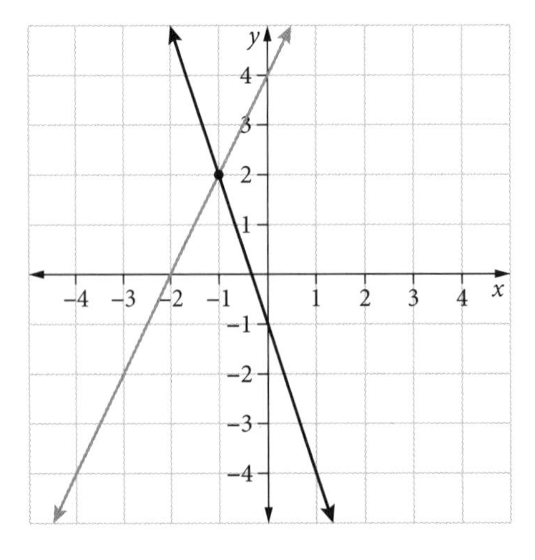

A $(0, -1)$ **B** $(0, 4)$ **C** $(2, -1)$ **D** $(-1, 2)$

Question 5

If $x^2 = 3.4$, what is the value of x correct to two significant figures?

A ± 1.7 **B** ± 1.8

C ± 1.84 **D** ± 2.0

Question 6

This is the graph of an equation of a line in the form $y = mx + c$.

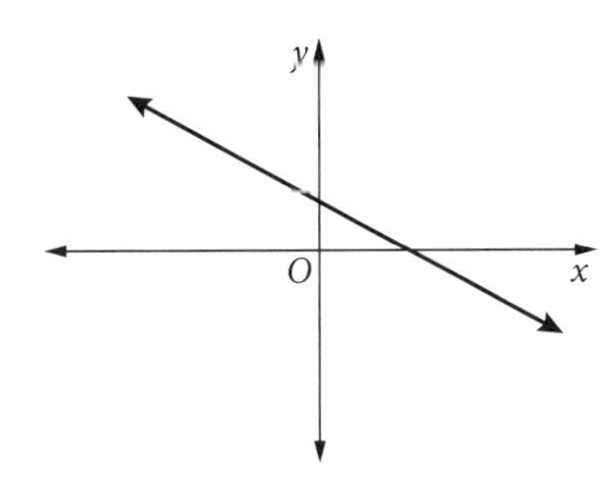

Which of the following statements is true?

A m and c are both positive. **B** m and c are both negative.

C m is positive and c is negative. **D** m is negative and c is positive.

Question 7 ©NESA 2020 HSC EXAM, QUESTION 1

Which of the following could represent the graph of $y = -x^2 + 1$?

A

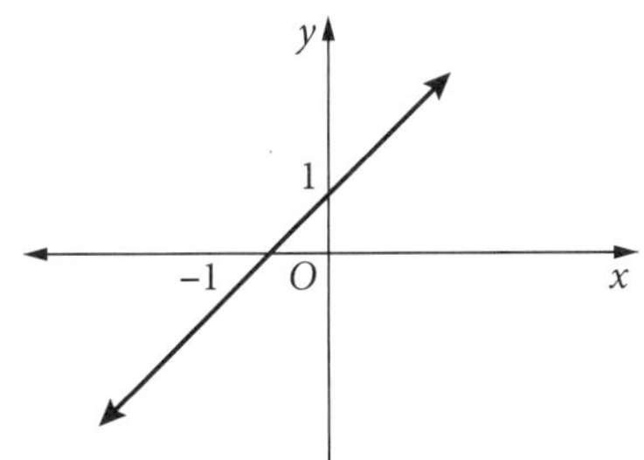

B

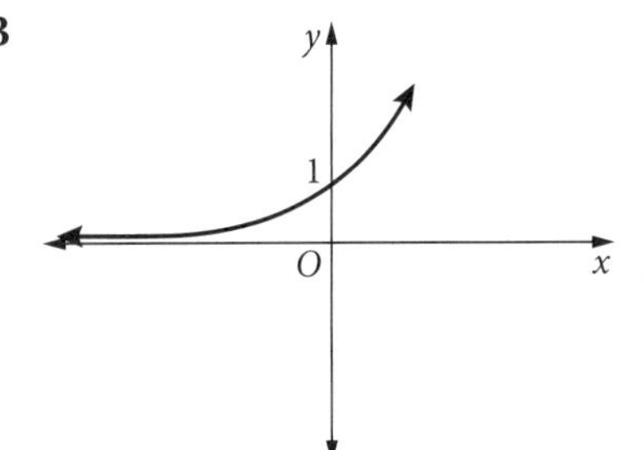

C

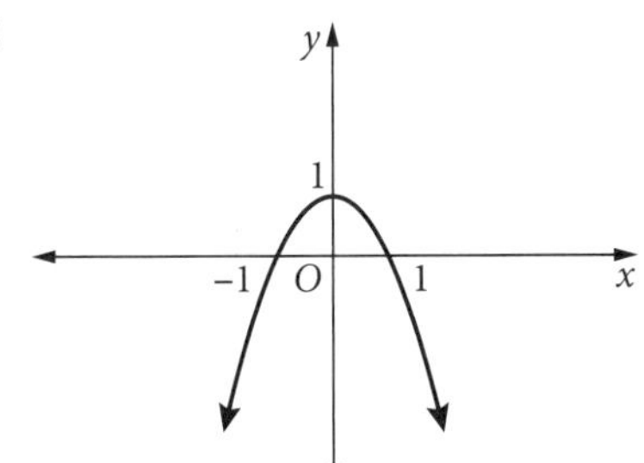

D

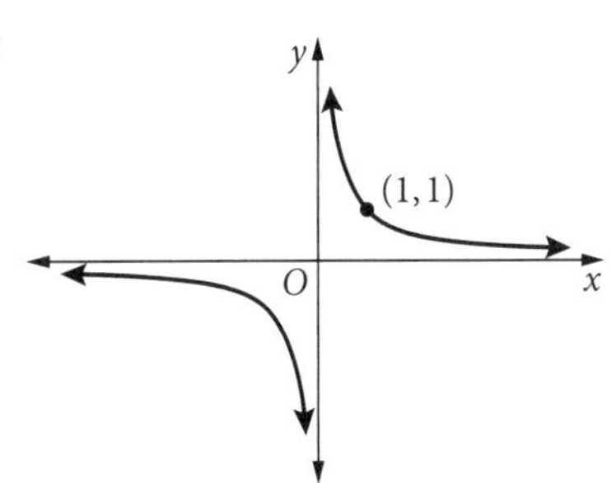

Question 8

Which of these functions has a graph with a curve that is decreasing?

A $y = 1.6^x$ **B** $y = 0.4(1.5)^x$

C $y = (0.5)^{-x}$ **D** $y = 0.8^x$

Question 9

The blood alcohol content (BAC) for a male adult is given by this formula.

$$\text{BAC}_{\text{male}} = \frac{10N - 7.5H}{6.8M}$$

N is the number of standard drinks consumed, H is the number of hours of drinking and M is the person's mass in kilograms.

Calculate the number of standard drinks consumed by a 68 kg male adult who drinks for 3.5 hours and has a BAC of 0.03.

A 2 **B** 4 **C** 6 **D** 8

Question 10

Which equation represents the relationship between x and y in this table?

x	0	1	2	3	4
y	0	1	4	9	16

A $y = x + 1$ **B** $y = x + 3$ **C** $y = x^2$ **D** $y = 2^x$

Question 11

Tyson tries to find the value of q when $p = 107$ in the equation $p = 7q^2 + 5$.

In which line has he made a mistake?

$p = 7q^2 + 5$
$107 = 7q^2 + 5$ Line A
$112 = 7q^2$ Line B
$16 = q^2$ Line C
$q = \pm 4$ Line D

A Line A **B** Line B **C** Line C **D** Line D

Question 12

The total cost, C, in dollars, of running a school function is given by $C = 75s + 450$, where s is the number of students attending.

If 7 students can no longer attend, how much cheaper is the function?

A \$443 **B** \$450 **C** \$525 **D** \$975

Question 13

Tavish hired a scooter while on holiday. The charges for hiring the scooter are shown in the table.

Time hired, h (h)	1	2	3
Cost, C (\$)	12	16	20

Which linear equation shows the relationship between cost and hire?

A $C = h + 4$ **B** $C = h + 12$ **C** $C = 4h + 8$ **D** $C = 4h + 12$

Question 14 ©NESA 2009 HSC EXAM, QUESTION 16

The time for a car to travel a certain distance varies inversely with its speed.

Which of the following graphs show this relationship?

A

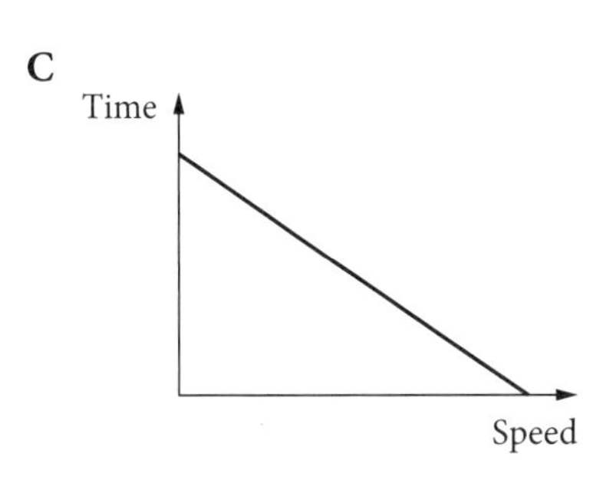

B

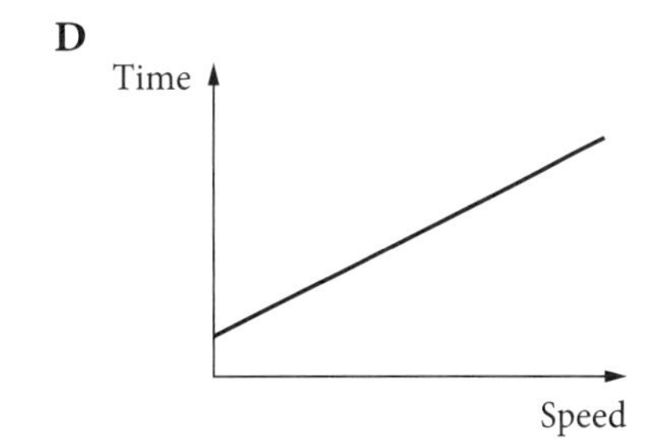

C

Time

Speed

D

Time

Speed

Question 15

Michaela draws a graph to solve the simultaneous equations $y = 2x - 3$ and $y = x + 5$.

Which graph is correct?

A

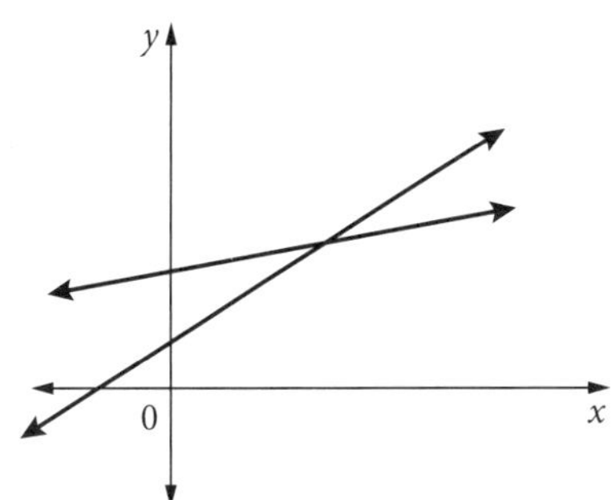

B

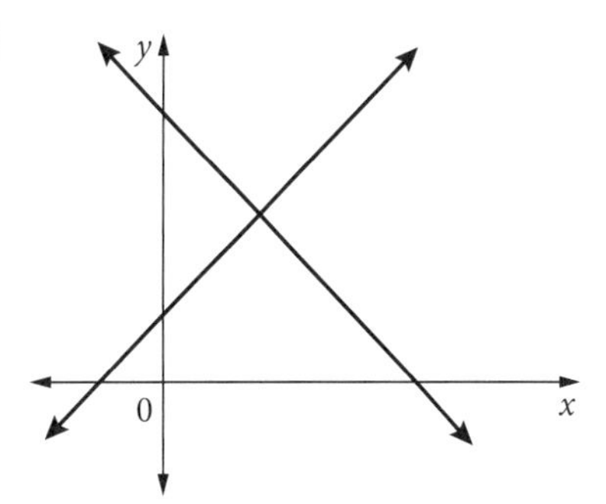

C

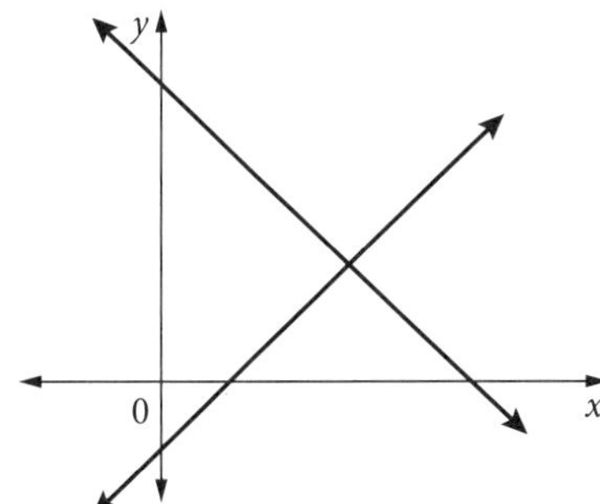

D

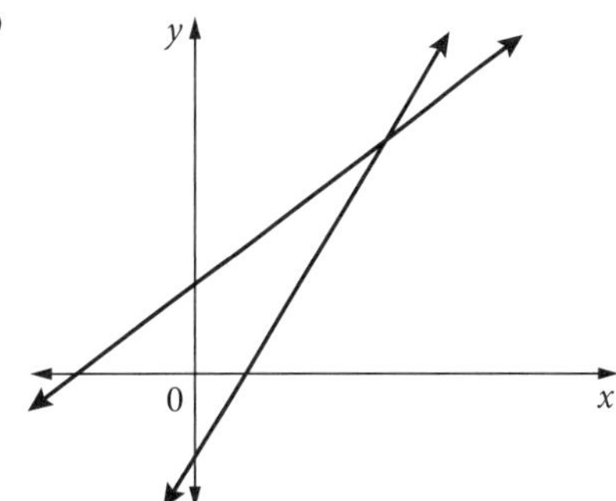

Question 16

The number of minutes, m, a tub of ice cream takes to melt varies inversely with the air temperature, T. It takes 10 minutes to melt at 28°C.

Which equation could be used to determine the number of minutes for the ice cream to melt at different temperatures?

A $m = 280T$

B $m = \frac{280}{T}$

C $m = \frac{T}{280}$

D $m = 280 + T$

Question 17

A car bought for \$35 000 depreciates in value each year according to the declining-balance formula $S = 35\,000(0.88)^n$, where S is salvage value and n is age of the car, in years.

What is the annual rate of depreciation?

A 0.12% **B** 0.88% **C** 12% **D** 88%

Question 18 ©NESA 2011 HSC EXAM, QUESTION 20

A function centre hosts events for up to 500 people. The cost C, in dollars, for the centre to host an event, where x people attend, is given by:

$$C = 10\,000 + 50x$$

The centre charges \$100 per person. Its income I, in dollars, is given by:

$$I = 100x$$

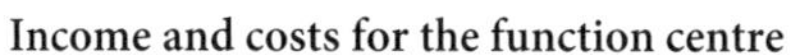

Income and costs for the function centre

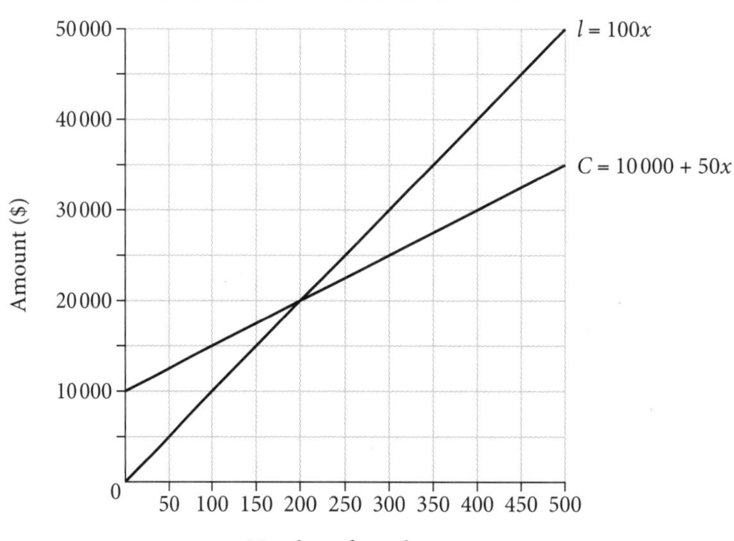

How much greater is the income of the function centre when 500 people attend an event than its income at the break-even point?

A \$15 000 **B** \$20 000 **C** \$30 000 **D** \$40 000

Question 19

Last Sunday, Jade had 705 followers on Instagram and Lisa had 473. On average, Jade gains 7 followers per day and Lisa loses 2 per day.

Which pair of equations models this situation if d represents number of days since last Sunday and f represents number of followers?

A $f = 7d + 705$, $f = 2d + 473$

B $f = 7d + 705$, $f = 473 - 2d$

C $f = 705d + 7$, $f = 473d - 2$

D $f = 705 - 7d$, $f = 473 - 2d$

Question 20

Tyrone wants to build a rectangular garden bed in his backyard. He is going to use a wall as one side of the garden and 17 m of fencing to border the remaining 3 sides.

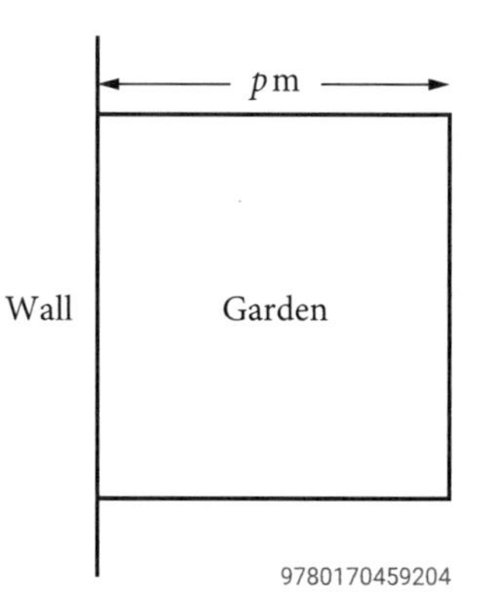

Which expression gives the area of the garden bed?

A $17 - 2p$ **B** $17p$

C $17p^2$ **D** $17p - 2p^2$

Practice set 2

Short-answer questions

Solutions start on page 34.

Question 1 (3 marks)

Match each equation with its graph.

3 marks

a $y = x^2 - 1$ **b** $y = \frac{1}{2}x + 1$ **c** $y = 2^x$ **d** $y = \frac{2}{x}$

A

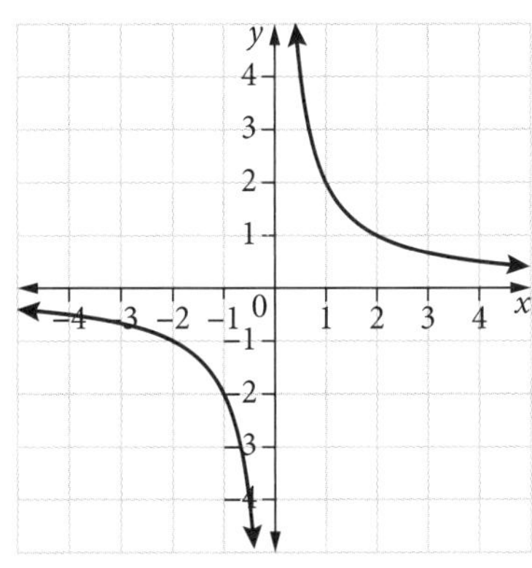

B

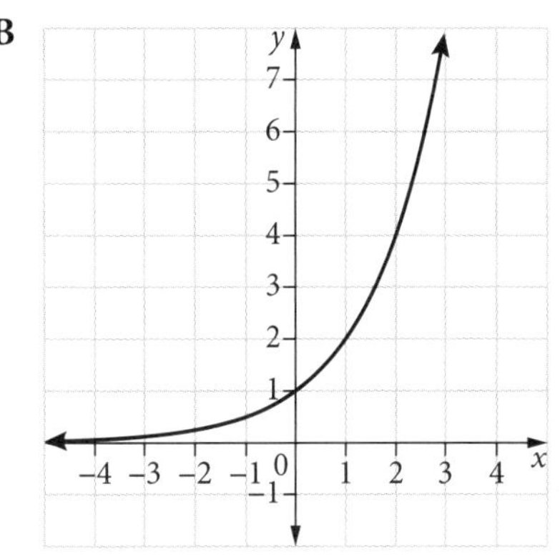

C

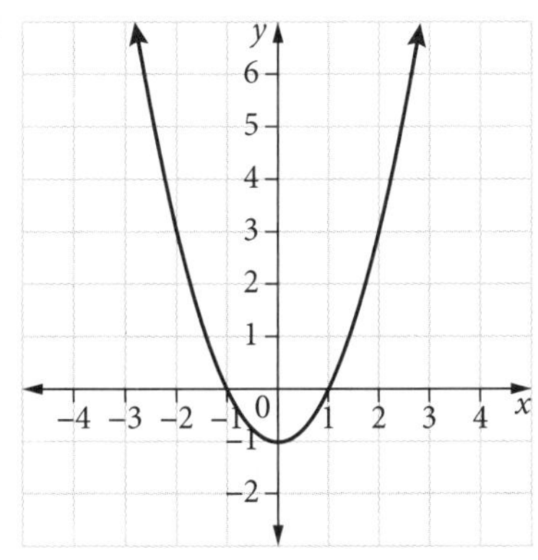

D

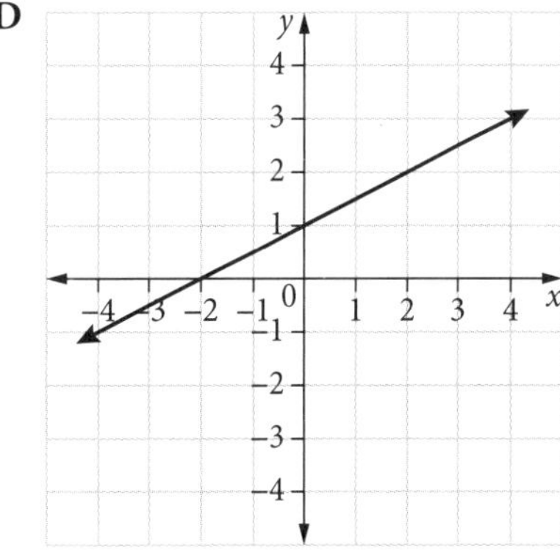

Question 2 (2 marks)

What is the equation of this line?

2 marks

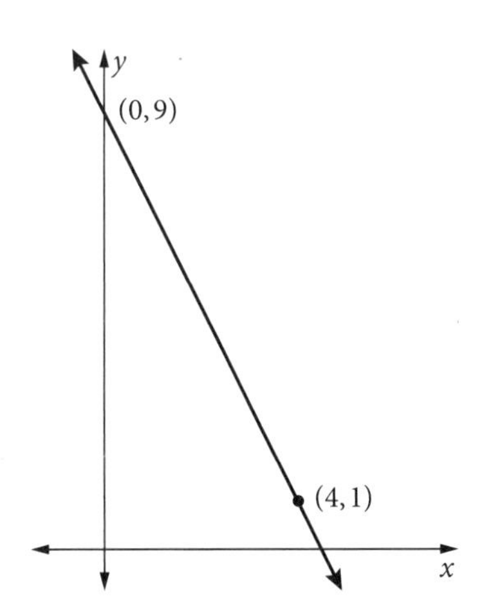

Question 3 (2 marks)

For babysitting, Louise charges $20 for the first hour and $15 for each hour after that.

The charge, C, in dollars, can be expressed using the formula $C = 20 + 15(h - 1)$, where h is time worked, in hours.

If Louise earns $95 for one day of babysitting, how many hours has she worked? 2 marks

Question 4 (5 marks)

The graph of the equation $y = 6 - 2x$ is shown.

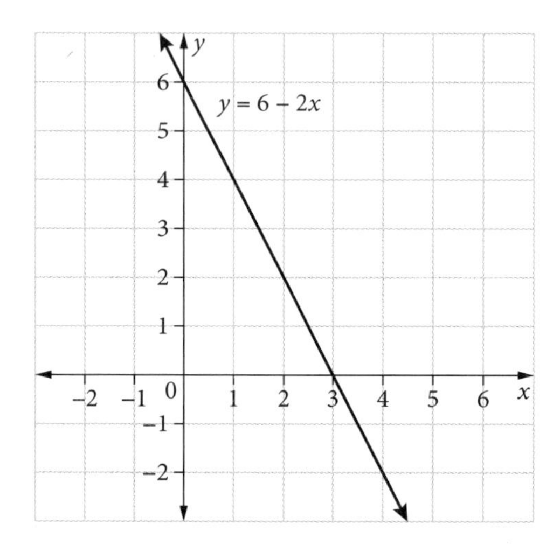

a Draw the line of the equation $y = 3x + 1$ on the number plane. 1 mark

b Write the coordinates of the point of intersection of the lines. 1 mark

c Write the solutions to the simultaneous equations $y = 3x + 1$ and $y = 6 - 2x$. 2 marks

d What is the connection between the answer to part **c** and the lines? 1 mark

Question 5 (5 marks)

This graph shows the linear relationship between the distance, d, in kilometres, travelled by a truck and the running costs, C, in dollars, for the trip.

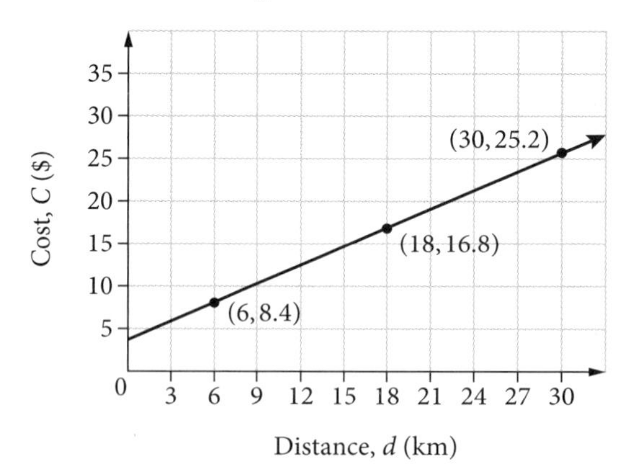

a What is the gradient of the line? 1 mark

b What does the gradient represent here? 1 mark

c Find an equation for this linear relationship in the form $C = md + c$. 2 marks

d If the length of the trip is 50 km, how much will the truck's running costs be? 1 mark

Question 6 (4 marks)

An electrician charges a call-out fee of $95 and $45 for every 30 minutes, or part thereof, worked.

a Write an equation that expresses the amount the electrician charges, C, in dollars, as a function of time, t, where t is each 30-minute interval. 1 mark

b How much would the electrician charge to work for 2 hours and 30 minutes? 1 mark

c For how long would the electrican have worked if the charge is $410? 2 marks

Question 7 (3 marks)

Vicky uses a microwave oven to heat a frozen pie. The time taken for heating varies inversely with the power setting in watts (W). The pie takes 6 minutes to heat at a power setting of 650 W.

How long would the pie take to heat at a power setting of 800 W? 3 marks

Question 8 (3 marks)

A local council is planning to build a rectangular dog park that has 4 hectares of space. The park will need to be fenced. The graph shows the possible dimensions of the park.

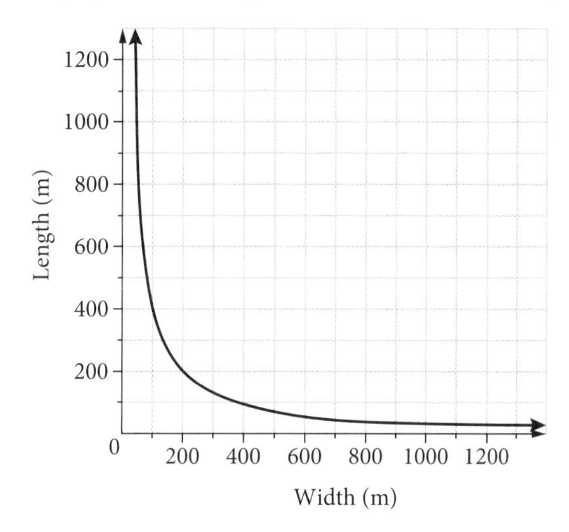

a How many square metres are in 4 hectares? 1 mark

b Is the curve an exponential curve, a parabola or a hyperbola? 1 mark

c If the council decides to have a park with a width of 100 m, what will be its length? 1 mark

Question 9 (3 marks)

A small rubber ball is dropped from a balcony. Its height, h metres above the ground, after each bounce is $h = 8(0.6)^b$, where b is number of bounces.

a What is the height from which the ball is first dropped? 1 mark

b Calculate, to the nearest centimetre, the height of the ball above the ground after 5 bounces. 1 mark

c What are the limitations of this exponential model for large values of b? 1 mark

Question 10 (3 marks)

Solve $\frac{5x + 2}{3} - 7 = 1 - x$. 3 marks

PRACTICE SET 2

Question 11 (2 marks)

The amount of Japanese yen, Y, varies directly with the Australian dollar, A, in dollars.

If A\$250 = ¥20 000, calculate the amount of yen for A\$1730. 2 marks

Question 12 (5 marks)

Rowan is comparing 2 student gym memberships for a 12-month contract. The cost of Gym Now, N dollars, includes a membership fee of \$30 per month, m, plus a \$99 joining fee. The cost of Fit Gym, F dollars, is \$40 a month with no joining fee. The 2 costs are represented by these formulas.

$$N = 30m + 99$$
$$F = 40m$$

a Graph both formulas on the same number plane. 2 marks

b After how many months are the costs the same for both gyms? 1 mark

c Which gym should Rowan sign up to? Explain your answer based on the formulas. 2 marks

Question 13 (3 marks)

The graph of the quadratic function $y = -x^2 + 6x - 5$ is shown.

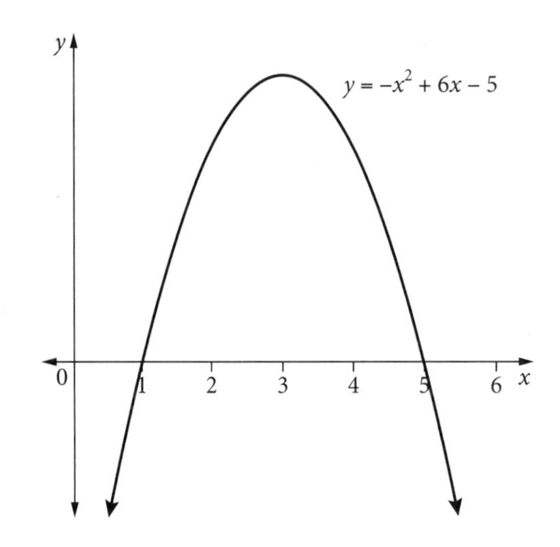

a What is the maximum value of the quadratic function? 2 marks

b What is the y-intercept of the graph? 1 mark

Question 14 (6 marks)

Ling and Andrew are planning a school social event to fundraise money for their school formal. They can use the school hall for free but have to hire the DJ for \$250 and photobooth for \$550. Refreshments per person cost \$10.

a Write an equation for the cost C, in dollars, of running the school social event for P people. 1 mark

b This graph shows the planned income, in dollars, when the ticket price is \$20 per person.

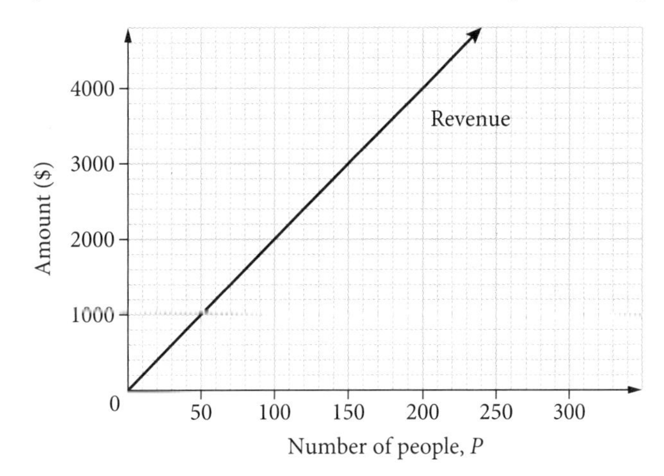

Draw the graph of the cost equation on the same axes. 2 marks

c How many people need to attend the event before it breaks even? 1 mark

d How much profit is made if 100 people attend the event? 1 mark

e Ling and Andrew wish to make a profit of \$1400. How many people need to attend the event for zthem to achieve their target? 1 mark

Question 15 (7 marks)

The formula $S = \frac{270}{t}$ can be used to calculate the average speed necessary to travel 270 km in a particular time, where S is speed, in km/h, and t is time, in hours.

a Use the formula to complete this table of values. 2 marks

t	1	2	3	4	5
S		135			

b Graph the formula on the axes. 2 marks

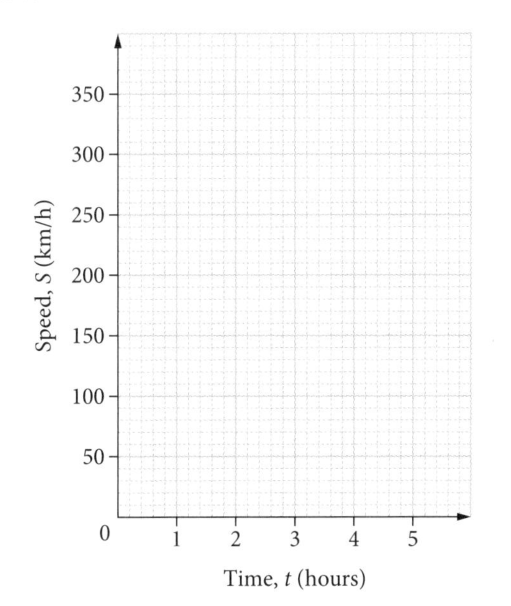

c What is the average speed required to complete the 270 km trip in 6 hours? 1 mark

d How long would the trip take at a speed of 80 km/h? 1 mark
Answer in hours and minutes, to the nearest minute.

e What is a limitation of this model? 1 mark

Question 16 (3 marks) ©NESA 2019 HSC EXAM, QUESTION 31

A rectangle has width w centimetres. The area of the rectangle, A, in square centimetres, is $A = 2w^2 + 5w$.

The graph of $A = 2w^2 + 5w$ is shown.

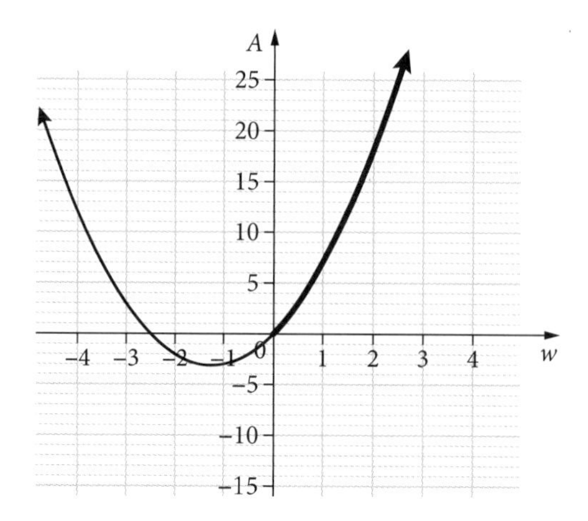

a Explain why, in this context, the model $A = 2w^2 + 5w$ only makes sense for the bold section of the graph. 1 mark

b The area of the rectangle is 18 cm^2. Calculate the perimeter of the rectangle. 2 marks

Question 17 (4 marks)

In 2020, the population of Australia was 25.39 million people.

a The growth rate is estimated to be 1.7% per annum. If y represents the estimated population of Australia at a time x years after 2020, write a formula for the population in the exponential form $y = k(a^x)$. Use appropriate values of k and a in your formula. 2 marks

b Using your formula, or otherwise, estimate the size of Australia's population in 2025. Express your answer to the nearest thousand. 2 marks

Question 18 (7 marks)

A farmer needs to fence part of her land for her lambs to graze. The grazing space of land will be fenced around 3 sides adjacent to her dam and she will be using 180 metres of fencing.

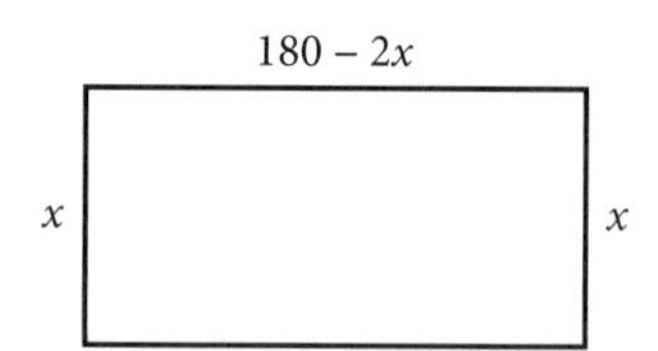

a Explain why the area of the land can be calculated using the formula $A = 180x - 2x^2$. 2 marks

b Use the formula $A = 180x - 2x^2$ to complete the table of values. 2 marks

x	0	15	30	45	60	75	90
A	0		3600		3600	2250	

9780170459204

c Use the values in the table to graph the formula on the number plane.  1 mark

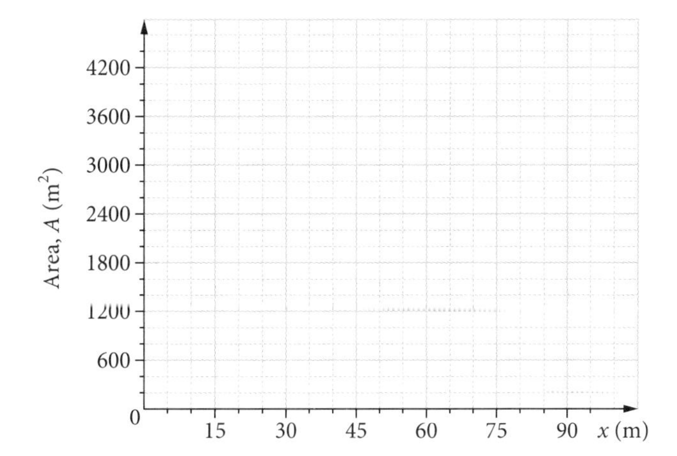

d What is the maximum area of the paddock and what are its dimensions? 2 marks

Question 19 (4 marks) ©NESA 2009 HSC EXAM, QUESTION 24(d) 

A factory makes boots and sandals. In any week:

- the total number of pairs of boots and sandals that are made is 200
- the maximum number of pairs of boots made is 120
- the maximum number of pairs of sandals made is 150.

The factory manager has drawn a graph to show the numbers of pairs of boots (x) and sandals (y) that can be made.

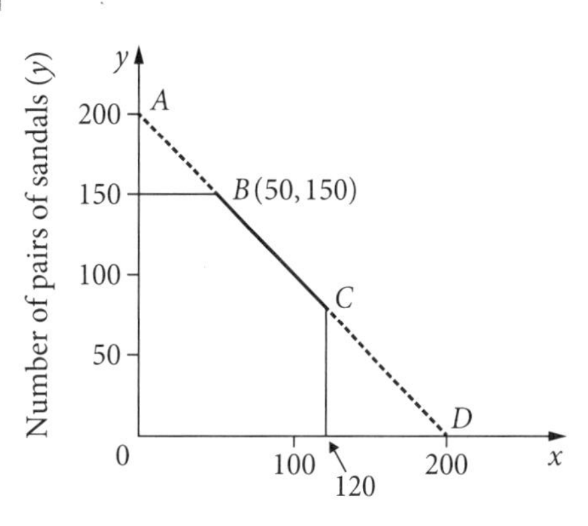

a Find the equation of the line AD. 1 mark

b Explain why the line is only relevant between B and C for this factory. 1 mark

c The profit per week, $\$P$, can be found by using this equation

$$P = 24x + 15y.$$

Compare the profits at B and C. 2 marks

Question 20 (8 marks) ©NESA 2006 HSC EXAM, QUESTION 28(b)

A new tunnel is built. When there is no toll to use the tunnel, 6000 vehicles use it each day. For each dollar increase in the toll, 500 fewer vehicles use the tunnel.

a Find the lowest toll for which no vehicles will use the tunnel. 1 mark

b For a toll of \$5.00, how many vehicles use the tunnel each day and what is the total daily income from tolls? 2 marks

c If d (dollars) represents the value of the toll, find an equation for the number of vehicles (v) using the tunnel each day in terms of d. 2 marks

d Anne says, 'A higher toll always means a higher total daily income'. 3 marks

Show that Anne is incorrect and find the maximum daily income from tolls. (Use a table of values, or a graph, or suitable calculations.)

Practice set 1

Worked solutions

1 C

$$\begin{aligned} C &= 1.6d + 5 \\ &= 1.6 \times 25 + 5 \\ &= 45 \end{aligned}$$

2 D

$$\begin{aligned} \frac{x-6}{7} &= 2 \\ x - 6 &= 14 \\ x &= 14 + 6 \\ &= 20 \end{aligned}$$

3 B

$$\begin{aligned} m &= \frac{2-(-1)}{0-1} \\ &= -3 \end{aligned}$$

$c = 2$

$y = -3x + 2$ or $y = 2 - 3x$

4 D

Point of intersection is $(-1, 2)$.

5 B

$$\begin{aligned} x^2 &= 3.4 \\ x &= \sqrt{3.4} \\ &= \pm 1.8439\ldots \\ &\approx \pm 1.8 \end{aligned}$$

6 D

The line is 'downhill', therefore the gradient is negative.

The y-intercept crosses the y-axis above the origin, therefore it is positive.

7 C

$y = -x^2 + 1$

The x^2 in the equation makes it a parabola.

8 D

$y = 0.8^x$ is a decreasing exponential function of the form $y = a^x$, where a is between 0 and 1.

9 B

$$\begin{aligned} \text{BAC} &= \frac{10N - 7.5H}{6.8M} \\ 0.03 &= \frac{10N - 7.5 \times 3.5}{6.8 \times 68} \\ 0.03 &= \frac{10N - 26.25}{462.4} \\ 13.872 &= 10N - 26.25 \\ 40.122 &= 10N \\ N &= 4.0122 \end{aligned}$$

$N \approx 4$ standard drinks

10 C

x	0	1	2	3	4
y	0	1	4	9	16

Only $y = x^2$ satisfies each point:

$0^2 = 0,\ 1^2 = 1,\ 2^2 = 4,\ 3^2 = 9,\ 4^2 = 16$

11 B

In line B, 5 should have been subtracted rather than added to each side. '112' should be '102'.

12 C

$75 \times 7 = 525$

13 C

Choosing $(1, 12)$ and $(2, 16)$:

$$\text{Gradient} = \frac{\text{rise}}{\text{run}} = \frac{16-12}{2-1} = \frac{4}{1} = 4$$

So, $C = 4h + c$

Substitute $(1, 12)$ to find c:

$$\begin{aligned} 12 &= 4 \times 1 + c \\ 12 &= 4 + c \\ c &= 8 \end{aligned}$$

So, $C = 4h + 8$

14 A

'Time varies inversely with speed' can be expressed as $T = \frac{k}{S}$, which means as speed increases, time decreases.

The answer cannot be C or D as the graphs are linear.

As S increases, T decreases, therefore the answer cannot be B.

15 D

Lines $y = 2x - 3$ and $y = x + 5$ both have positive gradients (lines going 'uphill'), and only graphs A and D have these.

The line of $y = 2x - 3$ has a negative y-intercept and only graph D has this.

16 B

$$y = \frac{k}{x}$$

$$10 = \frac{k}{28}$$

$$k = 280$$

$$m = \frac{280}{T}$$

17 C

$S = 35\,000(0.88)^n$

$S = V_0(1 - r)^n$, where $1 - r = 0.88 = 88\%$

Hence, depreciation rate, r, is $100\% - 88\% = 12\%$ per annum.

18 C

When $x = 500$, $I = 100 \times 500 = \$50\,000$.

Break-even point is $x = 200$ (from graph).

When $x = 200$, $I = 100 \times 200 = \$20\,000$.

Difference: $50\,000 - 20\,000 = \$30\,000$

19 B

Jade: Each day, followers are 'up' by 7:

$m = 7$

Initially ($d = 0$), there are 705 followers:

$c = 705$

Lisa: Each day, followers are 'down' by 2:

$m = -2$

Initially ($d = 0$), there are 473 followers:

$c = 473$

$f = 7d + 705$
$f = -2d + 473$ or $f = 473 - 2d$

20 D

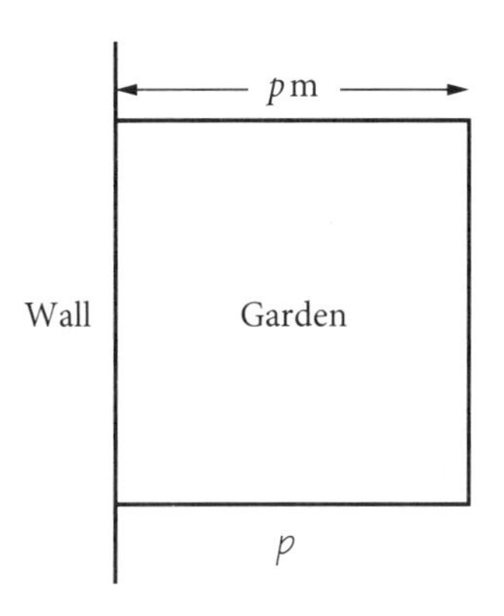

$p + p + \text{3rd side} = 17$

$2p + \text{3rd side} = 17$

$\text{3rd side} = 17 - 2p$

$A = p \times (17 - 2p)$
$= 17p - 2p^2$

WORKED SOLUTIONS

Practice set 2

Worked solutions

Question 1

a C. The graph of $y = x^2 - 1$ is a parabola, concave up with y-intercept at $(0, -1)$ or $y = -1$.

b D. The graph of $y = \frac{1}{2}x + 1$ is a straight line with gradient $\frac{1}{2}$ and y-intercept at $(0, 1)$ or $y = 1$.

c B. The graph of $y = 2^x$ is an increasing exponential curve with y-intercept at $(0, 1)$ or $y = 1$.

d A. The graph of $y = \frac{2}{x}$ is a hyperbola in the 1st and 3rd quadrants.

Question 2

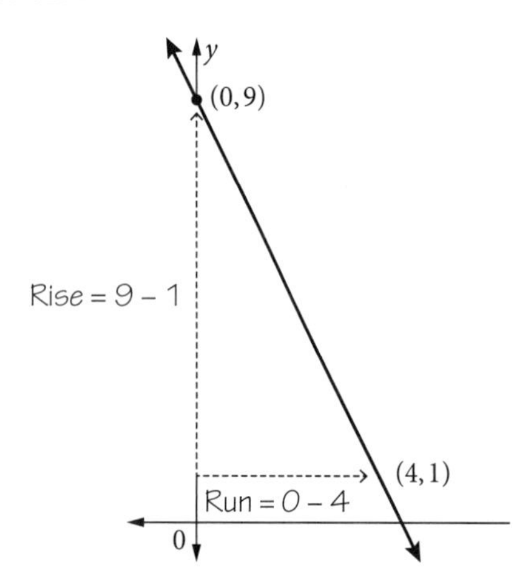

$m = \frac{9-1}{0-4} = \frac{8}{-1} = -2$

$c = 9$

$y = -2x + 9$

Question 3

$$\begin{aligned} C &= 20 + 15(h - 1) \\ 95 &= 20 + 15(h - 1) \\ 95 - 20 &= 15h - 15 \\ 75 &= 15h - 15 \\ 75 + 15 &= 15h \\ \frac{90}{15} &= h \\ h &= 6 \end{aligned}$$

Louise has worked for 6 hours.

Question 4

a

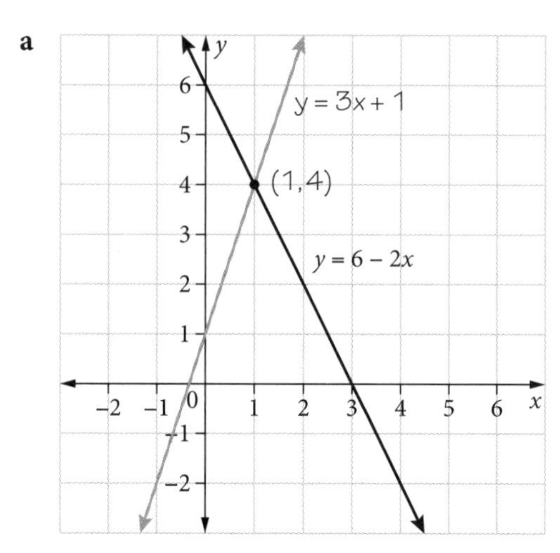

b $(1, 4)$

c $x = 1, y = 4$

d The solutions of the simultaneous equations give the coordinates of the point of intersection of the graphs of the simultaneous equations.

Question 5

a $m = \frac{8.4 - 16.8}{6 - 18}$

$= 0.7$

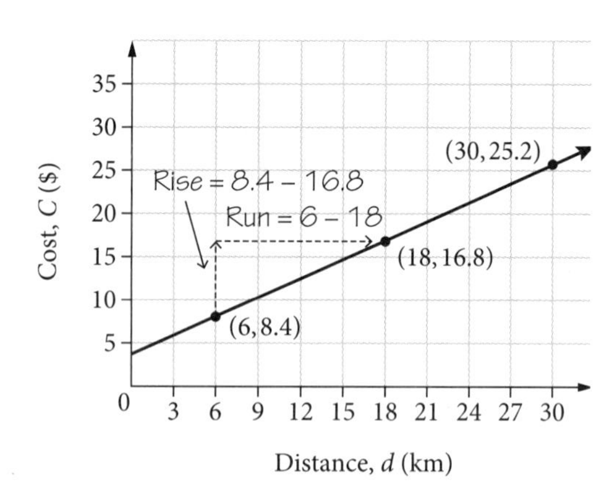

b The gradient of 0.7 is the truck's running cost per kilometre, that is, \$0.70/km or 70 cents/km.

c We know $m = 0.7$. To find the y-intercept, c, accurately, substitute one of the points from the line, such as (6, 8.4), into $C = 0.7d + c$.

$$8.4 = 0.7 \times 6 + c$$
$$8.4 = 4.2 + c$$
$$8.4 - 4.2 = c$$
$$c = 4.2$$

So, $C = 0.7d + 4.2$

d Substitute $d = 50$:

$$C = 0.7 \times 50 + 4.2$$
$$= \$39.20$$

Question 6

a $C = 45t + 95$

b 2 hours and 30 minutes
= 5 lots of 30-minute intervals

When $t = 5$:

$$C = 45 \times 5 + 95$$
$$= \$320$$

c When $C = 410$:

$$410 = 45t + 95$$
$$315 = 45t$$
$$t = 7$$

7 × 30 minutes = 3 hours and 30 minutes

Question 7

For the inverse equation, $t = \frac{k}{P}$, where t = time and P = power.

$$6 = \frac{k}{650}$$
$$6 \times 650 = k$$
$$k = 3900$$

$$t = \frac{3900}{800}$$
= 4.875 minutes (4 minutes and 52.5 seconds)

Question 8

a 1 ha = 10 000 m^2
4 × 10 000 = 40 000 m^2

b hyperbola

c The length of the paddock is 400 m.

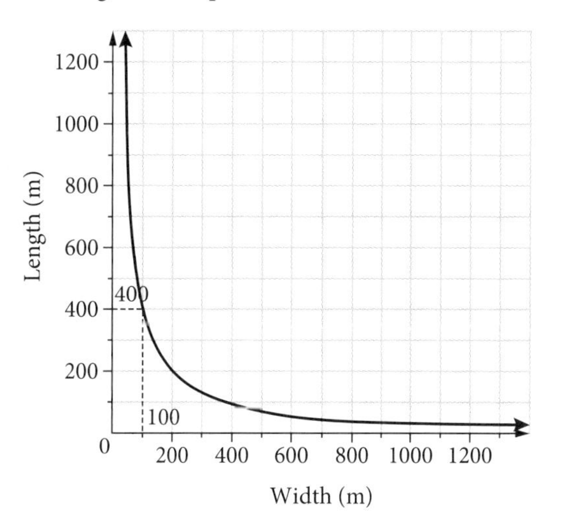

Question 9

$h = 8(0.6)^b$

a At the start, when $b = 0$, $h = 8$ m.

b $h = 8(0.6)^5$
$= 0.622\,08$
≈ 0.62 m
≈ 62 cm

c The exponential model assumes that the ball never reaches a height of 0 m, that is, it never stops bouncing.

Question 10

$$\frac{5x+2}{3} - 7 = 1 - x$$
$$\frac{5x+2}{3} = 8 - x$$
$$5x + 2 = 3(8 - x)$$
$$5x + 2 = 24 - 3x$$
$$8x = 22$$
$$x = \frac{22}{8}$$
$$= 2\frac{3}{4} \text{ or } 2.75$$

Question 11

$Y = kA$

When $A = 250$, $Y = 20\,000$

$$20\,000 = k \times 250$$
$$k = 80$$

$Y = 80A$

When $A = 1730$, $Y = 80 \times 1730$

$Y = 138\,400$

Amount is ¥138 400

Question 12

a

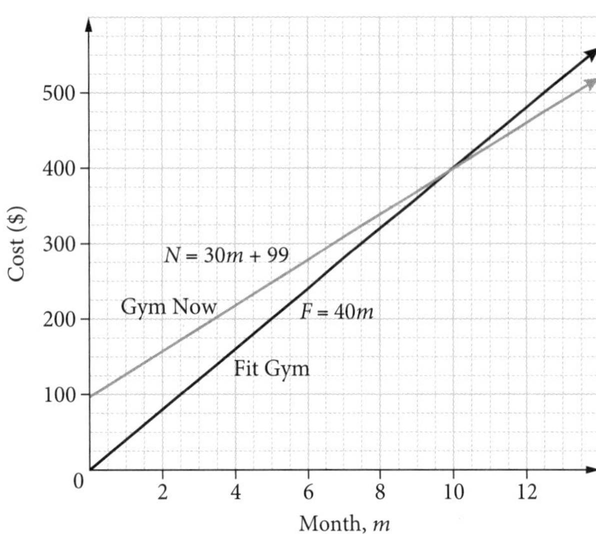

b 10th month (point of intersection)

c Rowan should sign up to Gym Now because it is cheaper over 12 months.

Question 13

a Maximum height is at $x = 3$

$$y = -3^2 + 6(3) - 5$$
$$= 4$$

b y-intercept occurs when $x = 0$

$$y = -0^2 + 6(0) - 5$$
$$= -5$$

Question 14

a C (y-intercept, c) $= 550 + 250$
$= 800$

Cost per person (gradient, m) $= 10$

Equation is $C = 10P + 800$.

b

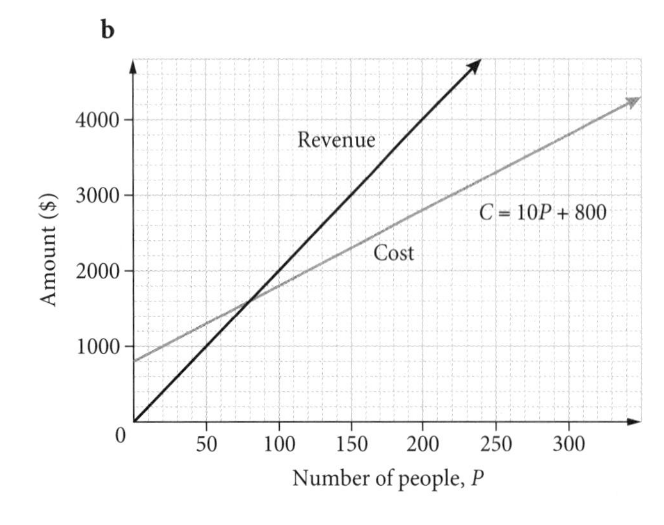

c Point of intersection of the 2 lines is (80, 1600).

80 people are needed to break even.

d When $P = 100$, revenue = \$2000, cost = \$1800, so profit = \$2000 – \$1800 = \$200.

e Need to find the value of P where the difference between revenue and cost is \$1400.

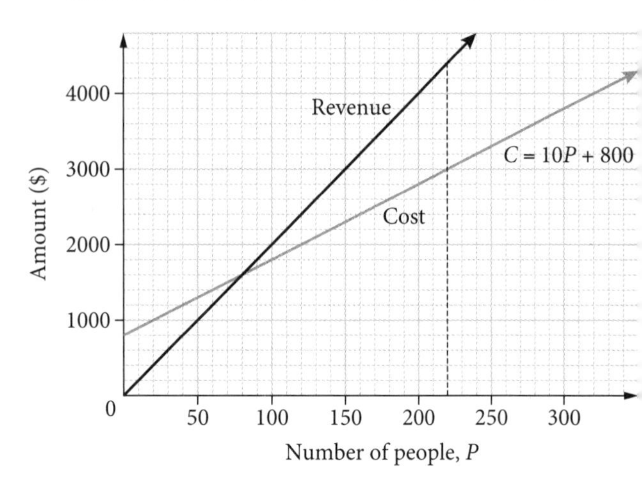

This occurs for 220 people, where profit = \$4400 – \$3000 = \$1400.

Question 15

a

t	1	2	3	4	5
S	270	135	90	67.5	54

b

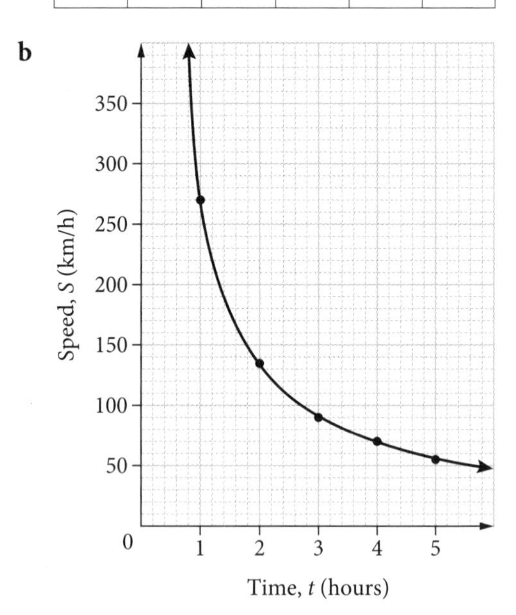

c $S = \dfrac{270}{t}$

When $t = 6$:

$$S = \frac{270}{6}$$
$$= 45\text{ km/h}$$

d $S = \frac{270}{t}$

When $S = 80$:

$80 = \frac{270}{t}$

$t = \frac{270}{80}$

$S = 3.375$
$= 3\text{ h } 22.5\text{ min}$
$\approx 3\text{ h } 23\text{ min}$

e It cannot take 0 hours to travel 300 km, nor can it take an infinite amount of time to travel 300 km.

Question 16

a The length, width and area of a shape cannot be negative, so the graph only exists in the 1st quadrant, where $w \geq 0$ and $A \geq 0$.

b From the graph, when $A = 18$, $w = 2$.

Area of rectangle = length × width

Then length is $18 \div 2 = 9$ cm

Perimeter is $2l + 2w = 2(9) + 2(2) = 22$ cm

Question 17

a In 2020, the population of Australia was 25.39 million = 25 390 000 people.

The growth rate is an increase of 100% + 1.7% = 1.017 (as a decimal).

Formula: $y = 25\,390\,000(1.017^x)$

b $y = 25\,390\,000(1.017^5)$
$= 27\,622\,785.15$
$\approx 27\,623\,000$

Question 18

a The formula for the area of a rectangle is $A = l \times w$.

In this case, $l = 180 - 2x$ and $w = x$.

Therefore, $A = (180 - 2x) \times x$
$= 180x - 2x^2$

b

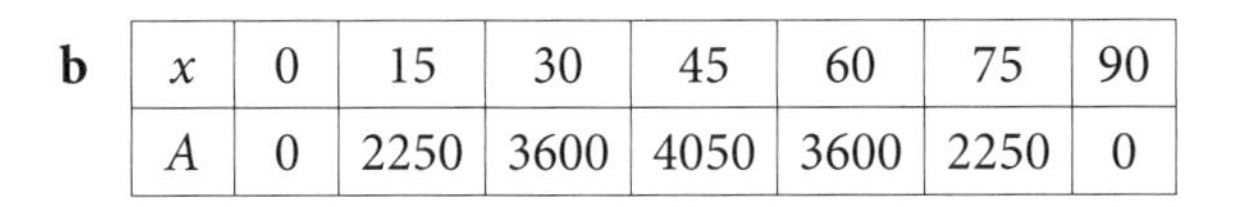

x	0	15	30	45	60	75	90
A	0	2250	3600	4050	3600	2250	0

c

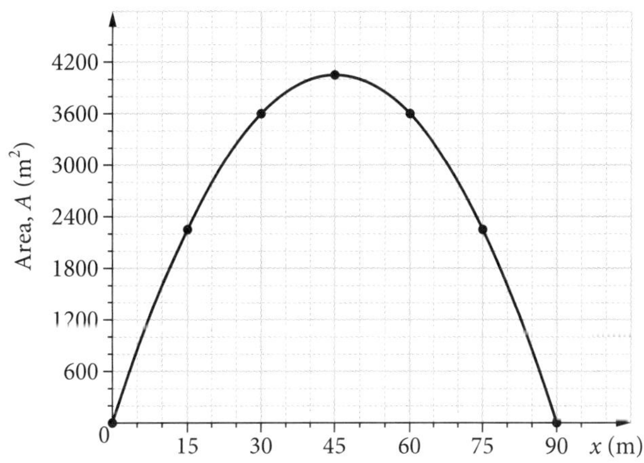

d Maximum y-value on the graph is 4050.

Therefore, the maximum area of the paddock is 4050 m^2 with the width 45 m.

Length is 180 m

$180 - 2x = 180 - 2(45)$
$= 90$

or $4050 \div 45 = 90$.

So, length is 90 m, width is 45 m and area is 4050 m^2.

Question 19

a $c = 200$ (from graph)

Line passes through points (0, 200) and (50, 150)

$m = \frac{200 - 150}{0 - 50} = -1$

$y = -x + 200$

b Maximum amount of boots is 120, so x cannot be greater than 120.

Maximum amount of sandals is 150, so y cannot be greater than 150.

Hence, the line only exists between B and C.

c For point B (50, 150), substitute $x = 50$, $y = 150$.

$P = 24 \times 50 + 15 \times 150$
$= \$3450$

For point C (120, 80), substitute $x = 120$, $y = 80$.

$P = 24 \times 120 + 15 \times 80$
$= \$4080$

So it is more profitable to make 120 pairs of boots and 80 sandals.

Question 20

a $6000 \div 500 = 12$

$12 toll is the lowest for which no vehicles will use the tunnel.

b $\$5 \times 500 = 2500$ fewer vehicles

Vehicles using tunnel: $6000 - 2500 = 3500$

Daily toll income: $\$5 \times 3500 = \$17\,500$

c 6000 vehicles initially (when $d = 0$)

$c = 6000$

For every dollar (x run) increase, 500 fewer vehicles (y rise).

$$m = -\frac{500}{1}$$
$$= -500$$

$v = -500d + 6000$

d Income = number of vehicles × toll

$$I = (-500d + 6000) \times d$$
$$= -500d^2 + 6000d$$
$$= 500d(12 - d)$$

Concave down parabola with intercepts 0 and 12

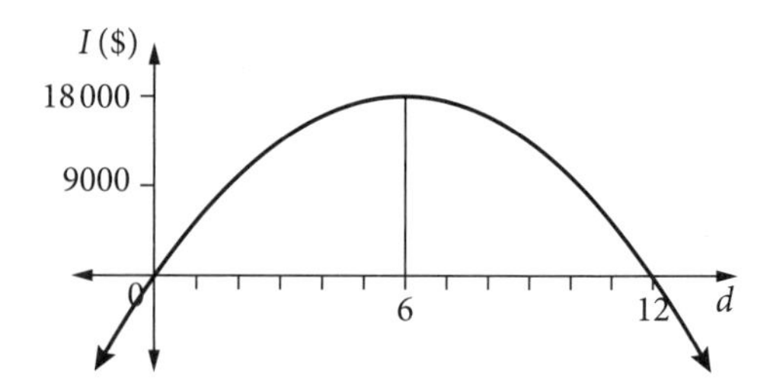

From the graph, the maximum income is when the toll is $6. Anne is incorrect.

HSC exam topic grid (2011–2020)

This grid shows the coverage of this topic in past HSC exams by question number. The past exams can be downloaded from the NESA website (www.educationstandards.nsw.edu.au) by selecting 'Year 11 – Year 12', 'HSC exam papers'. NESA marking feedback and guidelines can also be found there.

Before 2019, 'Mathematics Standard 2' was called 'Mathematics General 2' and, before 2014, 'General Mathematics'. For these exam papers, select 'Year 11 – Year 12', 'Resources archive', 'HSC exam papers archive'.

	Algebra and equations (Year 11)	Linear functions	Non-linear functions	Direct and inverse variation
2011	12, 18	**20***, 23(b), 28(b)	6#, 26(b)#	28(a)
2012	14, 21, 28(b)	5, 8, 13	16, 30(b)–(c)#	15
2013	5, 21, 29(a)	28(b)	22, 30(a)	
2014	4, 11, 26(a), 26(c), 29(b)	7, 26(d)*†, 29(a)	3	26(f)
2015	2, 23, 24, 26(b), 28(d)	13, 28(f)*	10#, 29(e)	
2016	2, 5, 24, 26(b)	4*, 14, 29(e)	29(b)#	
2017	7, 9, 19, 20, 28(a)(i), 28(d), 30(d)(i)	17*†	28(e)	
2018	16, 25, 26(b), 28(b), 28(e)	27(d)	4	29(c)
2019 new course	11, 28	14, 23(c), 36*	**31**	33, 34
2020	13	6, 10, 24*	**1**, 19, 33#	

Questions in **bold** can be found in this chapter.
* Simultaneous equations.
† Can be solved graphically.
\# Exponential function.

CHAPTER 2 TRIGONOMETRY

MS-M6 Non-right-angled trigonometry 44

TRIGONOMETRY

Except for compass radial surveys, all content for this topic is common with the Mathematics Advanced course.

Pythagoras' theorem

$c^2 = a^2 + b^2$

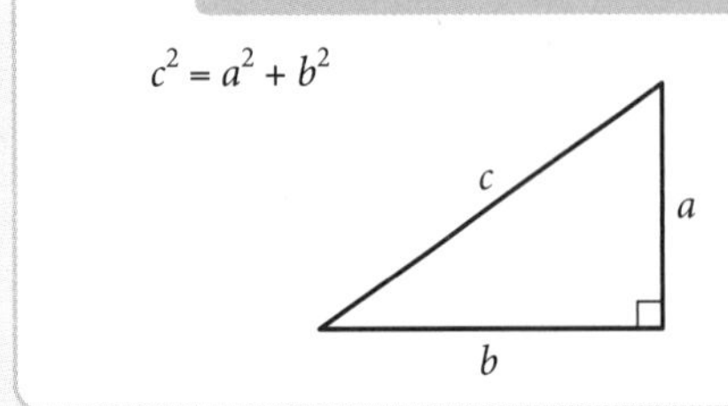

Right-angled trigonometry

$\sin\theta = \dfrac{\text{opposite}}{\text{hypotenuse}}$ (SOH)

$\cos\theta = \dfrac{\text{adjacent}}{\text{hypotenuse}}$ (CAH)

$\tan\theta = \dfrac{\text{opposite}}{\text{adjacent}}$ (TOA)

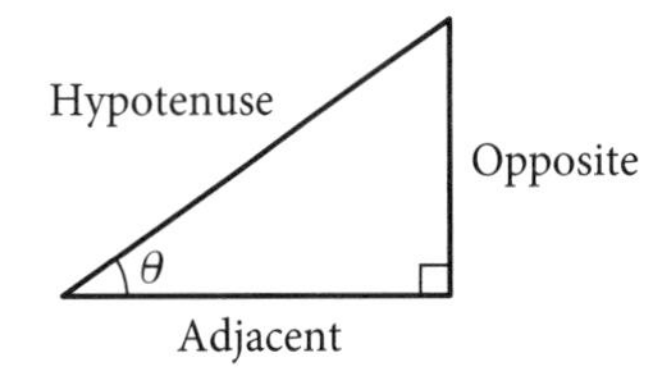

Angles of elevation and depression

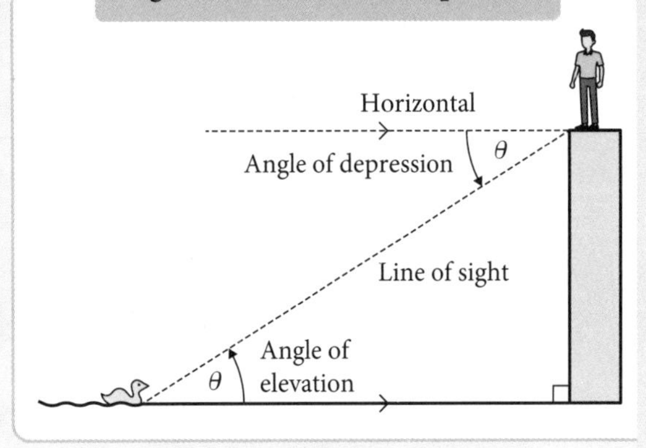

Compass radial surveys

A compass radial survey shows a field with a compass in the middle and the true bearings and distances to each corner.

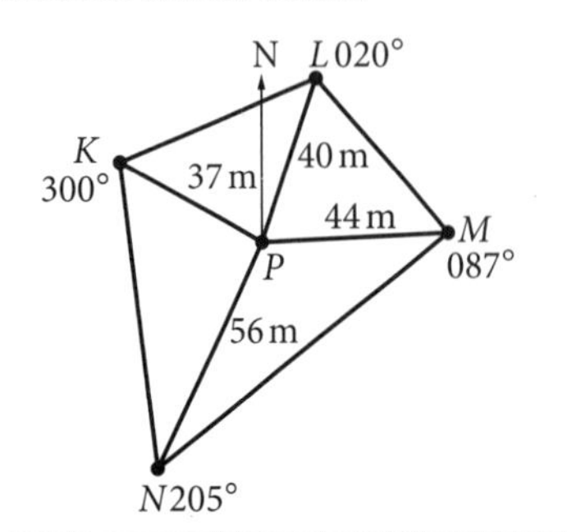

The sine and cosine rules

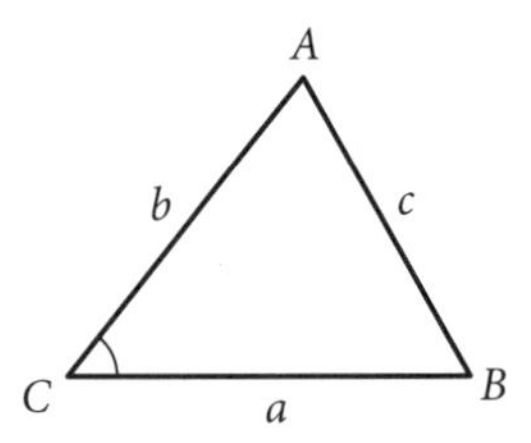

- **Sine rule**

  $\dfrac{a}{\sin A} = \dfrac{b}{\sin B} = \dfrac{c}{\sin C}$

- **Cosine rule**

  $c^2 = a^2 + b^2 - 2ab\cos C$ (to find a side)

  $\cos C = \dfrac{a^2 + b^2 - c^2}{2ab}$ (to find an angle)

Area of a triangle

$A = \dfrac{1}{2}ab\sin C$

Bearings

- **Compass bearings:** Acute angle measured from north or south towards east or west. For example, C is N 66° W from O (90° – 24° = 66°).

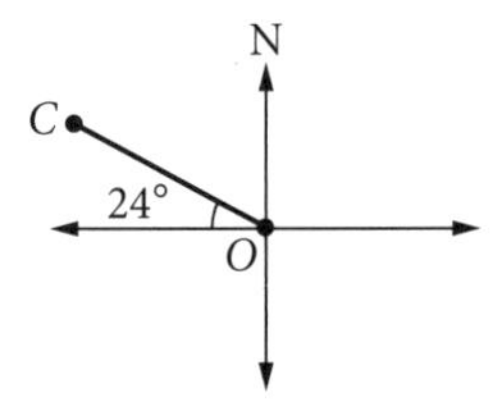

- **True bearings:** 3-digit angles from 000° to 360°, clockwise from north. For example, in the previous diagram, C is 294° from O (270° + 24° = 294°).

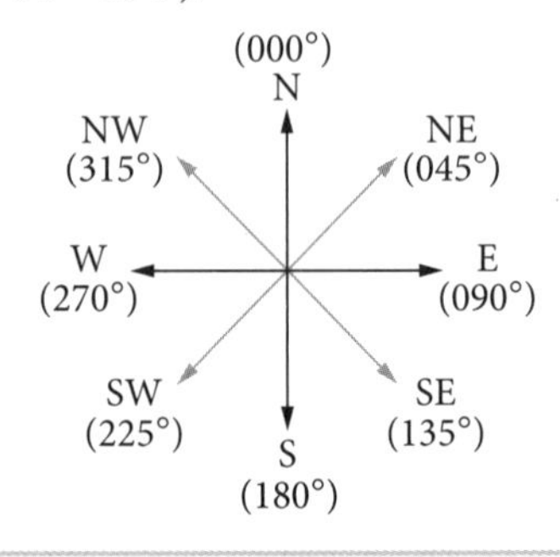

Except for Pythagoras' theorem, all formulas for this topic appear on the HSC exam reference sheet (and at the back of this book).

Glossary

A+ DIGITAL FLASHCARDS
Revise this topic's key terms and concepts by scanning the QR code or typing the URL into your browser.

https://get.ga/a-hsc-maths-standard-2

acute angle
An angle between 0° and 90°.

adjacent (side)
The side next to the given angle in a right-angled triangle. It is also the side that joins the given angle and the right angle.

angle of depression
The angle looking down at an object from the horizontal.

angle of elevation
The angle looking up at an object from the horizontal.

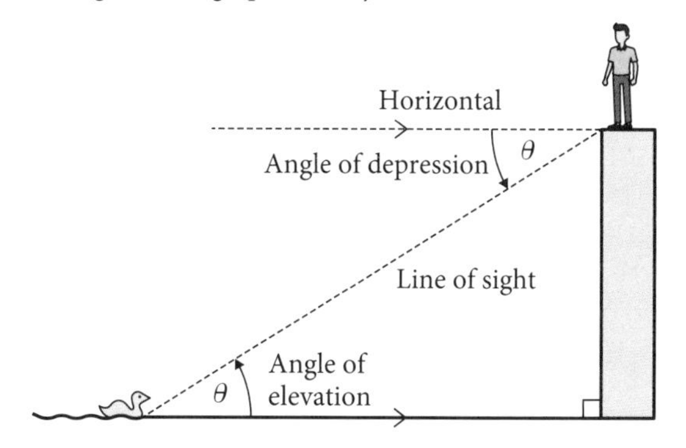

bearing
An angle that represents the direction from a given point to a certain location.

compass bearing
A bearing measured using acute angles and compass directions. It is measured from north or south towards east or west, for example, N 35° E.

compass radial survey
A survey of an irregular block of land where a compass is placed at its centre and the distance and bearing to each corner of the block is measured. It is used in surveying to find measurements such as perimeter and area of land.

cosine
A trigonometric ratio in a right-angled triangle.

$$\cos\theta = \frac{\text{adjacent}}{\text{hypotenuse}}$$

cosine rule
A trigonometric rule for triangle problems involving 3 sides and 1 angle.

$$c^2 = a^2 + b^2 - 2ab\cos C$$

$$\cos C = \frac{a^2 + b^2 - c^2}{2ab}$$

hypotenuse
The side opposite the right angle in a right-angled triangle; the longest side of a right-angled triangle.

minute (symbol ′)
A measure of angle size, $\frac{1}{60}$ of a degree. $1° = 60'$.

obtuse angle
An angle between 90° and 180°.

right-angled triangle
A triangle that has a 90° angle.

sine
A trigonometric ratio in a right-angled triangle.

$$\sin\theta = \frac{\text{opposite}}{\text{hypotenuse}}$$

sine rule
A trigonometric rule for triangle problems involving 2 sides and the 2 angles opposite them.

$$\frac{a}{\sin A} = \frac{b}{\sin B} = \frac{c}{\sin C}$$

tangent
A trigonometric ratio in a right-angled triangle.

$$\tan\theta = \frac{\text{opposite}}{\text{adjacent}}$$

true bearing
A three-digit angle from 000° and 360°, measured from north in a clockwise direction.

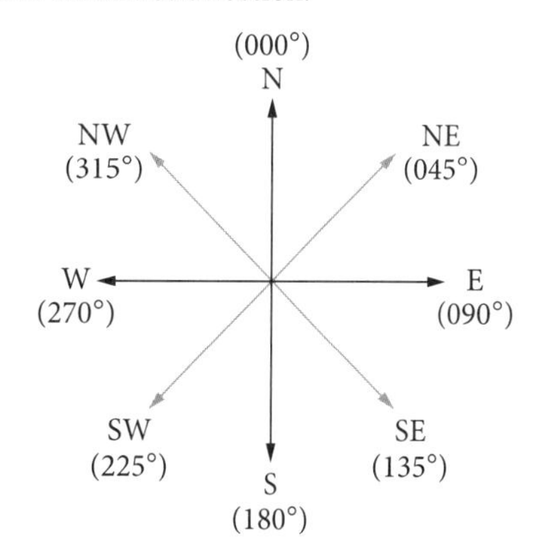

Topic summary

All content for this topic is common with the Mathematics Advanced course.

Review of right-angled trigonometry (Year 11)

Review of Pythagoras' theorem

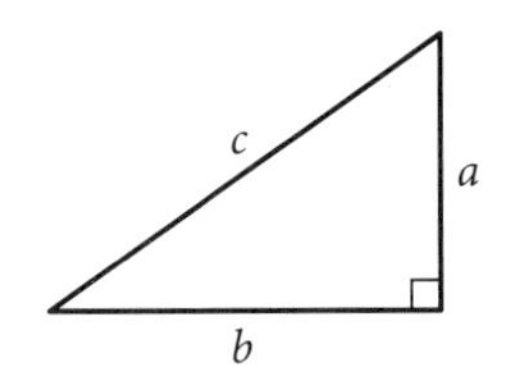

$$c^2 = a^2 + b^2$$

where c is the length of the **hypotenuse**, and a and b are the lengths of the shorter sides.

It does not matter which side you choose to be a or b.

Pythagoras' theorem is used to find a side length in a **right-angled** triangle when the other 2 sides are known. To find the length of a shorter side of the triangle, the formula may be rearranged to

$$a^2 = c^2 - b^2$$

Example 1

How high up the wall does the ladder reach? Give your answer to two decimal places.

Solution

$$\begin{aligned} h^2 &= 6^2 - 2^2 \\ &= 32 \\ h &= \sqrt{32} \\ &= 5.6568\ldots \\ &\approx 5.66\,\text{m} \end{aligned}$$

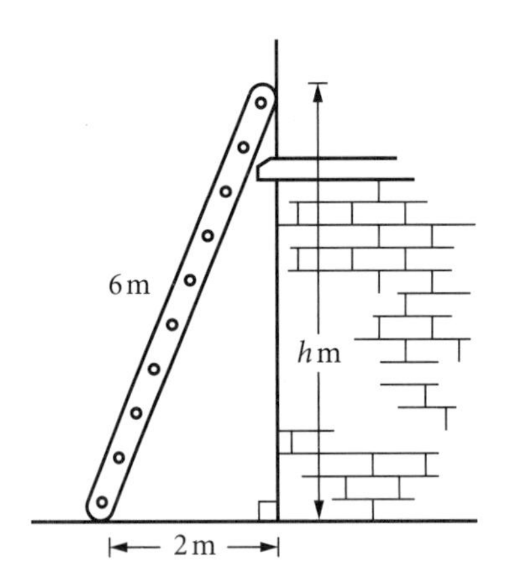

Review of right-angled trigonometry

Hint: These formulas are provided on the HSC exam reference sheet.

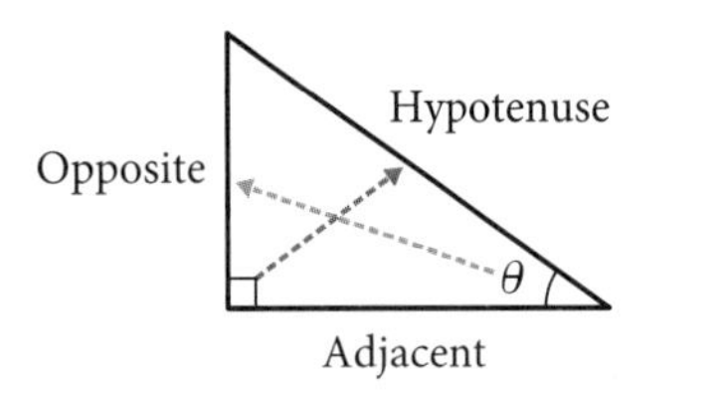

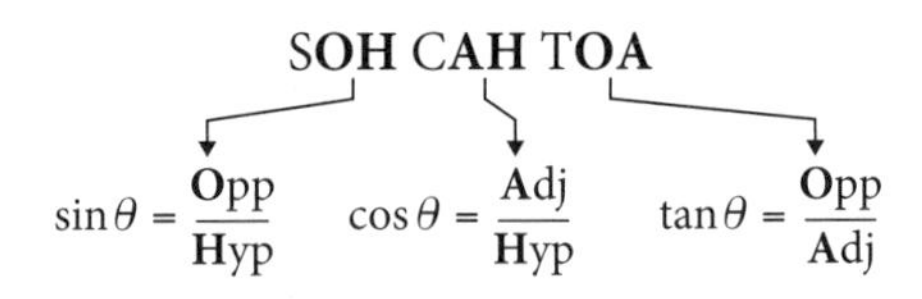

$$\sin\theta = \frac{\text{Opp}}{\text{Hyp}} \qquad \cos\theta = \frac{\text{Adj}}{\text{Hyp}} \qquad \tan\theta = \frac{\text{Opp}}{\text{Adj}}$$

When solving a trigonometry problem involving a right-angled triangle:

1. Label the sides opposite (O), **adjacent** (A) and hypotenuse (H).
2. Determine which trigonometric ratio to use: for example, if you have O and A you will use tan.
3. Substitute all known values into the formula.
4. Solve to find the unknown value.

Note: Make sure you check that your calculator is in DEGREE mode for trigonometry.

Example 2

Find the value of each variable.

a When finding an angle, press SHIFT on your calculator to 'undo the sin'.

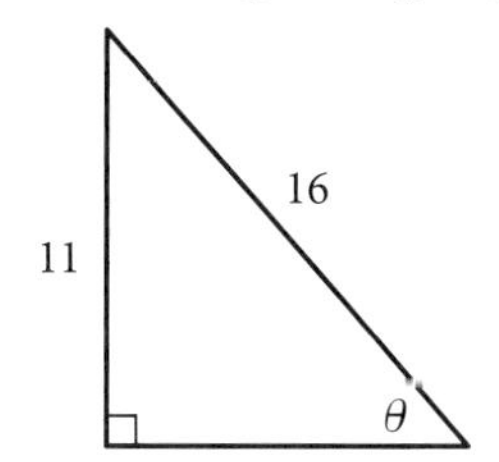

$$\sin \theta = \frac{11}{16}$$
$$\theta = \sin^{-1}\left(\frac{11}{16}\right)$$
$$\approx 43°$$

b To find an unknown side when the variable is on the top of the fraction, multiply both sides by the denominator.

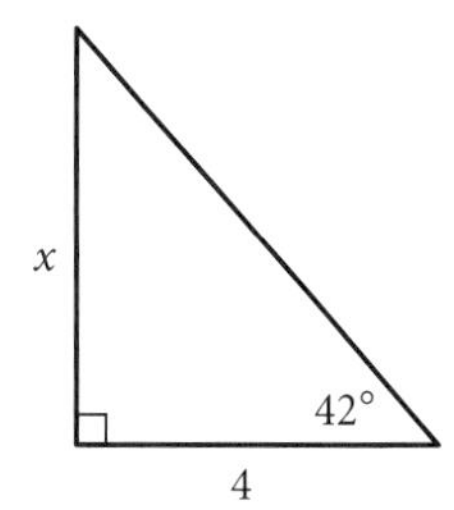

$$\tan 42° = \frac{x}{4}$$
$$x = 4 \times \tan 42°$$
$$\approx 3.6$$

c To find an unknown side when the variable is on the bottom of the fraction, 'BOTSWAP'.

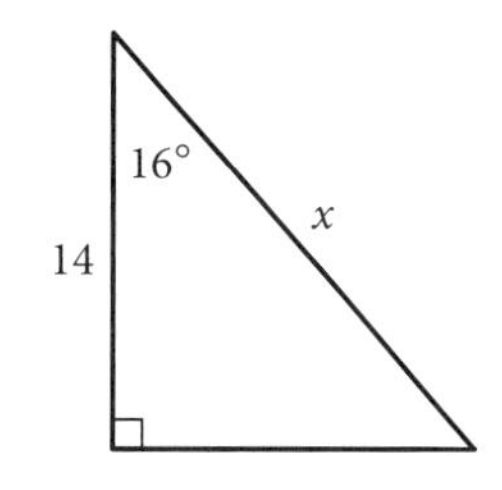

$$\cos 16° = \frac{14}{x}$$
$$x = \frac{14}{\cos 16°}$$
$$\approx 14.56$$

Hint
When the variable appears in the denominator, swap the positions of the variable and the sin, cos or tan value. Remember BOTSWAP for 'bottom swap'.

Review of angles of elevation and depression

Angle of elevation: The angle looking up to an object from the horizontal to your line of sight. 'Elevation' means 'up'.

Angle of depression: The angle looking down to an object from the horizontal to your line of sight. 'Depression' means 'down'.

The parallel line property of alternate angles means that:

angle of elevation = angle of depression

In the diagram, the angle of elevation from the duck to the person is equal to the angle of depression from the person down to the duck as the horizontal lines for both are parallel.

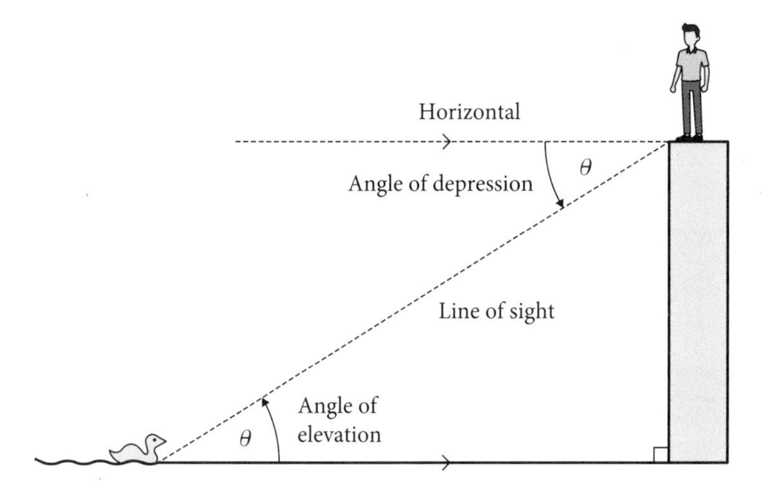

Hint
Most angle of elevation and depression problems use the tan ratio. Some problems are worded questions so *always* draw a diagram first.

Example 3

From the top of a building that is 90 m high, Milli sees a ferry on Sydney Harbour that is 400 m from the base of a building. What is the angle of depression, θ, of her line of sight? Answer correct to the nearest degree.

Solution

Let θ = angle of depression.

By alternate angles, it is also the angle of elevation of Milli from the ferry.

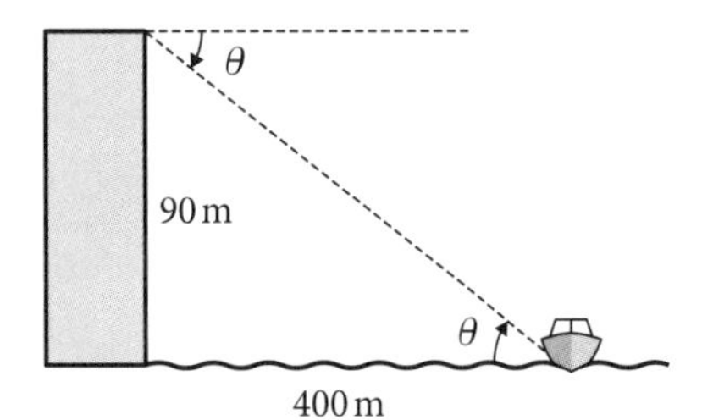

$$\tan\theta = \frac{90}{400}$$
$$\theta = 12.6803\ldots$$
$$\approx 13^\circ$$

Non-right-angled trigonometry (MS-M6)

To find the length of a side or the magnitude of an angle in non-right-angled triangles, use the sine or cosine rules. The sides and angles of a triangle are labelled so that sides a, b, c are *opposite* the angles A, B, C.

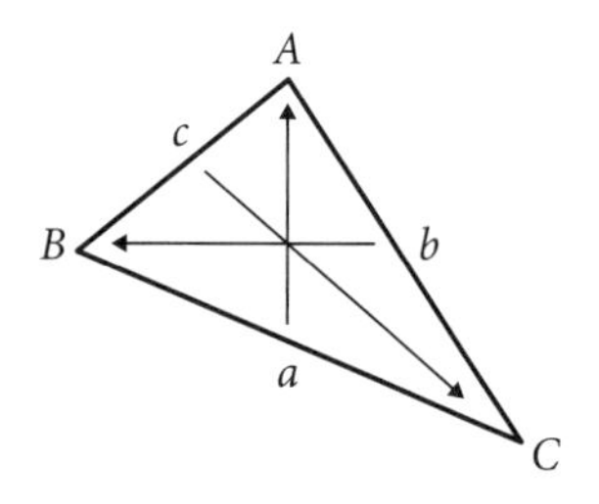

The sine rule

The **sine rule** is used when **2 sides and the 2 angles opposite** them are involved.

$$\frac{a}{\sin A} = \frac{b}{\sin B} = \frac{c}{\sin C}$$

Hint
This formula is provided on the HSC exam reference sheet.

Example 4

Find u, correct to three significant figures.

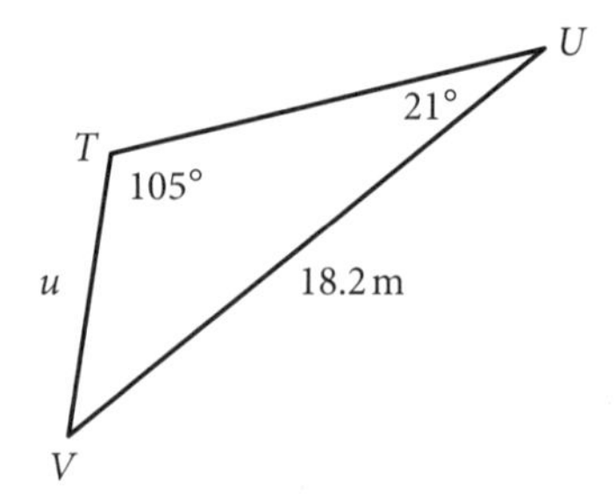

Solution

$$\frac{a}{\sin A} = \frac{b}{\sin B}$$
$$\frac{u}{\sin 21^\circ} = \frac{18.2}{\sin 105^\circ}$$
$$u = \frac{18.2 \times \sin 21^\circ}{\sin 105^\circ}$$
$$= 6.7523\ldots$$
$$\approx 6.75\text{ m}$$

The side labelled u m is opposite angle 21° and the side labelled 18.2 m is opposite angle 105°.

Hint
Although the sine rule formula shows 3 pairs of sides and angles, only 2 sides and 2 angles need to be involved to use it.

The cosine rule

The **cosine rule** is used when **3 sides and 1 angle** are involved.

To find the length of a side (c) when 2 sides and the included angle are known:

$$c^2 = a^2 + b^2 - 2ab\cos C$$

To find an angle (C) when all 3 sides are known:

$$\cos C = \frac{a^2 + b^2 - c^2}{2ab}$$

In the diagram, C is the **included angle** between sides a and b.

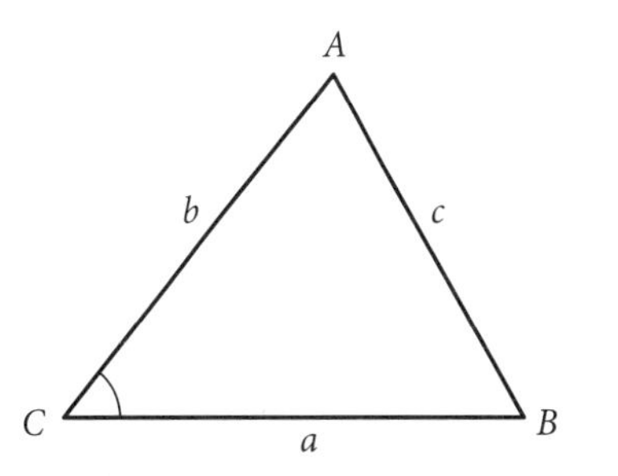

Hint
Both formulas are provided on the HSC exam reference sheet.

Example 5

Find d, correct to one decimal place.

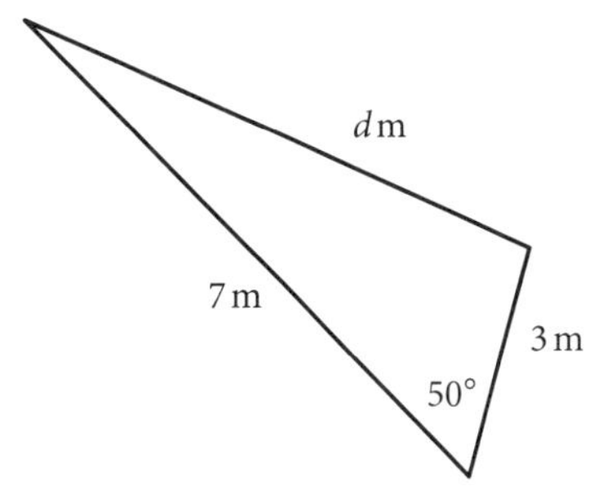

Solution

$c^2 = a^2 + b^2 - 2ab\cos C$, where unknown side c is opposite angle C

$$\begin{aligned} d^2 &= 7^2 + 3^2 - 2 \times 7 \times 3\cos 50^\circ \\ &= 31.0029\ldots \\ d &= \sqrt{31.0029\ldots} \\ &= 5.568\ldots \\ &\approx 5.6 \end{aligned}$$

Hint
Taking the square root will 'undo the square'.

So the side length, d, correct to one decimal place, is 5.6 m.

Area of a triangle sine formula

This formula is used to find the area of a triangle when **2 sides and the included angle** are known.

$$A = \frac{1}{2}ab\sin C$$

Hint

This formula is provided on the HSC exam reference sheet.

Example 6

A traffic island has sides of length 3.8 m, 4.0 m and 4.5 m.

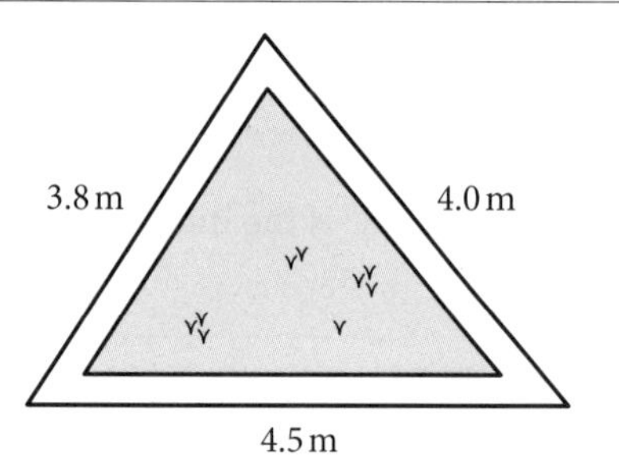

a Find the size of the smallest angle in the triangular island, correct to the nearest degree.

b Find the area of the island, correct to the nearest square metre.

Solution

a The smallest angle (C) is opposite the smallest side, 3.8 m.

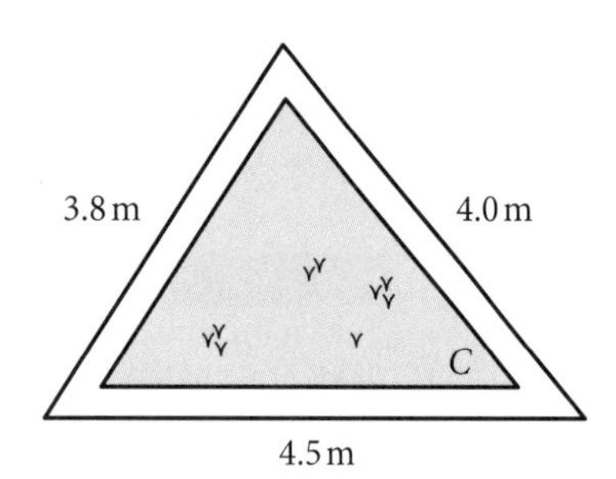

All 3 sides are given, so use the cosine rule:

$$\cos C = \frac{a^2 + b^2 - c^2}{2ab}$$

$$= \frac{4.5^2 + 4^2 - 3.8^2}{2 \times 4.5 \times 4}$$

$$C = 52.71\ldots$$

$$\approx 53°$$

b Using area formula:

$$A = \frac{1}{2}ab\sin C$$

$$= \frac{1}{2} \times 4.5 \times 4 \times \sin 52.71\ldots$$

$$= 7.1603\ldots$$

$$\approx 7\text{ m}^2$$

Hint

Remember to press SHIFT when finding an angle.

Trigonometry with obtuse angles

An **obtuse angle** is between 90° and 180°. It is possible to find the sine and cosine ratios of an obtuse angle. There is a special relationship between the sine and cosine ratios of supplementary angles such as 150° and 30°, that is, angles that add up to 180°.

Find the values of sin 150° and sin 30° on your calculator. Are they the same?

This can be expressed in the general form:

$$\sin \theta = \sin (180° - \theta)$$

sin (obtuse angle) = sin (supplementary **acute angle**)

Similarly, cos 150° is the same as −cos 30°. Notice that the cosine of an obtuse angle is negative.

This can be expressed in the general form:

$$\cos \theta = -\cos (180° - \theta)$$

cos (obtuse angle) = −cos (supplementary acute angle)

Example 7

Find the obtuse angle ϕ, correct to the nearest **minute**.

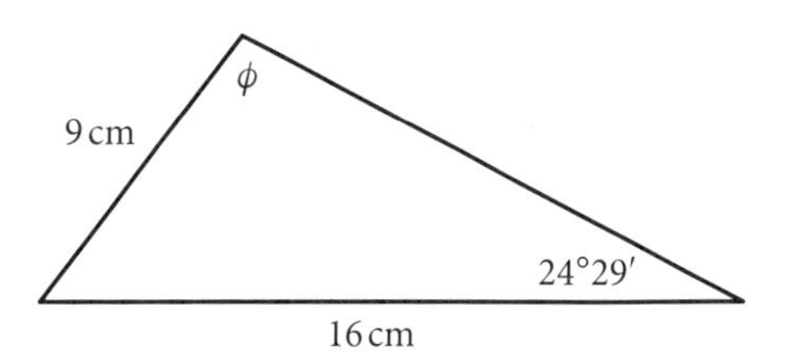

Solution

Two sides and the 2 opposite angles are involved, so use the sine rule.

$$\frac{a}{\sin A} = \frac{b}{\sin B}$$

$$\frac{16}{\sin \phi} = \frac{9}{\sin 24°29'}$$

$$\frac{\sin \phi}{16} = \frac{\sin 24°29'}{9}$$

$$\sin \phi = \frac{16 \times \sin 24°29'}{9}$$

$$= 0.73676\ldots$$

$$\phi = 47.456\ldots$$

Hint

When finding an unknown angle, flip the formula for simpler calculations:

$$\frac{\sin A}{a} = \frac{\sin B}{b} = \frac{\sin C}{c}$$

But ϕ is obtuse:

Obtuse $\phi = 180° - 47.456\ldots$

$= 132.54\ldots$

$= 132° 32' 37''$ Press ° ′ ″ or DMS

$\approx 132° 33'$

Bearings

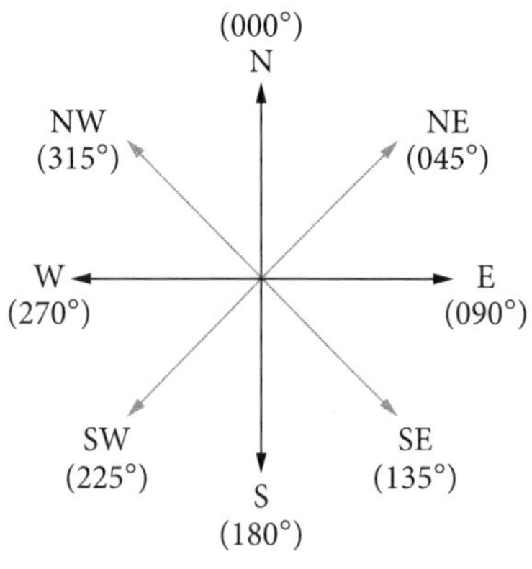

Bearings are used to give the direction of a location from a given reference point. They are expressed as an angle based on the 4 compass directions (N, S, E, W). There are 2 types of bearings: **compass bearings** and **true bearings**.

Compass bearings: acute angles measured from north or south and then to the east or west.

True bearings: three-digit angles measured from north in a clockwise direction from 000° to 360°.

If a direction or bearing is 'directly east' or 'due east', you can assume that it is 090° from north. If a direction or bearing is 'south-west', you can assume that it is 45° west of south, S 45° W or simply SW.

The word 'from' is important when describing bearings. For example, in the diagram, the bearing of '*A from B*' is 250°, whereas the bearing of '*A* to *B*' or '*B from A*' is 070°. Make sure you draw the north line at the reference point where the bearing is from, *not* where it is going to.

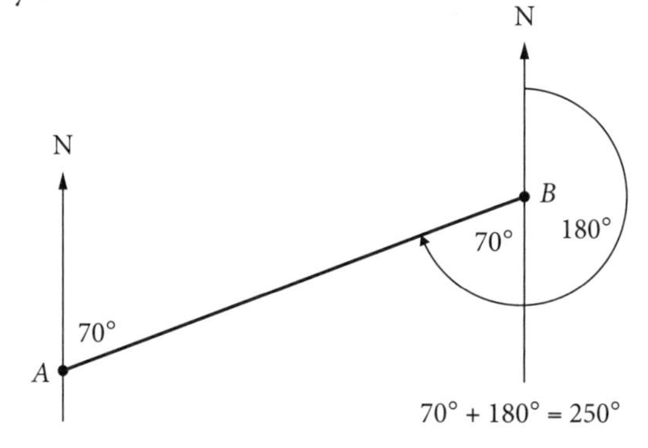

Example 8

What is the bearing of *A* from *O*, as a:

a true bearing?

b compass bearing?

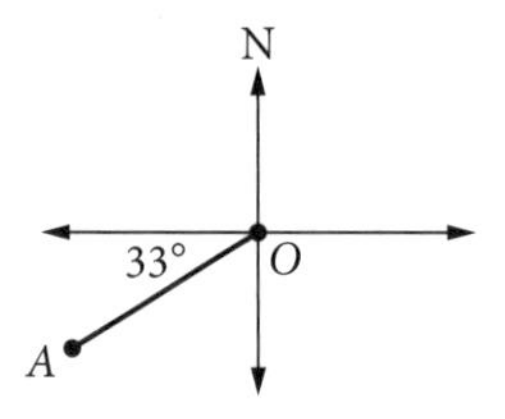

Solution

a True bearing = 270° − 33° (33° south of west)
= 237°

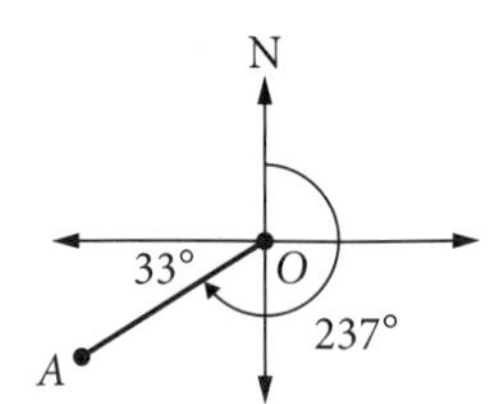

b *A* is west of south by an angle of 90° − 33° = 57°.

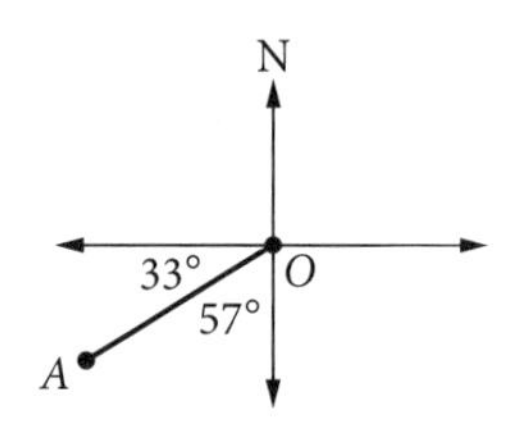

Compass bearing = S 57° W

Example 9

Robert paddles his kayak on a lake. He travels on a bearing of 125° for 7 km from A to B and then turns and paddles 10 km to C. The size of angle ABC is 112°. He then returns to A in a direct line from C.

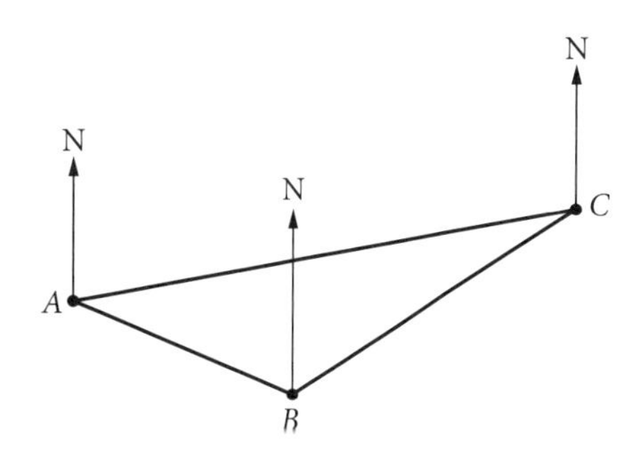

a Calculate the distance from C to A, correct to the nearest kilometre.

b What is the true bearing of A from C? Answer to the nearest degree.

Solution

a $c^2 = a^2 + b^2 - 2ab\cos C$, where the unknown side, c, is opposite angle C.

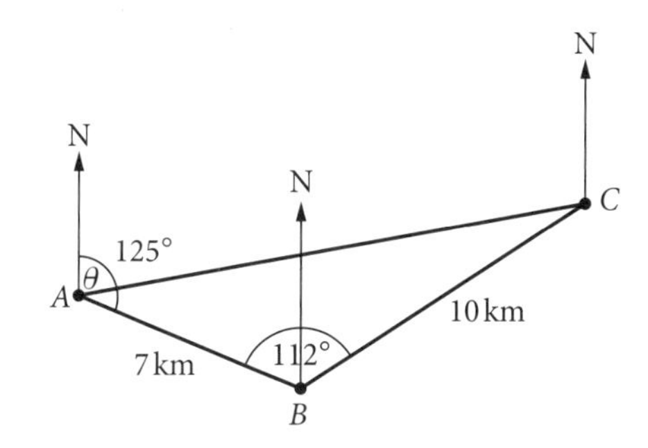

$$AC^2 = 7^2 + 10^2 - 2 \times 7 \times 10\cos 112°$$
$$= 201.4449\ldots$$
$$AC = \sqrt{201.4449\ldots}$$
$$= 14.1931\ldots$$
$$\approx 14\text{ km}$$

b To find the true bearing of A from C, use the north line at C to mark the angle showing the bearing.

Use the sine rule to find $\angle A$ in the triangle, which will help to calculate the bearing.

$$\frac{a}{\sin A} = \frac{b}{\sin B}$$
$$\frac{10}{\sin A} = \frac{14.1931\ldots}{\sin 112°}$$
$$\frac{\sin A}{10} = \frac{\sin 112°}{14.1931\ldots}$$
$$\sin A = \frac{10\sin 112°}{14.1931\ldots}$$
$$= 0.6532\ldots$$
$$A = 40.7881\ldots$$
$$\approx 41°$$

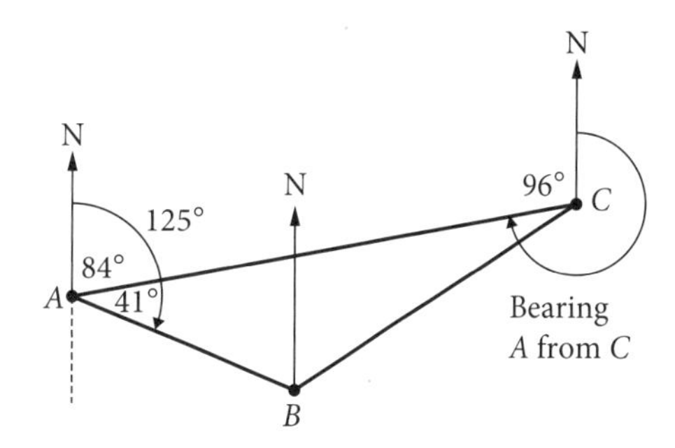

$$\angle NAC = 125° - 41°$$
$$= 84°$$

$$\angle NCA = 180° - 84°$$
$$= 96°$$
(co-interior angles on parallel lines are supplementary)

So bearing of A from $C = 360° - 96° = 264°$.

Compass radial surveys

Compass radial surveys are used to find the side lengths and area of an irregular field. A compass is placed at the centre, then the bearing and distance to each corner of the field are measured.

Example 10

This diagram shows the result of a compass radial survey of the field *LMNK*.

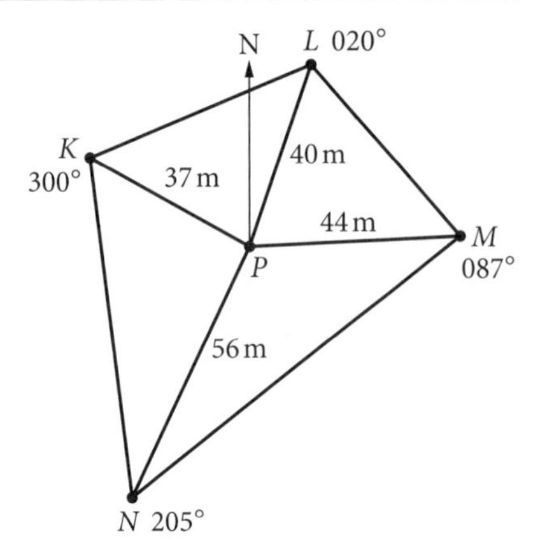

a Calculate the size of $\angle MPN$.

b Calculate the area of ΔMPN and the length of side *MN*, correct to two significant figures.

Solution

a $\angle MPN = 205 - 87 = 118°$

b Area $\Delta MPN = \frac{1}{2}ab\sin C$

$$A = \frac{1}{2} \times 56 \times 44\sin 118°$$
$$= 1087.7914\ldots$$
$$\approx 1100\,\text{m}^2$$

Use the cosine rule:

$c^2 = a^2 + b^2 - 2ab\cos C$, where $a = 56$, $b = 44$, $c = 118°$

$$MN^2 = 56^2 + 44^2 - 2 \times 56 \times 44 \times \cos 118°$$
$$= 7385.5558\ldots$$
$$MN = \sqrt{7385.5558\ldots}$$
$$= 85.9392\ldots$$
$$\approx 86\,\text{m}^2$$

Note: To find the area of the entire field *LMNK*, find the areas of all 4 triangles and add them together. To find the perimeter of the field, find the outer lengths of all 4 triangles and add them together.

Practice set 1

Multiple-choice questions

Solutions start on page 62.

All diagrams are not to scale.

Question 1

What is x, correct to the nearest whole number?

A 4 m

B 24 m

C 28 m

D 576 m

Question 2

What is the compass bearing of B from A?

A N 60° E

B N 60° W

C S 60° E

D S 60° W

Question 3

Which compass bearing is the same as the true bearing of 200°?

A S 20° E **B** S 20° W **C** S 70° E **D** S 70° W

Question 4

Which expression gives the length of the hypotenuse in this diagram?

A $18 \sin 36°$

B $18 \cos 36°$

C $18 \sin 54°$

D $\dfrac{18}{\sin 54°}$

Question 5

A tree of height 30 m casts a shadow of 14 m along the ground.

What is the angle of elevation, θ, of the sun, correct to the nearest degree?

A 25°

B 28°

C 62°

D 65°

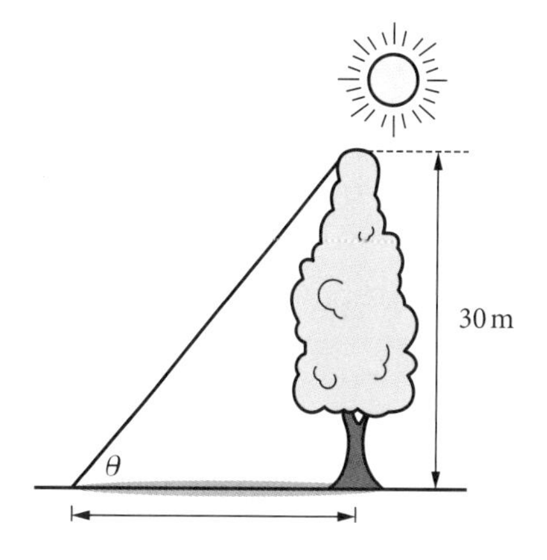

Question 6 ©NESA 2014 HSC EXAM, QUESTION 23

The following information is given about the locations of three towns, X, Y and Z.

- X is due east of Z.
- X is on a bearing of 145° from Y.
- Y is on a bearing of 060° from Z.

Which diagram best represents this information?

A

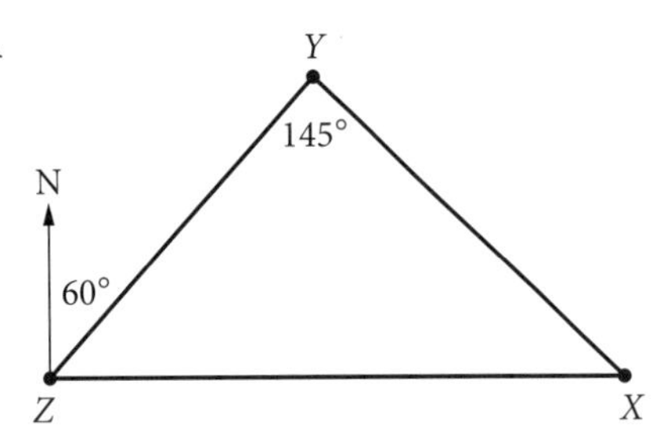

B

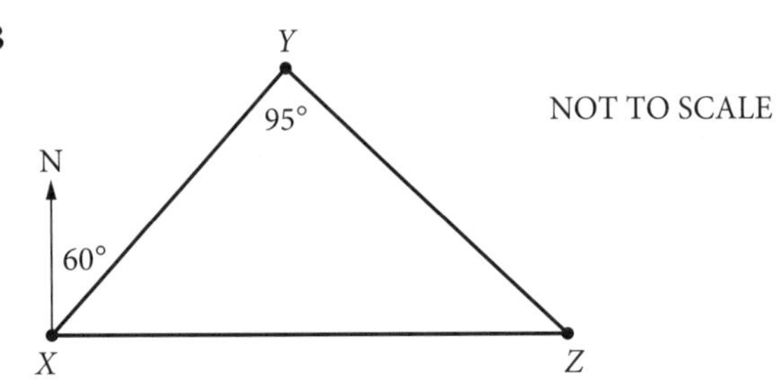

C

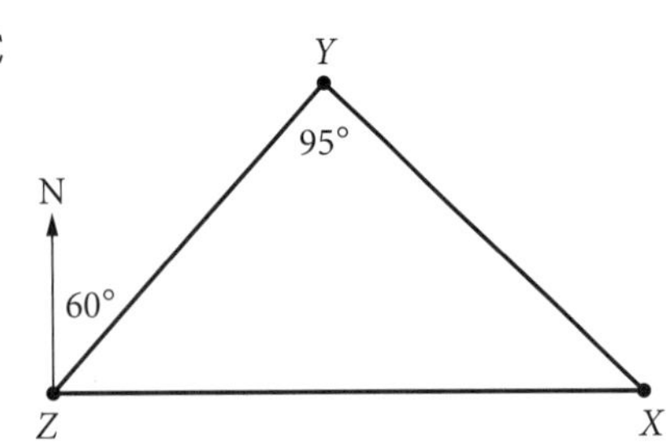

D

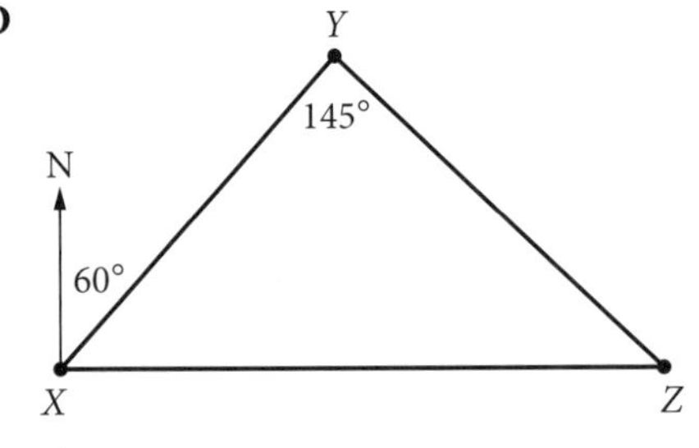

Question 7

What is the true bearing of P from Q?

A 060°

B 120°

C 240°

D 300°

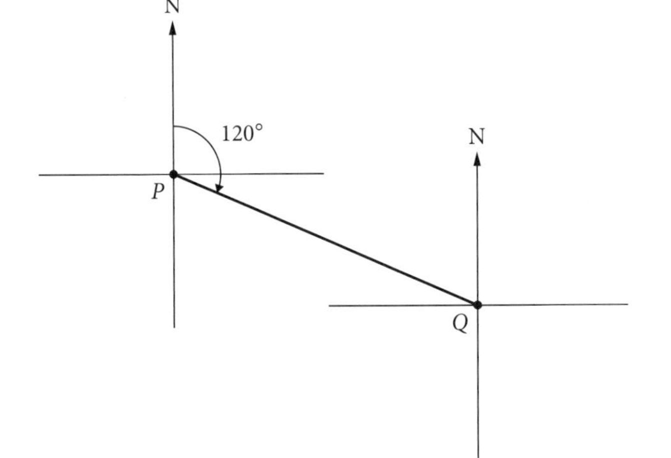

Question 8

What is the area of the triangle, correct to the nearest square metre?

A $101\,\text{m}^2$

B $153\,\text{m}^2$

C $166\,\text{m}^2$

D $306\,\text{m}^2$

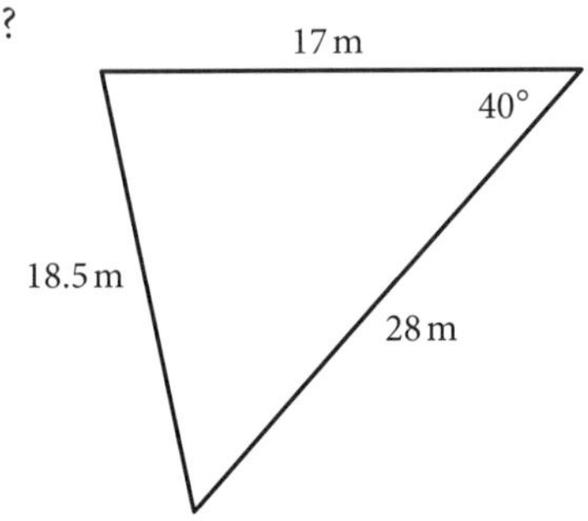

Question 9

Richard observes an airship at an angle of elevation of 52°.

If the airship is 220 m above the ground, what is its horizontal distance to Richard, correct to the nearest metre?

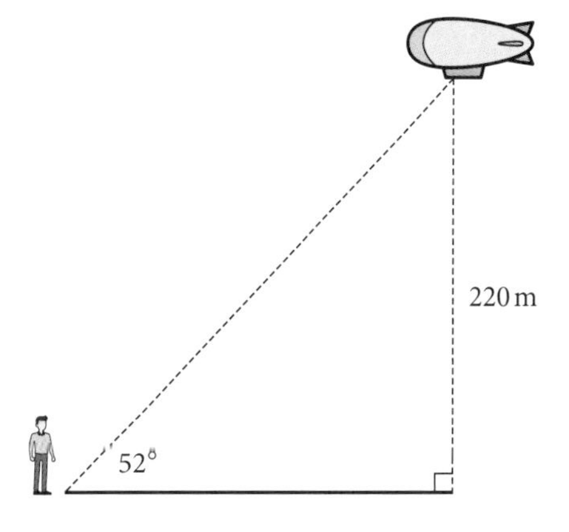

A 172 m

B 173 m

C 279 m

D 282 m

Question 10

The diagram shows a compass radial survey of a field *JKLM*.

What is the size of ∠*MOJ*?

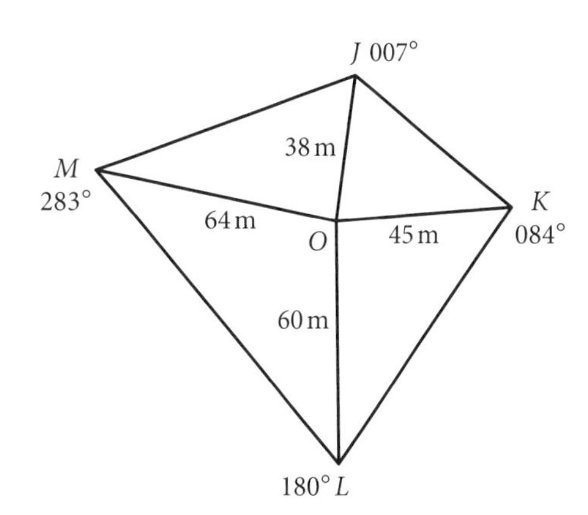

A 7°

B 77°

C 84°

D 154°

Question 11

What is the correct expression for x in this diagram?

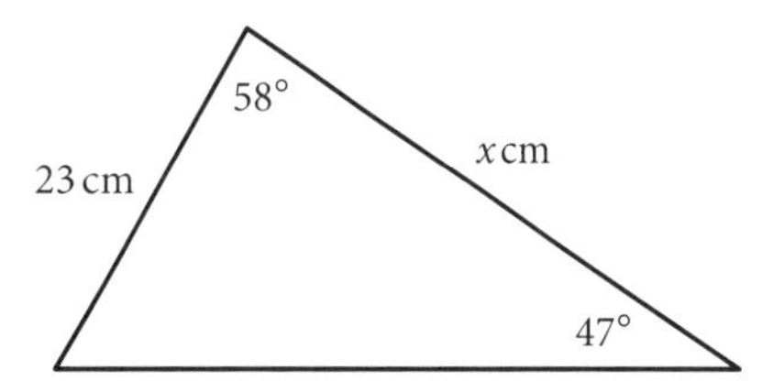

A $x = \frac{23\sin 75°}{\sin 47°}$ **B** $x = \frac{23\sin 47°}{\sin 58°}$

C $x = \frac{23\sin 47°}{\sin 75°}$ **D** $x = \frac{23\sin 58°}{\sin 47°}$

Question 12

Joshua walks 1.8 km due east from school, *S*, to the takeaway shop, *T*. He then turns and walks on a bearing of 218° towards his home, *H*, which is due south of the school.

How far does he live from the school, correct to the nearest 0.1 km?

A 1.1 km **B** 1.4 km **C** 2.3 km **D** 2.6 km

Question 13

Two vertical power poles of different heights are 15 m from each other. The angle of depression from the top of the taller pole to the top of the shorter pole is 41°. The shorter pole is 6 m high.

To the nearest metre, how high is the taller pole?

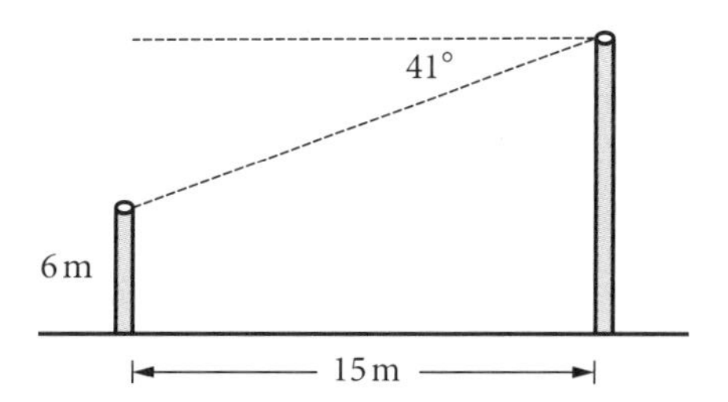

A 13 m

B 16 m

C 19 m

D 21 m

Question 14

What is the value of x, correct to the nearest centimetre?

A 45 cm

B 64 cm

C 85 cm

D 109 cm

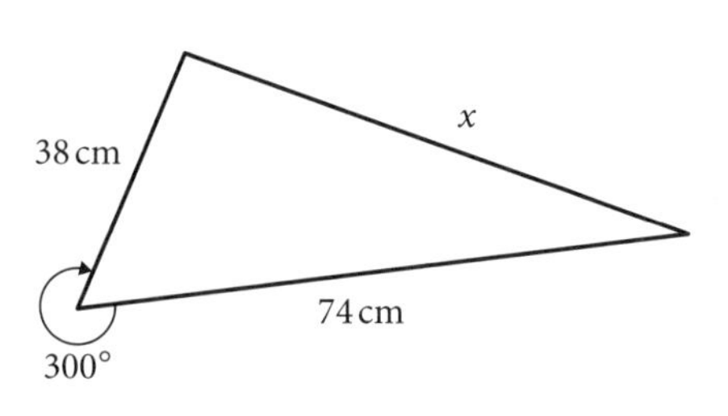

Question 15 ©NESA 2012 HSC EXAM, QUESTION 20

Town B is 80 km due north of Town A and 59 km from Town C. Town A is 31 km from Town C.

What is the true bearing of Town C from Town B?

A 019°

B 122°

C 161°

D 341°

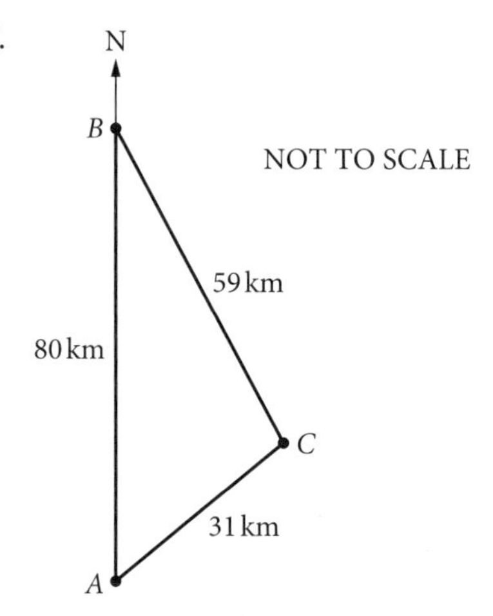

Question 16

What is the size of the smallest angle in this triangle?

A 38°

B 52°

C 59°

D 69°

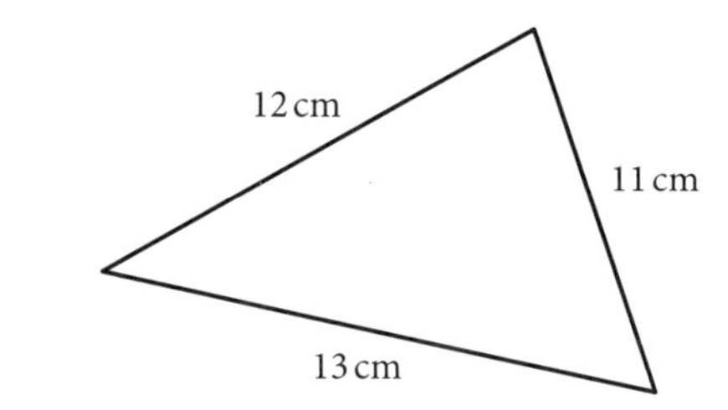

Question 17

A golfer hits a golf ball 3° off centre and lands it 16 m from a hole and 201 m from the tee.

What is the size of obtuse angle θ?

A 41°

B 124

C 139°

D 151°

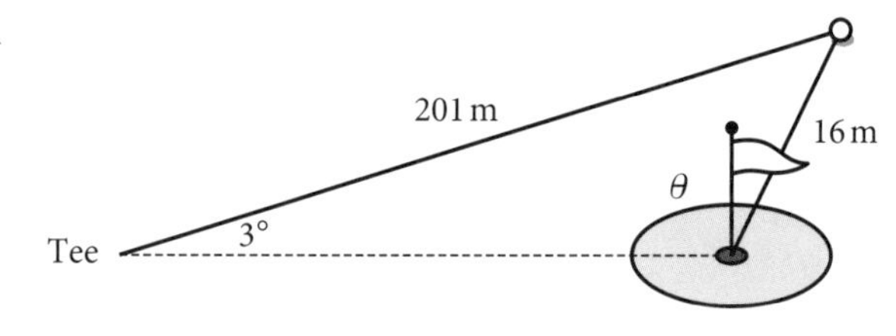

Question 18

What is the area of the field $QRST$, correct to the nearest square metre?

A 682 m^2

B 721 m^2

C 768 m^2

D 802 m^2

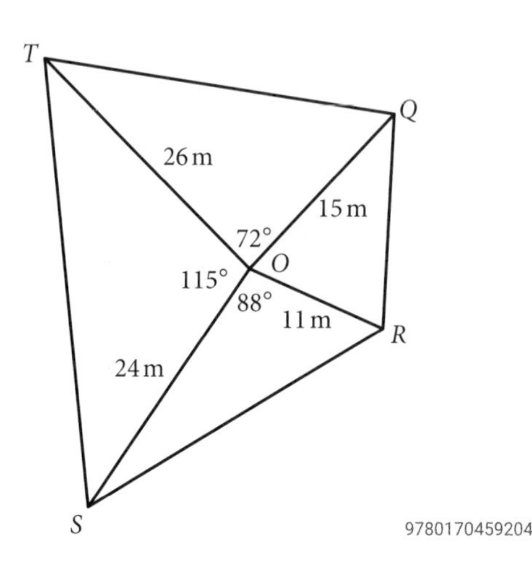

Question 19

What is the value of θ, correct to the nearest degree?

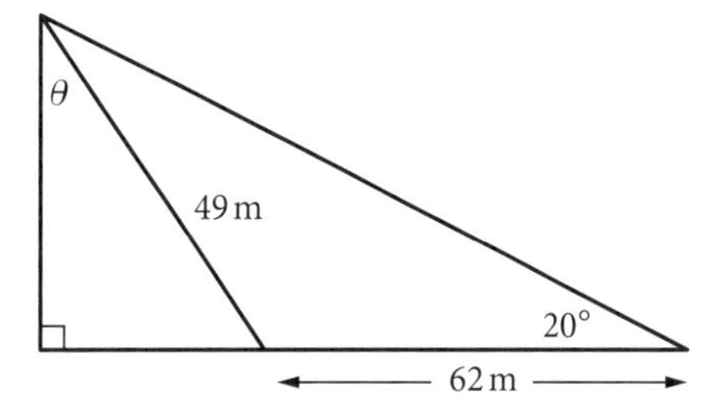

A 26°

B 35°

C 44°

D 53°

Question 20

The diagram shows the position of Q, R and T relative to P.

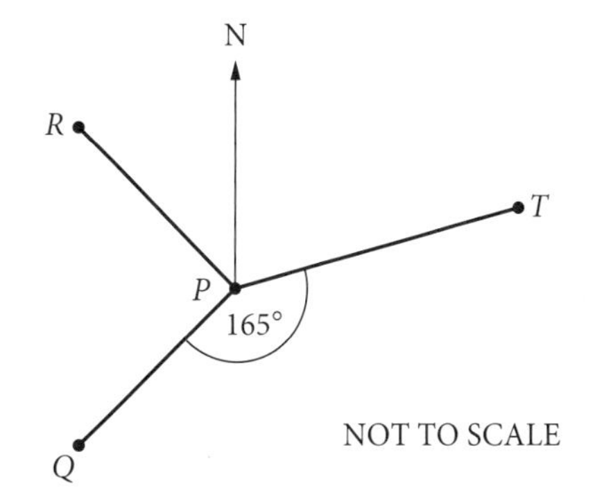

In the diagram:

Q is SW of P
R is NW of P
$\angle QPT$ is 165°

What is the bearing of T from P?

A 060° **B** 075°

C 105° **D** 120°

Practice set 2

Short-answer questions

Solutions start on page 65.

Question 1 (3 marks)

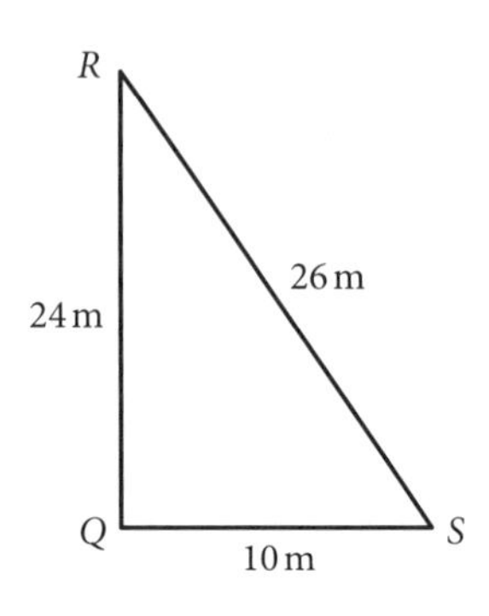

a Show that ΔQRS is a right-angled triangle. 1 mark

b Calculate the size of $\angle QRS$, correct to the nearest degree. 2 marks

Question 2 (2 marks) 

A pelican flying 8 m above the ground sees a fish in the water at an angle of depression of 35°.

Find the horizontal distance along the ground from the pelican to the fish. Give your answer correct to one decimal place. 2 marks

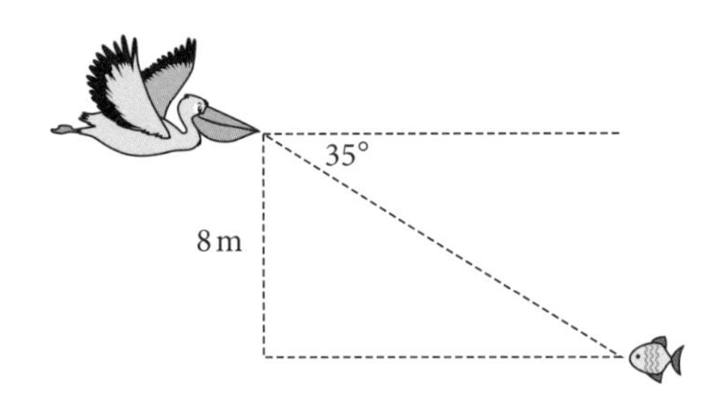

Question 3 (3 marks)

A plane flies from an airport, A, towards Brisbane, B, for 380 km on a bearing of 130°.

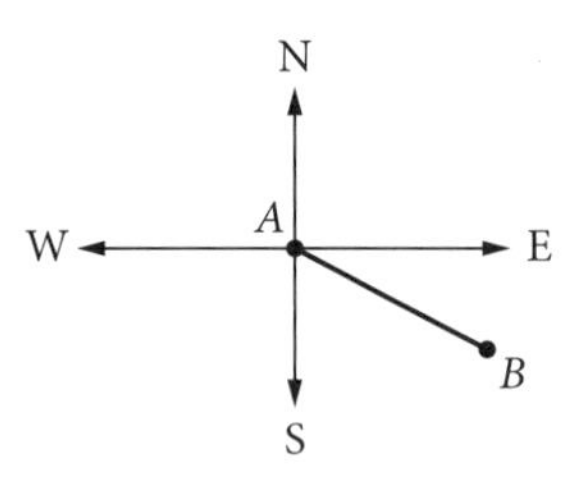

a What is the compass bearing of Brisbane from the airport? 1 mark

b How far east has the plane flown? Give your answer correct to the nearest kilometre. 2 marks

Question 4 (2 marks) 

To avoid a large reef, a boat sails 70 nautical miles due west from point A to point C and then 85 nautical miles due north to point B.

What is the true bearing from A to B, correct to the nearest degree? 2 marks

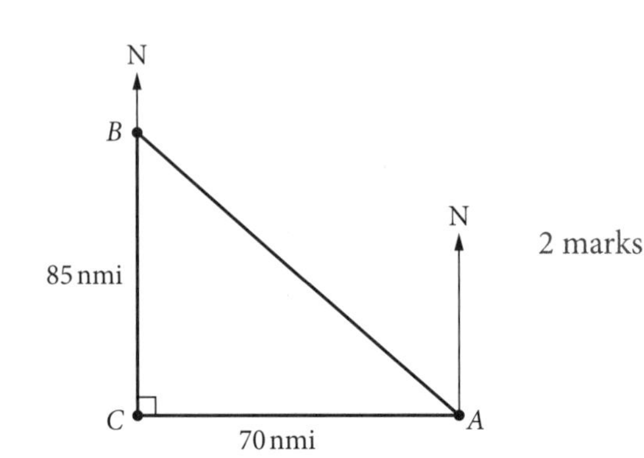

Question 5 (3 marks)

Calculate the length XY, correct to two significant figures. 3 marks

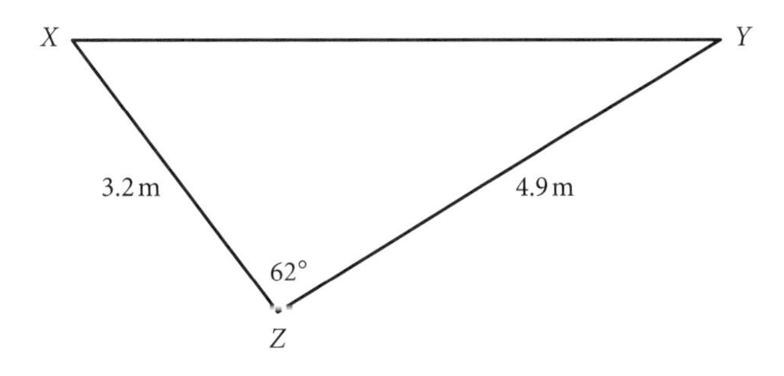

Question 6 (3 marks)

Find obtuse angle θ, correct to the nearest minute. 3 marks

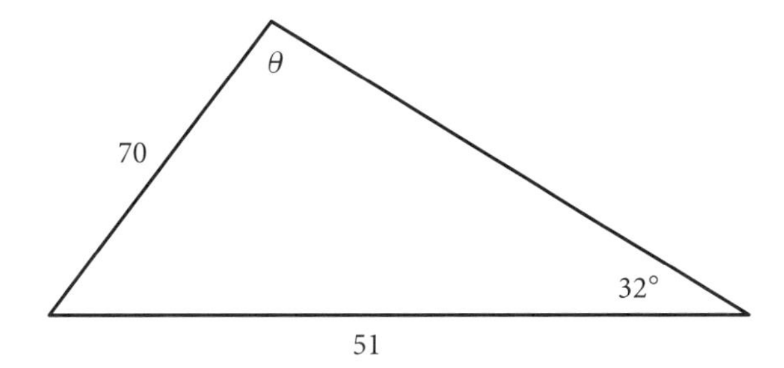

Question 7 (2 marks)

The area of triangle ABC is $46\ \text{cm}^2$.

Calculate the length AC, correct to the nearest centimetre. 2 marks

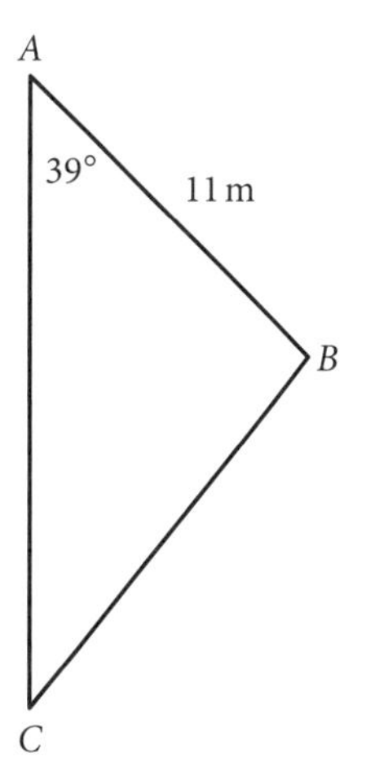

Question 8 (4 marks)

Tania looks up to the top of a 95 m tower with an angle of elevation of 10°.

After walking d m directly towards the foot of the tower, she finds that the angle of elevation increases to 18°.

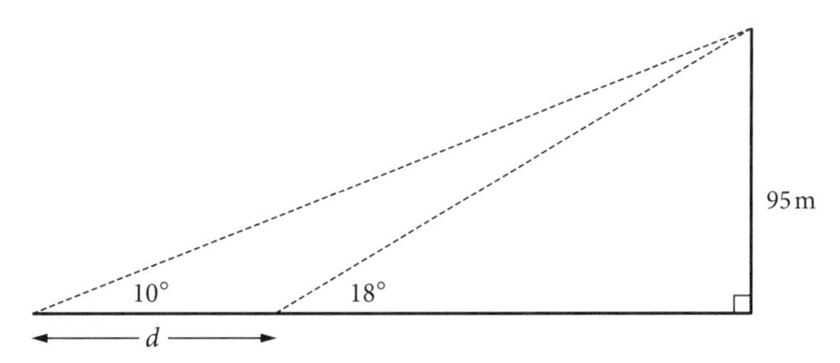

Calculate how far Tania has walked, correct to the nearest metre. 4 marks

Question 9 (6 marks)

A radial survey of a piece of land, $PQRS$, is shown.

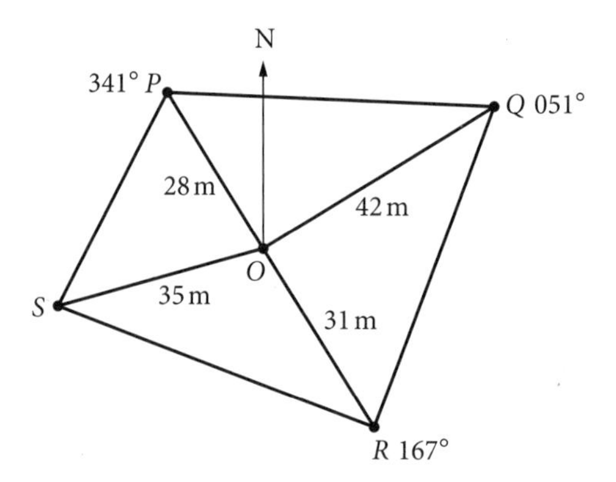

a What is the size of $\angle POQ$? 1 mark

b Find the area of ΔPOQ. Round your answer to the nearest square metre. 2 marks

c If the distance from R to S is 44 m, calculate the bearing of S from O. 3 marks

Question 10 (5 marks) ©NESA 2020 HSC EXAM, QUESTION 31

Mr Ali, Ms Brown and a group of students were camping at the site located at P. Mr Ali walked with some of the students on a bearing of 035° for 7 km to location A. Ms Brown, with the rest of the students, walked on a bearing of 100° for 9 km to location B.

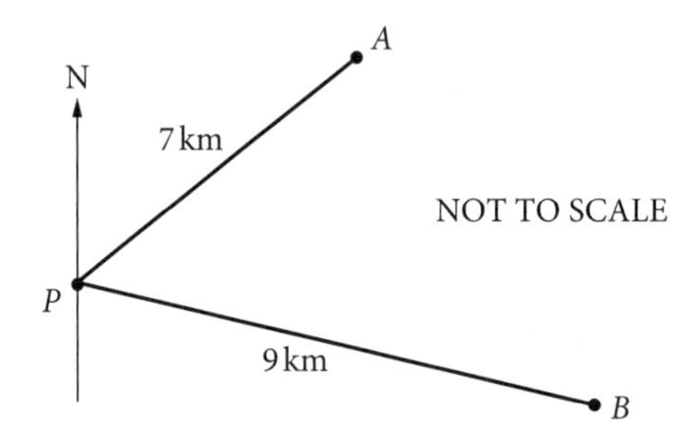

a Show that the angle APB is 65°. 1 mark

b Find the distance AB. 2 marks

c Find the true bearing of Ms Brown's group from Mr Ali's group.
Give your answer correct to the nearest degree. 2 marks

Question 11 (4 marks)

The diagram shows the triangular course of a sailing race, with $\angle XYZ = 65°$. The distance from Z to Y is 1.7 km with a true bearing of 061°. The race changes course at Y and requires the sailing boats to travel 3.1 km to X.

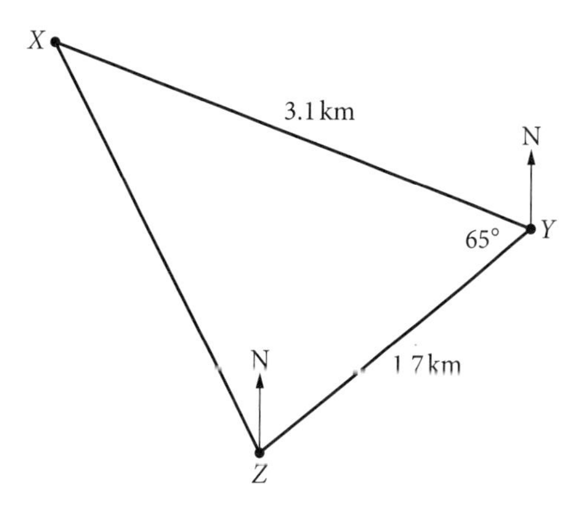

a What is the true bearing of Z from Y? 1 mark

b What is the distance X to Z, correct to one decimal place? 2 marks

c Other boats on the water are not allowed inside the area bounded by the course during the race. What is the size of the area that other boats are not allowed to use, correct to one decimal place? 1 mark

Question 12 (5 marks) ©NESA 2004 HSC EXAM, QUESTION 24(b)

The diagram shows a radial piece of land.

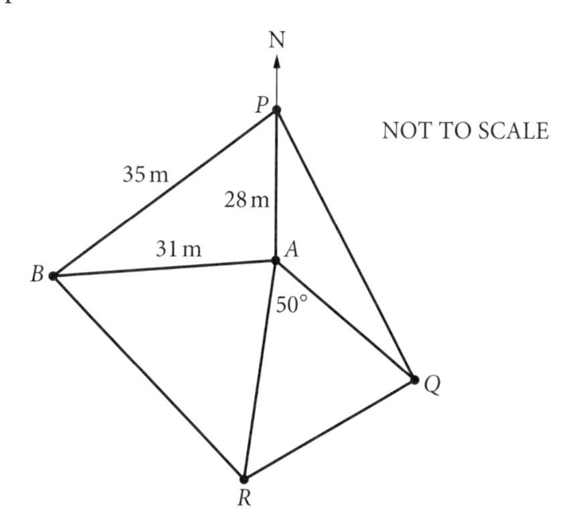

a Q is south-east of A. What is the size of $\angle PAQ$? 1 mark

b What is the bearing of R from A? 1 mark

c Find the size of angle PAB to the nearest degree. 3 marks

Question 13 (4 marks)

A cross-country running course follows a triangular track, ABC, around a bush reserve.

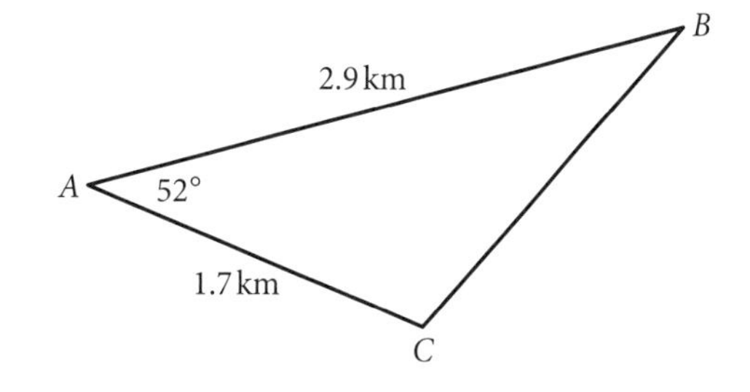

a Find the total length of the cross-country course ABC. Give your answer to the nearest kilometre. 2 marks

b Find the area of the bush reserve ABC. Give your answer in square kilometres, correct to two decimal places. 2 marks

9780170459204

PRACTICE SET 2

Question 14 (4 marks)

A racing speed boat sets off from a starting line, heading north for 7.5 km to marker *B*. It then turns and travels on a bearing of 130° for a further 16 km towards the next marker at *C*.

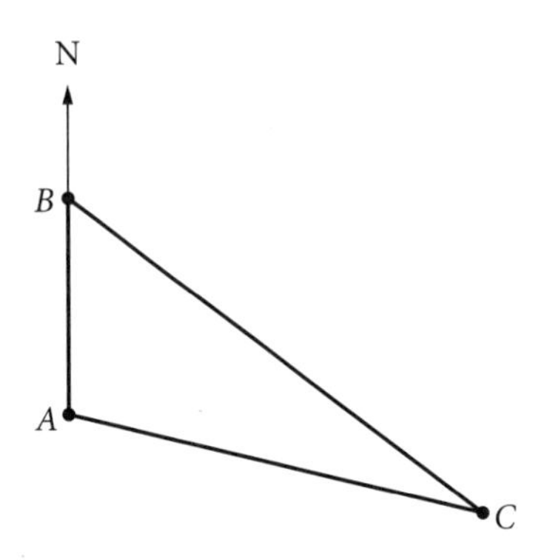

a Calculate the shortest distance, correct to one decimal place, for the remaining leg of the race, between *C* and *A*. 2 marks

b What is the true bearing of *A* from *C*? Give your answer to the nearest degree. 2 marks

Question 15 (2 marks)

Two soccer posts are 7.3 m apart.

If a striker is standing 9 m from one post and 12 m from the other, what is the angle, correct to the nearest minute, within which the ball must be kicked to score a goal? 2 marks

Question 16 (3 marks) ©NESA 2019 HSC EXAM, QUESTION 35

A compass radial survey shows the positions of four towns *A*, *B*, *C* and *D* relative to point *O*.

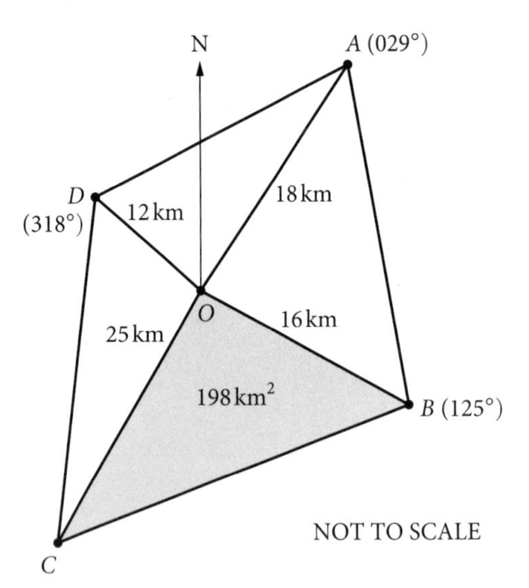

The area of triangle *BOC* is 198 km^2.

Calculate the true bearing of town *C* from point *O*, correct to the nearest degree. 3 marks

Question 17 (4 marks)

Tim cycles around a triangular course when training for a triathlon race.

He starts at A and passes through B, C and D, taking the shortest route back to his starting position at A.

The distances BC and CD are equal.

What is the distance of this course, correct to the nearest kilometre?

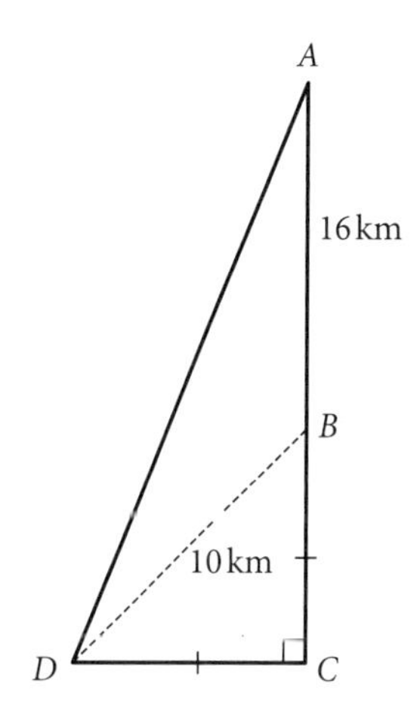

4 marks

Question 18 (4 marks) ©NESA 2020 HSC EXAM, QUESTION 32

The diagram shows a regular decagon (ten-sided shape with all sides equal and all interior angles equal). The decagon has centre O.

The perimeter of the shape is 80 cm.

By considering triangle OAB, calculate the area of the ten-sided shape. Give your answer in square centimetres, correct to one decimal place.

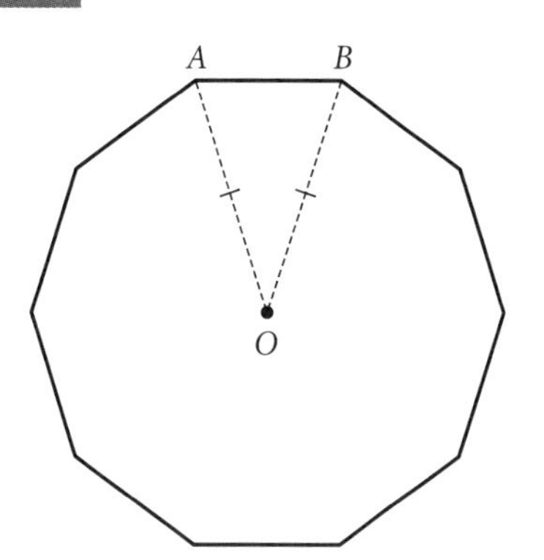

4 marks

Question 19 (3 marks)

A plane leaves Sydney Airport (A) and flies 800 km on a bearing of 280°. It then turns to fly 650 km on a bearing of 200°. How far is the plane from Sydney, correct to the nearest 10 km?

3 marks

Question 20 (5 marks) ©NESA 2015 HSC EXAM, QUESTION 30(e)

From point S, which is 1.8 m above the ground, a pulley at P is used to lift a flat object, F.

The lengths SP and PF are 5.4 m and 2.1 m respectively. The angle PSC is 108°.

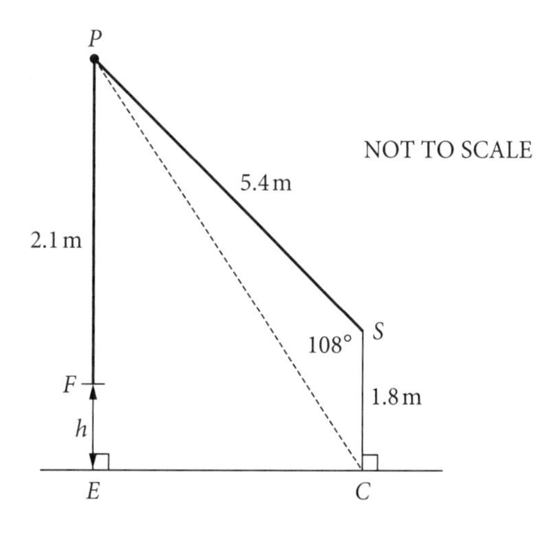

a Show that the length PC is 6.197 m, correct to three decimal places. 1 mark

b Calculate h, the height of the object above the ground, correct to two decimal places. 4 marks

Practice set 1

Worked solutions

1 B

$$x^2 = 26^2 - 10^2$$
$$= 576$$
$$x = 24\text{ m}$$

2 A

B is north and east of A: N 60° E

3 B

S 20° W

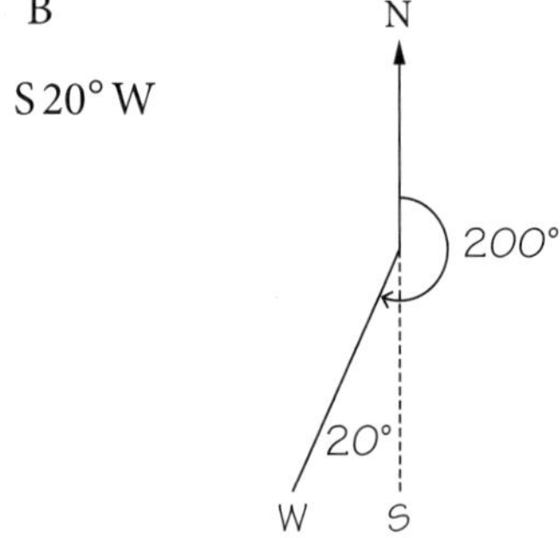

4 D

$$\sin 54° = \frac{18}{H}$$
$$H = \frac{18}{\sin 54°}$$

5 D

$$\tan\theta = \frac{30}{14}$$
$$\theta = 64°58'\ldots$$
$$\approx 65°$$

6 C

Not B or D because Z is west of X.

Not A because wrong angle inside triangle.

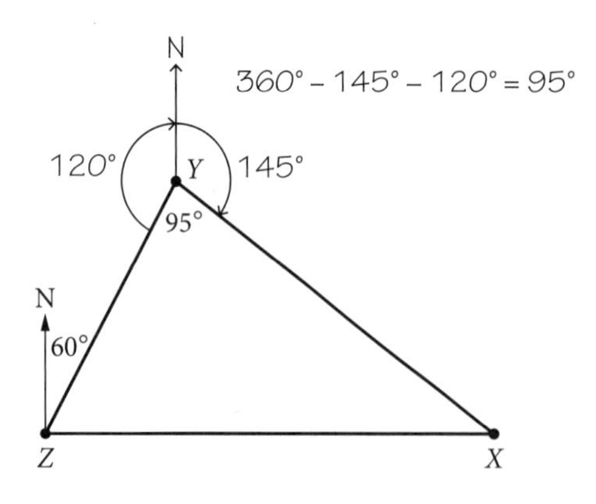

7 D

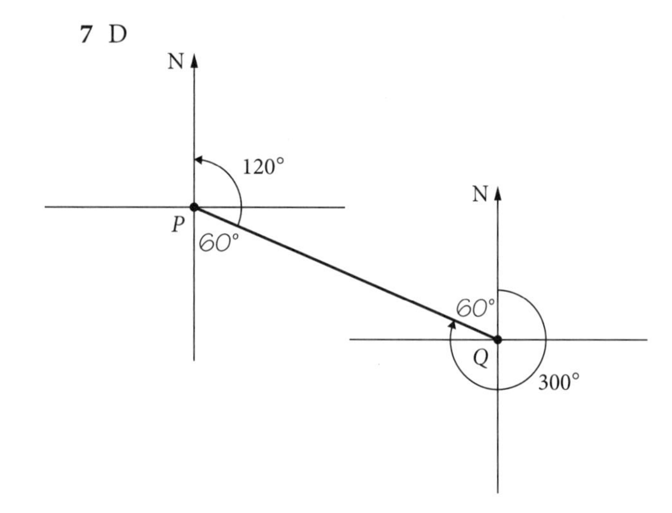

180° – 120° = 60° (straight angle)

Alternate angles are equal in parallel lines.

Bearing P from Q = 360° – 60°
= 300°

8 B

$$A = \frac{1}{2} \times 17 \times 28 \times \sin 40°$$
$$= 152.9834\ldots$$
$$\approx 153\text{ m}^2$$

9 A

Let x be the horizontal distance from airship to Richard.

$$\tan 52° = \frac{220}{x}$$
$$x = \frac{220}{\tan 52°}$$
$$= 171.88\ldots$$
$$\approx 172\text{ m}$$

10 C

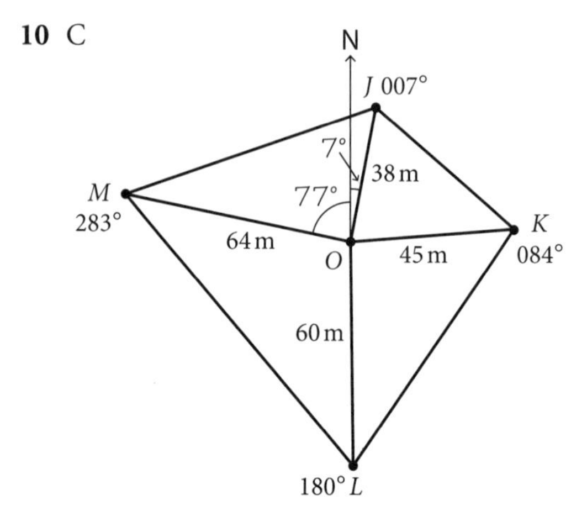

$\angle NOJ = 7°$ (given)

So $\angle NOM = 360° - 283°$ (angles at a point add to 360°)

$= 77°$

$\angle MOJ = \angle NOJ + \angle NOM$

$= 7° + 77°$

$= 84°$

11 A

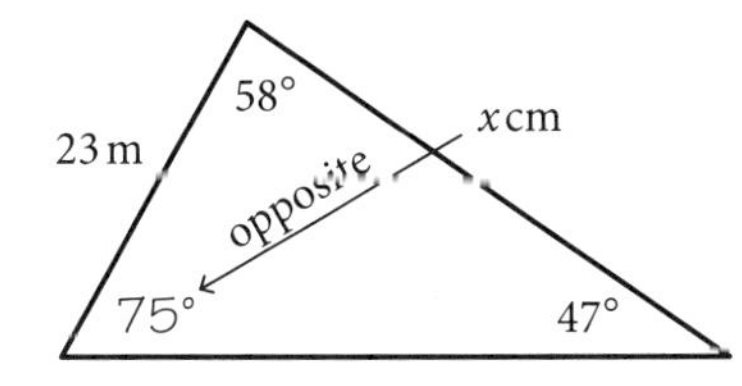

$180° - 58° - 47° = 75°$

$$\frac{x}{\sin 75°} = \frac{23}{\sin 47°}$$

$$x = \frac{23\sin 75°}{\sin 47°}$$

12 C

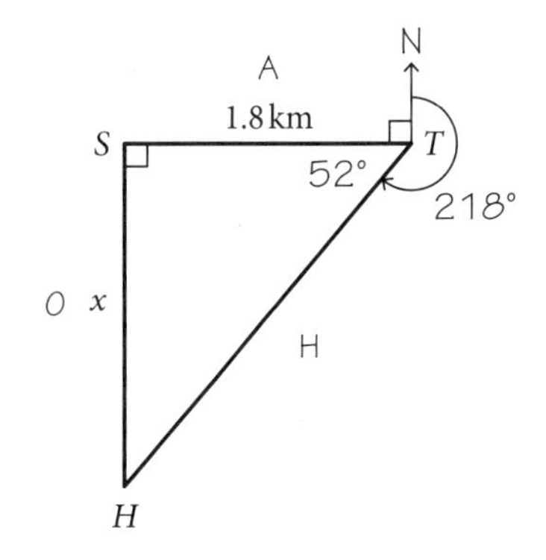

$\angle STH = 360° - 218° - 90°$ (angles at a point add to 360°)

$= 52°$

$$\tan 52° = \frac{x}{1.8}$$

$x = 1.8 \times \tan 52°$

$= 2.303\ldots$

$\approx 2.3\text{ km}$

13 C

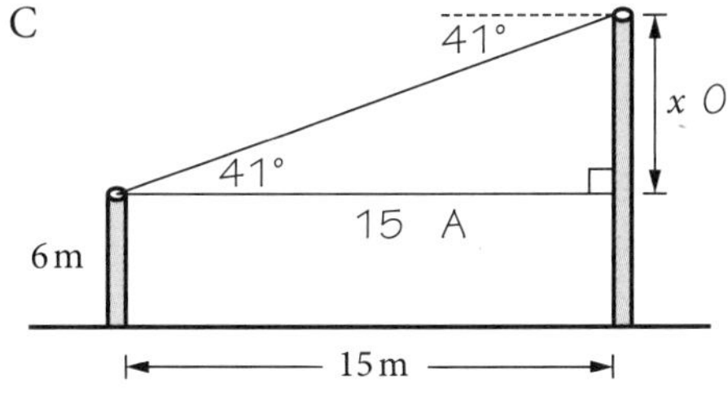

$$\tan 41° = \frac{x}{15}$$

$x = 15 \times \tan 41°$

$= 13.03\ldots$

$\approx 13\text{ m}$

Total height of taller pole $= 13 + 6$

$= 19\text{ m}$

14 B

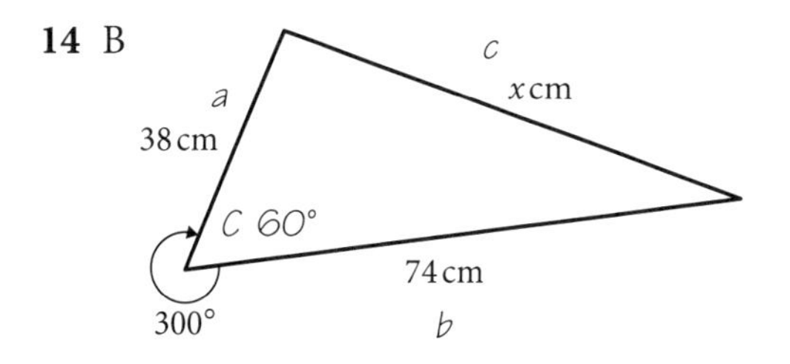

$\angle C = 360° - 300° = 60°$

$x^2 = 38^2 + 74^2 - 2 \times 38 \times 74 \times \cos 60°$

$= 4108$

$x = 64.09\ldots$

$\approx 64\text{ cm}$

15 C

$$\cos B = \frac{59^2 + 80^2 - 31^2}{2 \times 59 \times 80}$$

$$= \frac{8920}{9440}$$

$B = 19°$

Bearing of C from B

$= 180° - 19°$ (straight angles add to 180°)

$= 161°$

So bearing of C from B is 161°.

16 B

Smallest angle is opposite smallest side.
Let the smallest angle be θ.

$$\cos\theta = \frac{12^2 + 13^2 - 11^2}{2 \times 12 \times 13}$$

$$\theta = \cos^{-1}\left(\frac{12^2 + 13^2 - 11^2}{2 \times 12 \times 13}\right)$$

$\approx 52°$

17 C

$$\frac{\sin\theta}{201} = \frac{\sin 3°}{16}$$

$$\sin\theta = \frac{\sin 3°}{16} \times 201$$

$$\theta = \sin^{-1}\left(\frac{\sin 3°}{16} \times 201\right)$$

$\approx 41°$

Obtuse angle $= 180° - 41°$

$= 139°$

18 A

$$\text{Area } OTS = \frac{1}{2} \times 26 \times 24 \times \sin 115^\circ$$
$$\approx 282.768$$

$$\text{Area } OTQ = \frac{1}{2} \times 26 \times 15 \times \sin 72^\circ$$
$$\approx 185.456$$

$$\text{Area } OSR = \frac{1}{2} \times 24 \times 11 \times \sin 88^\circ$$
$$\approx 131.920$$

$$\text{Area } OQR = \frac{1}{2} \times 15 \times 11 \times \sin 85^\circ$$
(angles at a point add to 360°)
$$\approx 82.186$$

Total area $QRST$
$= 282.768 + 185.456 + 131.920 + 82.186$
$\approx 682 \text{ m}^2$

19 C

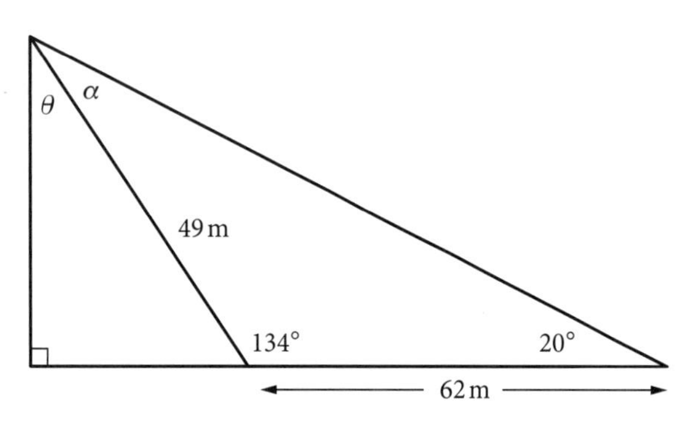

$$\frac{\sin \alpha}{62} = \frac{\sin 20^\circ}{49}$$
$$\sin \alpha = \frac{62 \sin 20^\circ}{49}$$
$$\alpha = 25.64\ldots$$
$$\approx 26^\circ$$

Unknown angle in α triangle:

$180^\circ - 20^\circ - 26^\circ$ (angle sum of a triangle)
$= 134^\circ$

So $\theta = 180^\circ - 134^\circ$ (angle sum of a triangle)
$= 44^\circ$

20 A

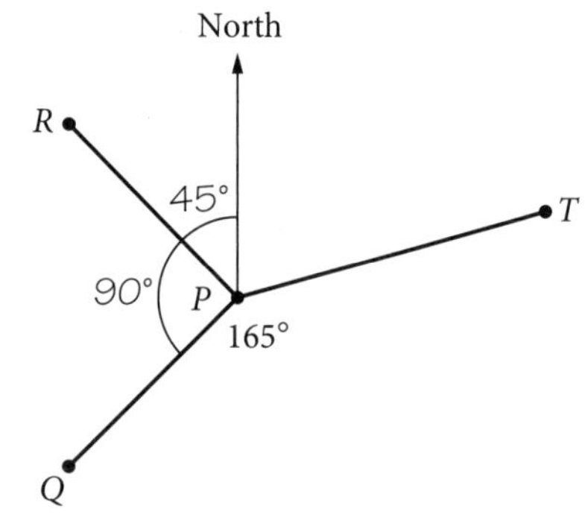

$\angle NPR = 45^\circ$ (R is NW of P)
$\angle RPQ = 90^\circ$ (R is NW of P and Q is SW of P)

So bearing T from P
$= 360^\circ - 45^\circ - 90^\circ - 165^\circ$ (angles at a point add to 360°)
$= 060^\circ$

Practice set 2

Worked solutions

Question 1

a Right-angled triangle if it satisfies $c^2 = a^2 + b^2$

$26^2 = 24^2 + 10^2$

$676 = 676$

So it is a right-angled triangle.

b $\sin\theta = \frac{10}{26}$

$\theta = 22.619\ldots$

$\approx 23°$

Question 2

Let x be the horizontal distance from the pelican to the fish.

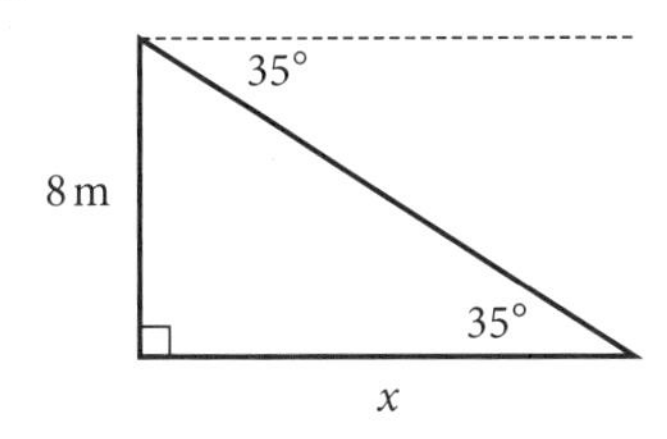

$\tan 35° = \frac{8}{x}$

$x = \frac{8}{\tan 35°}$

$= 11.435\ldots$

$\approx 11.4\,\text{m}$

Question 3

a S 50° E

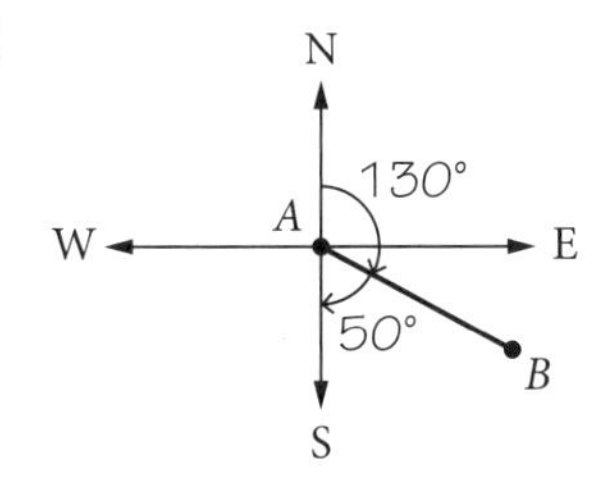

b

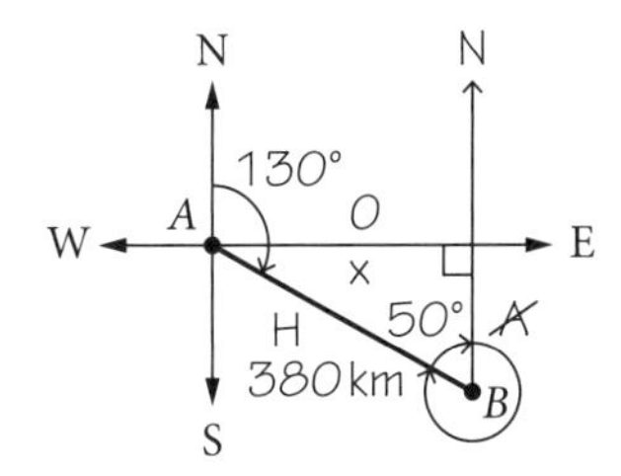

$\sin 50° = \frac{x}{380}$

$x = 380 \times \sin 50°$

$= 291.069\ldots$

≈ 291 km (nearest kilometre)

Question 4

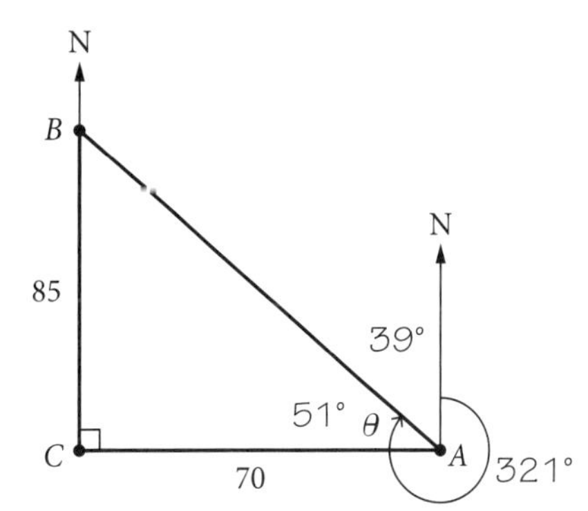

$\tan\theta = \frac{85}{70}$

$\theta = 50.527\ldots$

$\approx 51°$

True bearing of B from A is $270° + 51° = 321°$ or compass bearing is N 39° W.

Question 5

$c^2 = a^2 + b^2 - 2ab\cos C$

$XY^2 = 3.2^2 + 4.9^2 - 2 \times 3.2 \times 4.9\cos 62°$

$XY = \sqrt{19.52737...}$

$= 4.4189\ldots$

$\approx 4.4\,\text{m}$

Question 6

$\frac{\sin\theta}{51} = \frac{\sin 32°}{70}$

$\sin\theta = \frac{51\sin 32°}{70}$

$\theta = 22.711\ldots$

$= 22°42'39.8''$

$\approx 22°43'$

Hence, obtuse angle $\theta \approx 180° - 22°43'$

$\approx 157°17'$

Question 7

$A = \frac{1}{2}ab\sin C$

$46 = \frac{1}{2} \times a \times 11 \times \sin 39°$

$46 = a \times 3.4612\ldots$

$a = \frac{46}{3.4612\ldots}$

$= 13.289\ldots$

$AC \approx 13\,\text{cm}$

Question 8

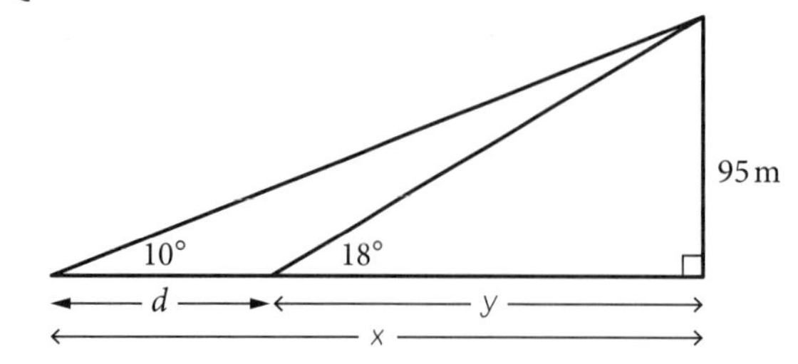

$$\tan 10° = \frac{95}{x}$$

$$x = \frac{95}{\tan 10°}$$

$$= 538.771 \text{ m}$$

$$\tan 18° = \frac{95}{y}$$

$$y = \frac{95}{\tan 18°}$$

$$= 292.3799\ldots$$

$$\approx 293 \text{ m}$$

$$d \approx 538.771 - 292.3799$$

$$\approx 246.39 \text{ m}$$

$$\approx 24\,639 \text{ cm}$$

Question 9

a $\angle POQ = \angle PON + \angle NOQ$

$= 360° - 341° + 51°$

$= 70°$

b Area of ΔPOQ:

$$A = \frac{1}{2}bc\sin A$$

$$= \frac{1}{2} \times 28 \times 42 \times \sin 70°$$

$$= 552.539\ldots$$

$$\approx 553 \text{ m}^2$$

c Let $\angle SOR = \theta$

$$\cos\theta = \frac{a^2 + b^2 - c^2}{2ab}$$

$$= \frac{35^2 + 31^2 - 44^2}{2 \times 35 \times 31}$$

$$\theta = 83.38\ldots$$

$$\approx 83°$$

So bearing of S from O is $167° + 83° = 250°$

Question 10

a $\angle APB = 100° - 35°$ (difference between the bearings)

$= 65°$

b Using the cosine rule:

$$AB^2 = 7^2 + 9^2 - 2 \times 7 \times 9\cos 65°$$

$$= 76.7500\ldots$$

$$AB = 8.760\,07\ldots$$

$$\approx 8.8 \text{ km}$$

c

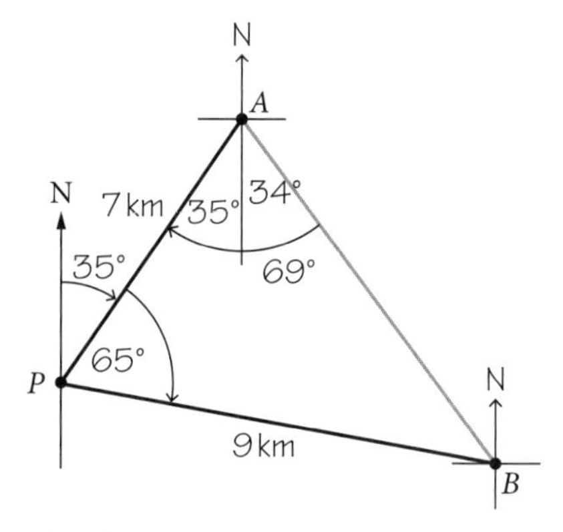

Must find $\angle PAB$:

Using the sine rule:

$$\frac{\sin\angle PAB}{9} = \frac{\sin 65°}{AB}$$

$$\sin\angle PAB = \frac{9\sin 65°}{8.7601\ldots}$$

$$= 0.9310\ldots$$

$$\angle PAB = 68.601\ldots$$

$$\approx 69°$$

OR using the cosine rule:

$$\cos\angle PAB = \frac{7^2 + AB^2 - 9^2}{2 \times 7 \times AB}$$ (from part **b**, $AB \approx 8.76$ km)

$$= \frac{44.75009\ldots}{122.6499\ldots}$$

$$\angle PAB = 68.601\ldots$$

$$\approx 69°$$

Angle east of south in $\angle PAB = 69° - 35°$

$= 34°$

Bearing is $180° - 34° = 146°$

The bearing of Ms Brown's group from Mr Ali's group is 146°.

Question 11

a

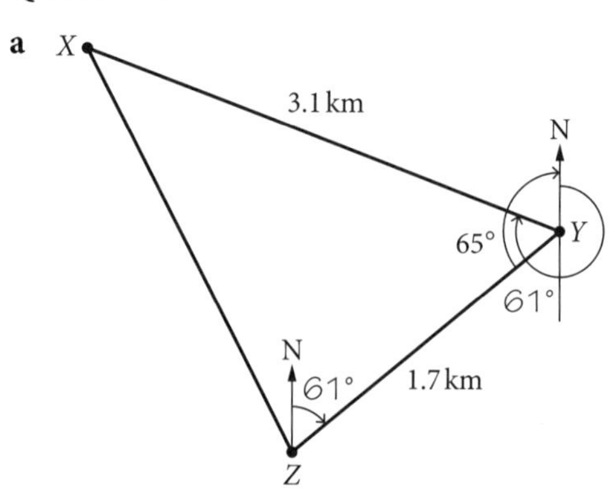

Alternate angles are equal (61°), north lines are parallel:

Bearing to Z from Y

$= 180° - 61° + 65°$

$= 306°$

b $c^2 = a^2 + b^2 - 2ab\cos C$

$$XZ^2 = 1.7^2 + 3.1^2 - 2 \times 1.7 \times 3.1 \times \cos 65°$$
$$XZ = \sqrt{1.7^2 + 3.1^2 - 2 \times 1.7 \times 3.1 \times \cos 65°}$$
$$= 2.836\ldots$$
$$\approx 2.8\text{ km}$$

c $A = \frac{1}{2} \times 1.7 \times 3.1 \times \sin 65°$

$$= 2.388\ldots$$
$$\approx 2.4\text{ km}^2$$

Question 12

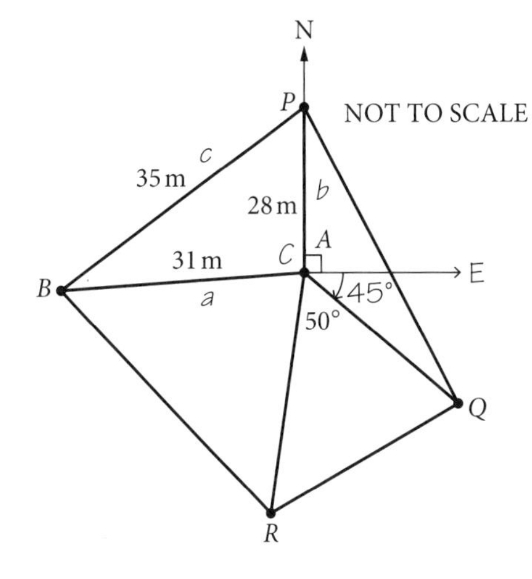

a Q is SE of A.

So $\angle PAQ = 90° + 45°$
$= 135°$

b Bearing of R from $A = 90 + 45 + 50 = 185°$

c $\cos C = \dfrac{a^2 + b^2 - c^2}{2ab}$

$$\cos C = \frac{31^2 + 28^2 - 35^2}{2 \times 31 \times 28}$$
$$C = 72.57\ldots$$

$\angle PAB \approx 73°$

Question 13

a $c^2 = a^2 + b^2 - 2ab\cos C$

$$BC^2 = 1.7^2 + 2.9^2 - 2 \times 1.7 \times 2.9 \times \cos 52°$$
$$BC = \sqrt{1.7^2 + 2.9^2 - 2 \times 1.7 \times 2.9 \times \cos 52°}$$
$$= 2.286\ldots$$
$$\approx 2.3\text{ km}$$

Length of course $= 2.9 + 2.3 + 1.7$
$= 6.9$
≈ 7 km

b $A = \frac{1}{2} \times 2.9 \times 1.7 \times \sin 52°$

$$= 1.942\ldots$$
$$\approx 1.94\text{ km}^2$$

Question 14

a

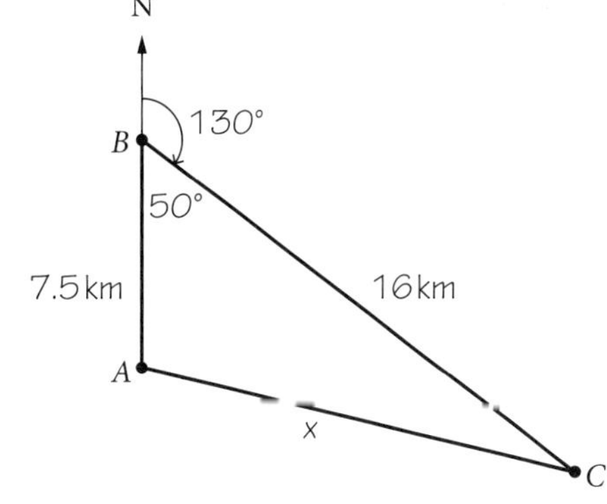

$c^2 = a^2 + b^2 - 2ab\cos C$

$$x^2 = 7.5^2 + 16^2 - 2 \times 7.5 \times 16 \times \cos 50°$$
$$AC = \sqrt{7.5^2 + 16^2 - 2 \times 7.5 \times 16 \times \cos 50°}$$
$$= 12.569\ldots$$
$$\approx 12.6\text{ km}$$

b

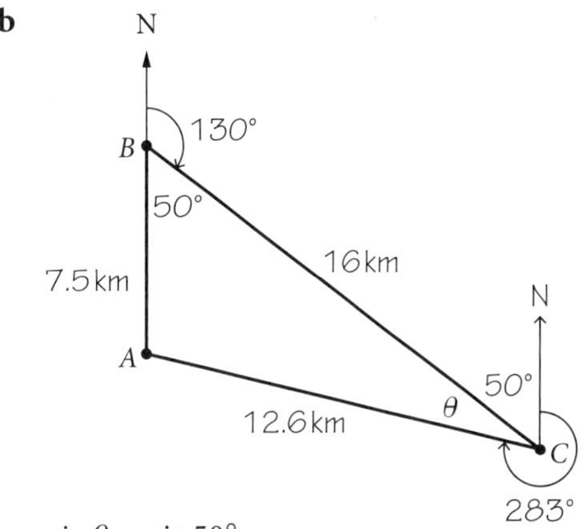

$$\frac{\sin\theta}{7.5} = \frac{\sin 50°}{12.6}$$
$$\sin\theta = \frac{7.5\sin 50°}{12.6}$$
$$\theta = 27.127\ldots$$
$$\approx 27°$$

The bearing of A from C is:
$360° - 50° - 27° = 283°$

Question 15

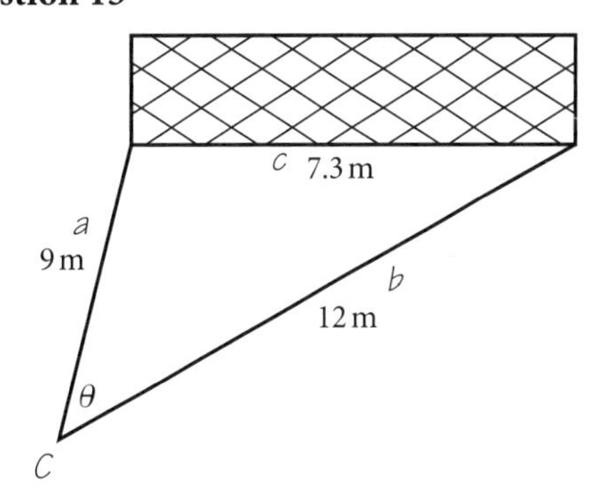

$$\cos\theta = \frac{9^2 + 12^2 - 7.3^2}{2 \times 9 \times 12}$$
$$\theta = 37°20'82''$$
$$\approx 37°21'$$

Question 16

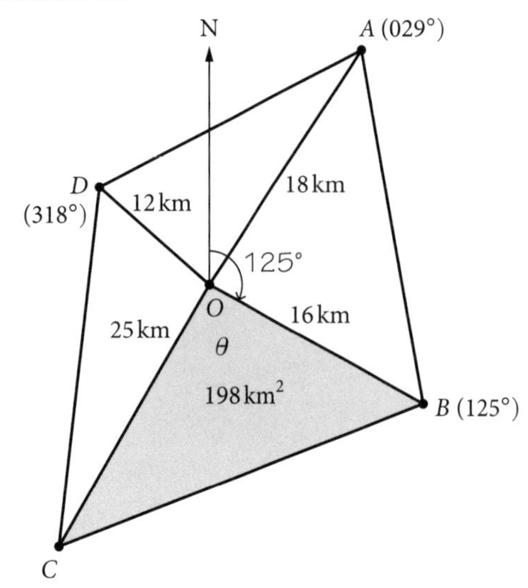

$$A = \frac{1}{2}ab\sin C$$

$$198 = \frac{1}{2} \times 25 \times 16 \times \sin\theta$$

$$198 = 200 \times \sin\theta$$

$$\sin\theta = \frac{198}{200}$$

$$\theta = 81.89\ldots$$

$$\approx 82^\circ$$

So bearing of C from $O = 125^\circ + 82^\circ$

$= 207^\circ$

Question 17

Let BC and $CD = x$.

Using Pythagoras' theorem:

$$x^2 + x^2 = 10^2$$

$$2x^2 = 100$$

$$x^2 = 50$$

$$x = \sqrt{50}$$

$$\approx 7\text{ km}$$

Distance A to $C = 16 + 7 = 23$ km

Using Pythagoras' theorem for distance D to A:

$$DA^2 = 23^2 + 7^2$$

$$DA = \sqrt{578}$$

$$= 24\text{ km}$$

So total distance cycled $= 16 + 7 + 7 + 24$

≈ 54 km

Question 18

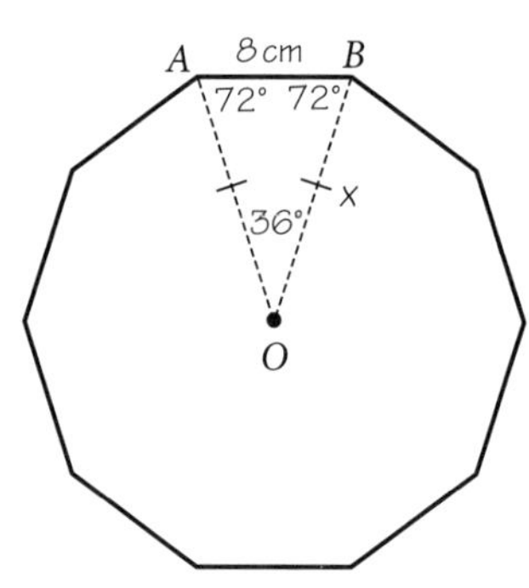

$$AB = 80 \div 10$$

$$= 8\text{ cm}$$

$$\angle AOB = 360^\circ \div 10$$

$$= 36^\circ$$

$$\angle OAB = \angle OBA = \frac{180 - 36}{2}$$

$$= 72^\circ$$

$$\frac{x}{\sin 72^\circ} = \frac{8}{\sin 36^\circ}$$

$$x = \frac{8}{\sin 36^\circ} \times \sin 72^\circ$$

$$= 12.94\ldots$$

Area of $\Delta OAB = \frac{1}{2} \times 8 \times 12.94\ldots \times \sin 72^\circ$

$= 49.24\ldots$

Total area $= 49.24 \times 10$

$= 492.429\ldots$

$\approx 492.4\text{ cm}^2$

Question 19

$180^\circ - 80^\circ = 100^\circ$

(cointerior angles are supplementary)

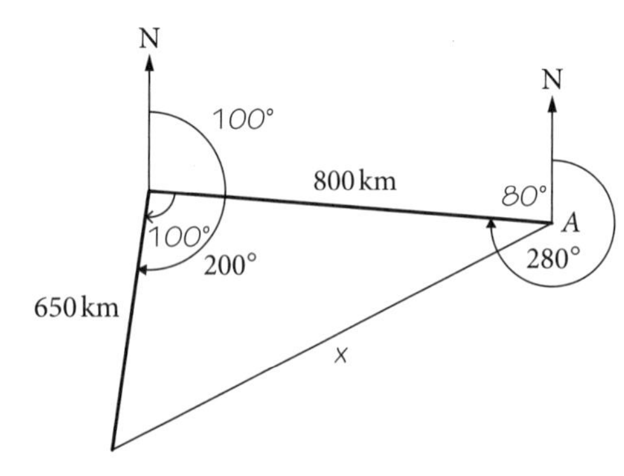

$$c^2 = a^2 + b^2 - 2ab\cos C$$

$$x^2 = 650^2 + 800^2 - 2 \times 650 \times 800 \times \cos 100^\circ$$

$$x = \sqrt{650^2 + 800^2 - 2 \times 650 \times 800 \times \cos 100^\circ}$$

$$= 1114.941\ldots$$

$$\approx 1110\text{ km}$$

The plane is 1110 km from Sydney.

Question 20

a $PC^2 = 5.4^2 + 1.8^2 - 2 \times 5.4 \times 1.8 \times \cos 108°$

$PC = \sqrt{5.4^2 + 1.8^2 - 2 \times 5.4 \times 1.8 \times \cos 108°}$

$PC = 6.19736$

$\approx 6.197\text{ m}$

b $\dfrac{\sin C}{5.4} = \dfrac{\sin 108°}{6.197}$

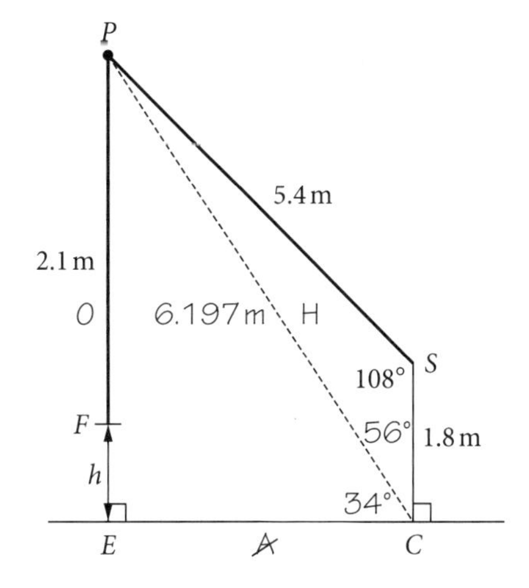

$\sin C = \dfrac{\sin 108°}{6.197} \times 5.4$

$C = 55.969\ldots$

$\angle PCS \approx 56°$

So $\angle ECP = 90° - 56°$

$= 34°$

$\sin 34° = \dfrac{PE}{6.197}$

$PE = 6.197 \times \sin 34°$

$= 3.465\ldots$

$h = 3.468\ldots - 2.1$

$= 1.368\ldots$

$\approx 1.37\text{ m}$

Alternative calculation:

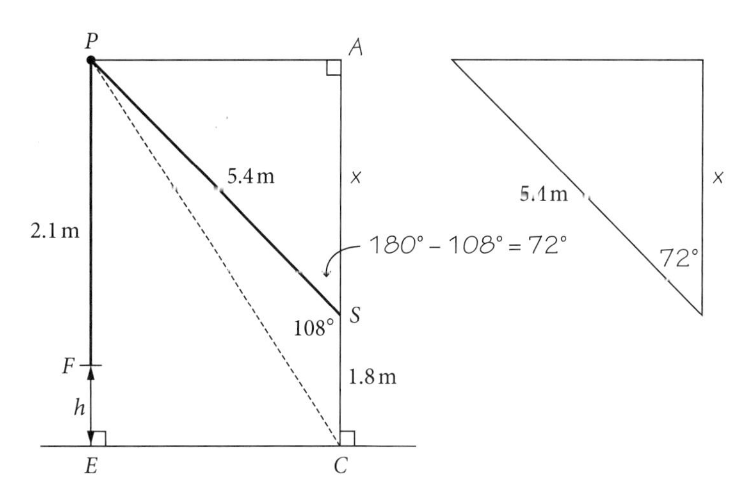

$\cos 72° = \dfrac{x}{5.4}$

$5.4 \times \cos 72° = x$

$x = 1.6686\ldots$

$AC = PE = 1.6686 + 1.8 = 3.468\ldots$

$h = 3.468\ldots - 2.1$

$= 1.368\ldots$

$\approx 1.37\text{ m}$

WORKED SOLUTIONS

HSC exam topic grid (2011–2020)

This grid shows the coverage of this topic in past HSC exams by question number. The past exam papers can be downloaded from the NESA website (www.educationstandards.nsw.edu.au) by selecting 'Year 11 – Year 12', 'HSC exam papers'. NESA marking feedback and guidelines can also be found there.

Before 2019, 'Mathematics Standard 2' was called 'Mathematics General 2' and, before 2014, 'General Mathematics'. For these exams, select 'Year 11 – Year 12', 'Resources archive', 'HSC exam papers archive'.

	Right-angled trigonometry	The sine and cosine rules	Area of a triangle	Bearings
2011	4, 9	24(c)		24(c)
2012	4, 27(d)	**20**, 29(c)	10	**20**
2013	4	24, 26(a)	28(a)*	28(a)*
2014	2(b)	28(b)*	28(b)*	**23**, 28(b)*
2015	9	**30(e)**	22	7*
2016	26(d)	25, 30(c)	30(c)	25
2017	8, 26(d)	30(c)		30(c)
2018	22, 30(c)	12	30(c)	7
2019 new course	12, 22	17	**35***	4, **35***
2020	16	**31**, **32**	**32**	**31**

Questions in **bold** can be found in this chapter.
* Radial survey.

CHAPTER 3
RATES AND RATIOS

MS-M7 Rates and ratios 76

RATES AND RATIOS

Rate problems

- Best buys
- Speed
- Heart rates
- Electricity and energy
- Fuel consumption

Ratio problems

- Simplifying ratios
- Building materials, mixtures and costs
- Dividing a quantity in a given ratio
- Capture–recapture method
- Scales on maps

Scale drawings

- Building plans, elevation views
- Perimeter, area, surface area and volume
- The trapezoidal rule,
  $A \approx \frac{h}{2}(d_f + d_l)$,
  to approximate irregular areas

Glossary

A+ DIGITAL FLASHCARDS
Revise this topic's key terms and concepts by scanning the QR code or typing the URL into your browser.

https://get.ga/a-hsc-maths-standard-2

best buy
The cheapest item per quantity or item. It is determined by comparing unit prices.

capture–recapture method
A sampling method commonly used in ecology to estimate the size of an animal population by capturing a portion of the population, marking and releasing them, then capturing another sample.

elevation view
A scale drawing showing what a building looks like from the front, back or sides, or from the north, south, east or west.

fuel consumption
The rate at which an engine uses fuel; measured in litres per 100 kilometres (L/100 km).

heart rate
The speed of a heartbeat in beats per minute (bpm); the average number of contractions of a person's heart in 1 minute.

kilowatt-hour (kWh)
A measure of energy usage (usually electrical) by a 1000-watt load drawing power for 1 hour.

rate
A measurement that compares two quantities with different units, such as pay rate ($/h), fuel consumption (L/100 km) and heart rate (beats/min).

ratio
A proportion of 2 or more quantities with the same units that compares parts or shares. For example, the ratio of the length of a side of a square to its perimeter is 1 : 4.

scale drawing
A drawing that is a reduction or enlargement of the real object it represents. The scale compares the lengths in the drawing to the actual lengths. For example, if the scale is 1 : 200, the actual lengths are 200 times the size of the scaled lengths. Scale scan be written with units (1 cm to 1 m) or without units (1 : 100).

scale ratio
A ratio or scale factor that enlarges or reduces lengths in scale drawing and corresponding side lengths in similar figures. It is a measure of magnification or reduction. A scale ratio of 1 : 2 indicates that lengths on the actual object are 2 times the matching lengths on the similar figure or scale drawing.

speed
A rate that indicates how fast an object is moving; the rate at which an object covers distance in a given amount of time.

$$\text{Average speed } (S) = \frac{\text{distance } (D)}{\text{time } (T)}$$

trapezoidal rule
The rule and formula $A \approx \frac{h}{2}(d_f + d_l)$, which uses trapeziums to approximate the area of an irregular shape, often with a curved boundary.

unit price
The cost of 1 unit, which is used to compare prices and find the best buy. Unit costs are expressed as a rate such as $/kg, cents/L and are calculated using:

Unit price = total price ÷ number of items or units

Topic summary

Review of metric units (Year 11)

Length

Metric multiple	Unit name	Unit symbol	Conversion
$\frac{1}{1000}$ m	millimetre	mm	
$\frac{1}{100}$ m	centimetre	cm	1 cm = 10 mm
1 m	metre	m	1 m = 100 cm 1 m = 1000 mm
1000 m	kilometre	km	1 km = 1000 m

millimetre (mm) ⟶ centimetre (cm) ⟶ metre (m) ⟶ kilometre (km)

Area

square centimetre (cm^2)

1 cm

Area = 1 cm × 1 cm
= 10 mm × 10 mm
= 100 mm^2

square metre (m^2)

1 m

Area = 1 m × 1 m
= 100 cm × 100 cm
= 10 000 cm^2

Unit name	Unit symbol	Conversion
square millimetre	mm^2	
square centimetre	cm^2	1 cm^2 = 100 mm^2
square metre	m^2	1 m^2 = 10 000 cm^2 1 m^2 = 1 000 000 mm^2
hectare	ha	1 ha = 10 000 m^2
square kilometre	km^2	1 km^2 = 100 ha 1 km^2 = 1 000 000 m^2

square millimetre (mm^2) ⟶ square centimetre (cm^2) ⟶ square metre (m^2) ⟶ hectare (ha) ⟶ square kilometre (km^2)

Volume

cubic centimetre (cm^3)

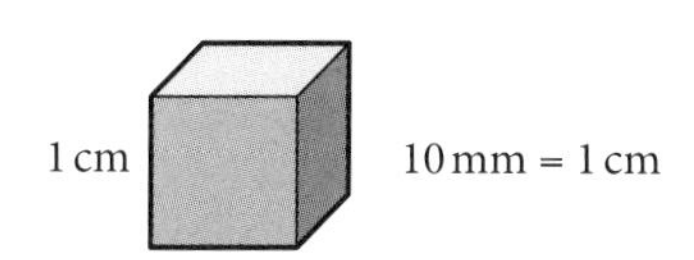

cubic metre (m^3)

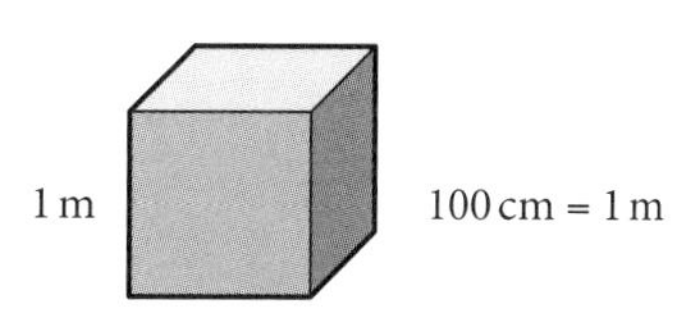

$$\text{Volume} = 1\text{ cm}^3 = 1\text{ cm} \times 1\text{ cm} \times 1\text{ cm} = 10\text{ mm} \times 10\text{ mm} \times 10\text{ mm} = 1000\text{ mm}^3$$

$$\text{Volume} = 1\text{ m}^3 = 1\text{ m} \times 1\text{ m} \times 1\text{ m} = 100\text{ cm} \times 100\text{ cm} \times 100\text{ cm} = 1\,000\,000\text{ cm}^3$$

Unit name	Unit symbol	Conversion
cubic millimetre	mm^3	
cubic centimetre	cm^3	$1\text{ cm}^3 = 1000\text{ mm}^3$
cubic metre	m^3	$1\text{ m}^3 = 1\,000\,000\text{ cm}^3$

cubic millimetre (mm^3) ⟶ cubic centimetre (cm^3) ⟶ cubic metre (m^3)

Capacity

Metric multiple	Unit name	Unit symbol	Conversion
$\frac{1}{1000}$ L	millilitre	mL	
1 L	litre	L	1 L = 1000 mL
1000 L	kilolitre	kL	1 kL = 1000 L
1 000 000 L	megalitre	ML	1 ML = 1000 kL 1 ML = 1 000 000 L

millilitre (mL) ⟶ litre (L) ⟶ kilolitre (kL) ⟶ megalitre (ML)

Volume and capacity

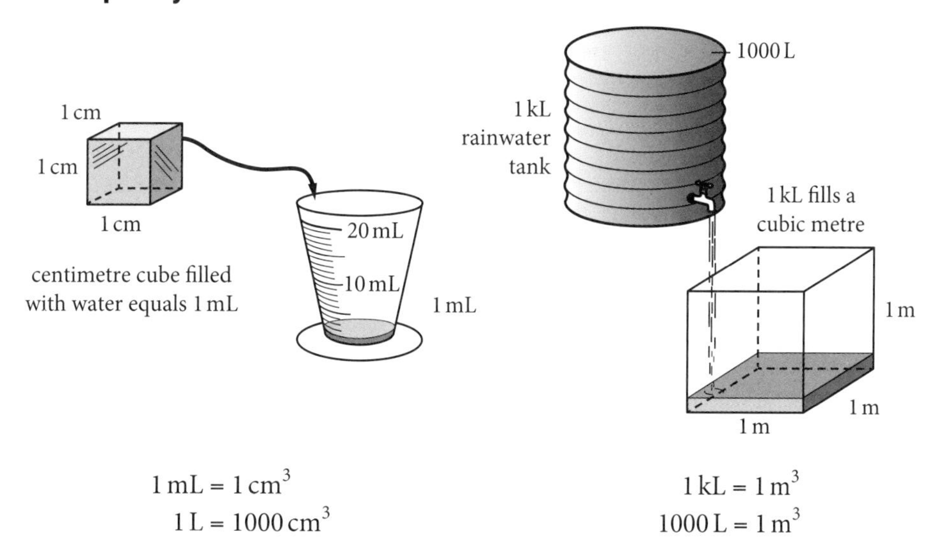

$1\text{ mL} = 1\text{ cm}^3$
$1\text{ L} = 1000\text{ cm}^3$

$1\text{ kL} = 1\text{ m}^3$
$1000\text{ L} = 1\text{ m}^3$

Mass

Metric multiple	Unit name	Unit symbol	Conversion
$\frac{1}{1000}$ g	milligram	mg	
1 g	gram	g	1 g = 1000 mg
1000 g	kilogram	kg	1 kg = 1000 g
1 000 000 g	tonne	t	1 t = 1000 kg

milligram (mg) ⟶ gram (g) ⟶ kilogram (kg) ⟶ tonne (t)

Time

Unit name	Unit symbol	Conversion
second	s	
minute	min	1 min = 60 s
hour	h	1 h = 60 min 1 h = 3600 s
day		1 day = 24 h
year		1 year = 365 days 1 leap year = 366 days

second (s) ⟶ minute (min) ⟶ hour (h) ⟶ day ⟶ year

Note: 1 h = 60 × 60 s = 3600 s

Rates and ratios (MS-M7)

Rates

A **rate** is a measurement that compares 2 quantities that have different units.

Simplifying rates

Example 1

a A car travels 160 km in 2 hours. What is its average speed in km/h?

b A cook works for 3 hours and earns $75. What is the chef's hourly pay rate?

Solution

a $\frac{160\text{ km}}{2\text{ h}} = 80\text{ km/h}$

b $\frac{\$75}{3\text{ h}} = \$25\text{/h}$

Converting rates

Look at both units in the rate a/b.

- For the 'a' (numerator) part of the rate, if converting:
  - a smaller unit to a larger unit, then DIVIDE
  - a larger unit to a smaller unit, then MULTIPLY.
- The opposite applies for the 'b' (denominator) part of the rate. If converting:
  - a smaller unit to a larger unit, then MULTIPLY
  - a larger unit to a smaller unit, then DIVIDE.

Example 2

Convert \$8.20/kg to c/g.

Solution

Numerator: dollars, \$8.20 to cents, ? c

large unit ⟶ small unit, so MULTIPLY by 100

× 100

\$8.20/kg = ? **c**/g

$8.20 \times 100 = 820$c

So \$8.20/kg is 820c/kg.

Denominator: per kilogram (1 kg) to per gram (1000 g)

large unit ⟶ small unit, so DIVIDE by 1000

\$8.20/**kg** = ? c/**g**

÷ 1000

$820 \div 1000 = 0.82$c

So \$8.20/kg is 0.82c/g.

Speed

Speed is a rate that compares distance travelled with time taken. It is measured in km/h or m/s.

$$\text{Average speed } (S) = \frac{\text{distance } (D)}{\text{time } (T)}$$

Hint
The speed formula is not given on the HSC exam reference sheet.

Placing your thumb to cover one of the variables in the 'speed triangle' shown gives you the formula for that variable.

$$S = \frac{D}{T}$$

$$D = S \times T$$

$$T = \frac{D}{S}$$

D

S × T

Example 3

a An athlete runs 100 m in 11.23 s. What is her average speed? Give your answer in m/s, correct to one decimal place.

b How long will a horse take to run 2400 m if it runs at an average speed of 60 km/h? Give your answer to the nearest second.

Solution

a $S = \frac{D}{T}$

$= \frac{100\,\text{m}}{11.23\,\text{s}}$

$\approx 8.9\,\text{m/s}$

b $T = \frac{D}{S} = \frac{2400\,\text{m}}{60\,\text{km/h}}$

Distance unit, m, and speed unit, km/h, need to include the same unit of length, so convert 2400 m to km:

$2400\,\text{m} = 2400 \div 1000 = 2.4\,\text{km}$

$T = \frac{2.4\,\text{km}}{60\,\text{km/h}} = 0.04\,\text{h}$

Next, convert hours to minutes:
$0.04\,\text{h} = 0.04 \times 60 = 2.4\,\text{min}$

Now convert 2.4 min to minutes and seconds by using the degrees–minutes–seconds key (° ′ ″ or DMS) on your calculator or by calculating 0.4×60 for the number of seconds in 2.4 minutes:

$T = 2.4\,\text{min} = 2\,\text{min}\ 24\,\text{s}$

Converting units of speed

Some questions in HSC exams involve conversions of speed.

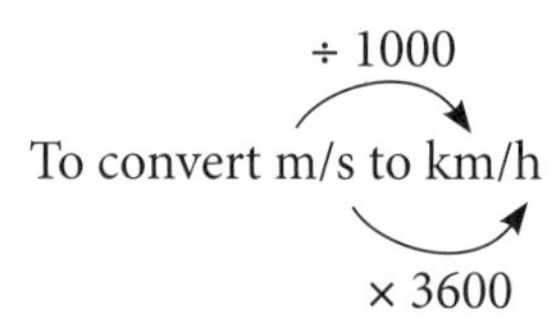

Numerator: Convert metres (m) to kilometres (km) by dividing by 1000 (small unit to large unit).

Denominator: Convert per 1 second (s) to per 1 hour (h) by multiplying by 3600 (small unit to large unit).

Note: There are 3600 seconds in 1 hour and 1000 m in 1 km.

Example 4

Convert 15 m/s to km/h.

Solution

$15\,\text{m/s} = 15 \div 1000 \times 3600$

$= 54\,\text{km/h}$

Example 5

Convert 60 km/h to m/s, correct to two decimal places.

Solution

$60\text{ km/h} = 60 \times 1000 \div 3600$
$\approx 16.67\text{ m/s}$

Heart rates

Example 6

The average **heart rate** in adults is 72 beats per minute. How many times will a heart beat in 1 hour at this rate?

Solution

$72\text{ beats/min} \times 60 = 4320\text{ beats/h}$

Energy and power

Energy, such as electricity, is measured in joules (J).

Power is the rate that energy is used or transferred per unit time. It is measured in watts (W), which is equal to 1 joule per second.

The units of power are shown in the table.

Metric multiple	Unit name	Unit symbol	Conversion
$\frac{1}{1000}$ W	milliwatt	mW	
1 W	watt	W	1 W = 1000 mW
1000 W	kilowatt	kW	1 kW = 1000 W
1 000 000 W	megawatt	MW	1 MW = 1000 kW
1 000 000 000 W	gigawatt	GW	1 GW = 1000 MW

milliwatt (mW) ⟶ watt (W) ⟶ kilowatt (kW) ⟶ megawatt (MW) ⟶ gigawatt (GW)

Electrical energy usage for households is measured in **kilowatt-hours** (kWh).

The cost of electricity usage is given in cents/kWh.

Energy consumption = power (kW) × time (h) = kWh

Hint
Express power in kilowatts (kW) and express time in hours (h).

Example 7

Calculate the energy consumed by a dishwasher of power 1400 W during a washing cycle of 90 minutes.

Solution

Convert 1400 W to kilowatts: 1400 W ÷ 1000 = 1.4 kW

Convert 90 min to hours: 90 ÷ 60 = 1.5 h

$$\begin{aligned}\text{Energy consumption} &= 1.4\,\text{kW} \times 1.5\,\text{h}\\ &= 2.1\,\text{kWh}\end{aligned}$$

Example 8

Calculate the cost of running a 2400 W fan heater for 6 hours per day for 1 week. Assume the cost of electricity is $0.29/kWh.

Solution

Convert 2400 W to kW: 2400 W ÷ 1000 = 2.4 kW

Total energy consumption = 2.4 kW × 6 h × 7 days = 100.8 kWh.

$$\begin{aligned}\text{Running cost of appliance} &= \text{energy consumption (kWh)} \times \text{electricity cost (\$/kWh)}\\ &= 100.8\,\text{kWh} \times \$0.29/\text{kWh}\\ &= \$29.23\end{aligned}$$

Fuel consumption

Fuel consumption (or fuel economy) is the rate at which a vehicle uses fuel, such as petrol, measured in litres per 100 kilometres (L/100 km). This is the average number of litres of fuel used to travel a distance of 100 km.

$$\text{Fuel consumption rate (L/100 km)} = \frac{\text{amount of fuel used (L)}}{\text{distance travelled (km)}} \times 100$$

The lower this rate of L/100 km, the better the fuel consumption for the vehicle, because fewer litres of fuel are used to travel 100 km.

Example 9

A car travels 520 km and uses 40 L of petrol. Calculate its fuel consumption rate in L/100 km, correct to one decimal place.

Solution

$$\text{Fuel consumption rate} = \frac{40\,\text{L}}{520\,\text{km}} \times 100 \approx 7.7\,\text{L/100 km}$$

Example 10

If a car has an average fuel economy rate of 8.6 L/100 km, how far can it travel on a full tank of 70 L of petrol, correct to the nearest kilometre?

Solution

$$\text{Distance travelled (km)} = \frac{\text{amount of fuel used (L)}}{\text{fuel consumption (L/100 km)}} \times 100$$

$$\text{Distance} = \frac{70\,\text{L}}{8.6\,\text{L/100 km}} \times 100 \approx 814\,\text{km}$$

9780170459204

Best buys (unit pricing)

When comparing prices of similar items, the **best buy** is the cheapest item per quantity or item, calculated by comparing unit prices.

Unit price = total price ÷ number of items or units

The best buy is the item with the lowest **unit price**.

> **Hint**
> Before comparing items, make sure the units are the same.

Example 11

Which is the better buy for fruit juice: \$1.50 for 250 mL or \$7 for 1.25 L?

Solution

Find each cost as a rate in \$/L.

Convert 250 mL to litres: 250 mL = 0.25 L

Product 1:

Unit price $= \dfrac{\$1.50}{0.25\text{L}} = \$6/\text{L}$

Product 2:

Unit price $= \dfrac{\$7}{1.25\text{L}} = \$5.60/\text{L}$

Hence, \$7 for 1.25 L is the better buy because the unit price is less.

Ratios

A **ratio** is a proportion of 2 or more quantities with the same units that compares parts or shares.

$a:b$ ('a to b') compares the first quantity, a, to the second quantity, b.

The terms in a ratio must have the same units.

$$1\text{ g}:1\text{ kg} = 1\text{ g}:1000\text{ g} = 1:1000$$

1 : 2 is equivalent to 1 mL : 2 mL, or 10 mL : 20 mL, or 1 L : 2 L.

Example 12

a A class contains 7 tables and 10 chairs. What is this as a ratio?

b Write a ratio that compares 1 mm to 1 m.

Solution

a Ratio of tables to chairs = 7 : 10

b 1 mm : 1 m

1 m is equal to 1000 mm

1 mm : 1000 m

So ratio is 1 : 1000.

Simplifying ratios

A ratio is simplified by dividing all terms in the ratio by their highest common factor (HCF).

If the terms are fractions or decimals, multiply each term in the ratio by a common factor to make them whole numbers.

Example 13

Express the following ratios in simplest form.

a $10:15$ **b** $1.2:2.4$ **c** 50 cents : $2

Solution

a

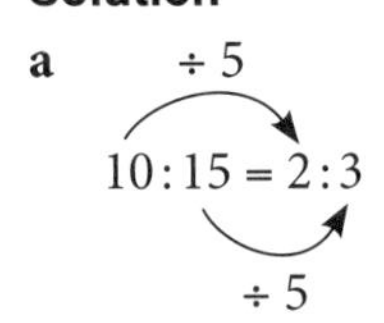

b

$\times 10$ $\div 12$

$$1.2:2.4 = 12:24 = 1:2$$

$\times 10$ $\div 12$

c

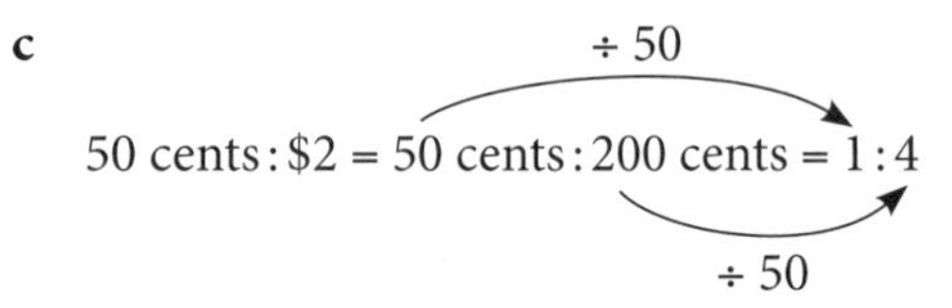

> **Hint**
> If there are 2 terms in the ratio, then you can use the fraction key on your calculator to simplify: $\frac{50}{200} = \frac{1}{4} = 1:4$.

Ratio problems

Example 14

The ratio of jackets to jumpers worn by students in a class is $2:3$. If there are 8 jackets, how many jumpers are there?

Solution

Let x be the number of jumpers.

jackets : jumpers

$2:3$

$8:x$

Method 1

Form an equation and solve it:

$$\frac{2}{3} = \frac{8}{x}$$

$$\frac{3}{2} = \frac{x}{8}$$

$$\frac{x}{8} = \frac{3}{2}$$

$$x = \frac{3}{2} \times 8$$

$$= 12$$

So there are 12 jumpers.

Method 2

Solve by equal ratios:

$\times 4$

$$2:3 = 8:x$$

$\times 4$

$$2 \times 4 = 8$$

$$x = 3 \times 4 = 12$$

So there are 12 jumpers.

Dividing a quantity in a given ratio

Example 15

Divide \$40 in the ratio 2 : 3.

Solution

Total parts = 2 + 3 = 5

$$\begin{aligned} 1 \text{ part} &= \text{quantity} \div \text{total parts} \\ &= \$40 \div 5 \\ &= \$8 \end{aligned}$$

Multiply the value of 1 part by each number of parts in the ratio (2 : 3):

$$\begin{aligned} 2 \text{ parts} &= 2 \times \$8 = \$16 \\ 3 \text{ parts} &= 3 \times \$8 = \$24 \end{aligned}$$

Check by adding these values to get the original quantity:

$$\$16 + \$24 = \$40$$

So \$40 divided in the ratio 2 : 3 is \$16 : \$24.

Capture–recapture method

The **capture–recapture method** is commonly used to estimate an animal population's size. A small portion of the population is captured, marked and then released. Later, another sample is captured and the marked animals are counted. The ratio of marked animals is used to estimate the total population.

Example 16

Sonja uses the following method to estimate a population of fish in a lake. She catches 65 fish, tags them and releases them. Later, she catches 100 fish at random. Of these 100 fish, 20 fish are tagged. Estimate the population of fish in the lake.

Solution

Let P be the estimated population of fish.

Compare the proportion of tagged fish in the population and the recaptured sample:

Population: tagged = 65, total = P
Sample: tagged = 20, total = 100

$$\begin{aligned} \frac{65}{P} &= \frac{20}{100} \\ \frac{P}{65} &= \frac{100}{20} \\ \frac{P}{65} &= 5 \\ P &= 65 \times 5 \\ &= 325 \end{aligned}$$

So there are an estimated 325 fish in the lake.

Alternative formula:

$$\begin{aligned} \text{Estimated population } (P) &= \frac{\text{number tagged in population}}{\text{number tagged in sample}} \times \text{total sample} \\ &= \frac{65}{20} \times 100 \\ &= 325 \end{aligned}$$

Scale drawings

A **scale drawing** is a representation of an object at a size larger or smaller than the object's actual size. For example, the drawing could be a reduced-size representation of a house or an enlarged representation of a computer chip. All dimensions of a scale drawing are in the same proportion as the original object.

The **scale ratio** compares scale lengths (the length on the drawing) to actual lengths.

For example, if the scale ratio is 1 : 50, the actual lengths are 50 times the scaled lengths, which indicates every 1 cm measured on a scale drawing is equal to 50 cm on the real object.

Example 17

By measurement and calculation, find the actual length (in centimetres) of the tennis racquet.

Tennis racquet 1 : 12

Solution

Scaled length of racquet = 5.5 cm

Actual length of racquet = 12 × 5.5 cm = 66 cm

Example 18

Write each scale as a ratio.

a (scale bar) 0 5 10 15 20 km

b 10 mm to 1 m

Solution

a Scale = 1 cm : 5 km
= 1 cm : 500 000 cm
= 1 : 500 000

b Scale = 10 mm : 1 m
= 10 mm : 1000 mm
= 1 : 100

Building plans

A plan or floor plan is a diagram showing the layout of rooms and features in a house or building.

An **elevation view** is a scale drawing of the front, back or sides of a building, or the north, south, east or west side.

Measurements on plans are usually given in millimetres to avoid using decimal points.

For area and volume problems, it is often best to first convert dimensions from millimetres to metres.

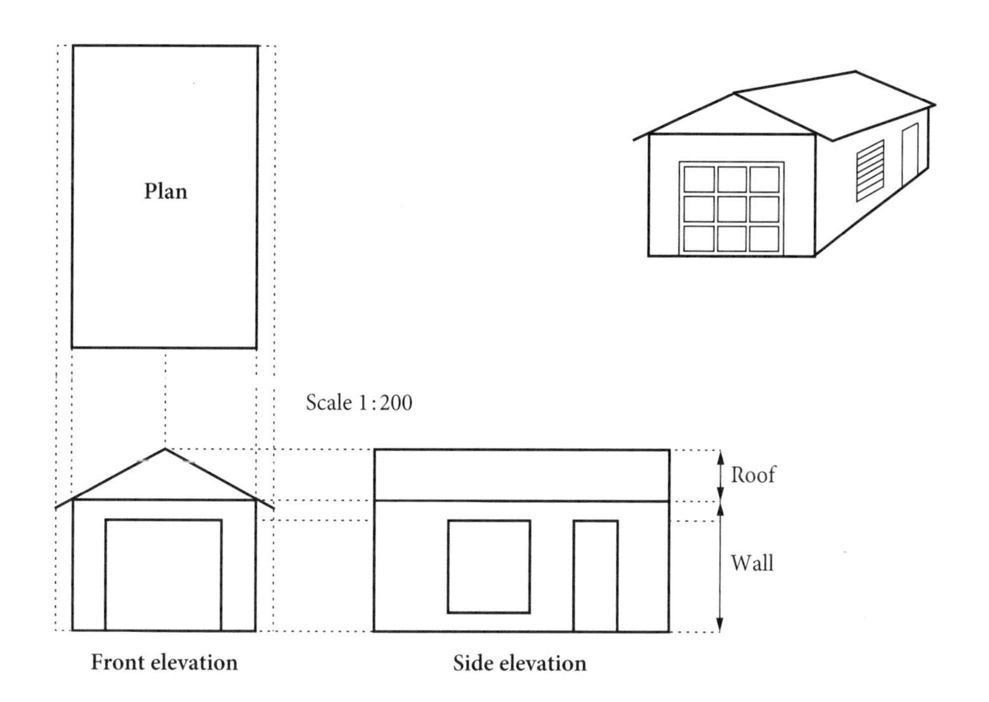

Symbols on house plans

	Hinged door	Ens	Ensuite (bathroom)
	Toilet (water closet) or WC		Bath
	Shower		Vanity
	Stove		Kitchen sink
Ptry	Pantry		Laundry tub
WM	Washing machine		Linen cupboard
Robe	Wardrobe	WIR	Walk-in (ward)robe
FL	Floor level	GL	Ground level
CL	Ceiling level		Window

Example 19

This is the floor plan of the ground level of a split-level home. Measurements are in metres.

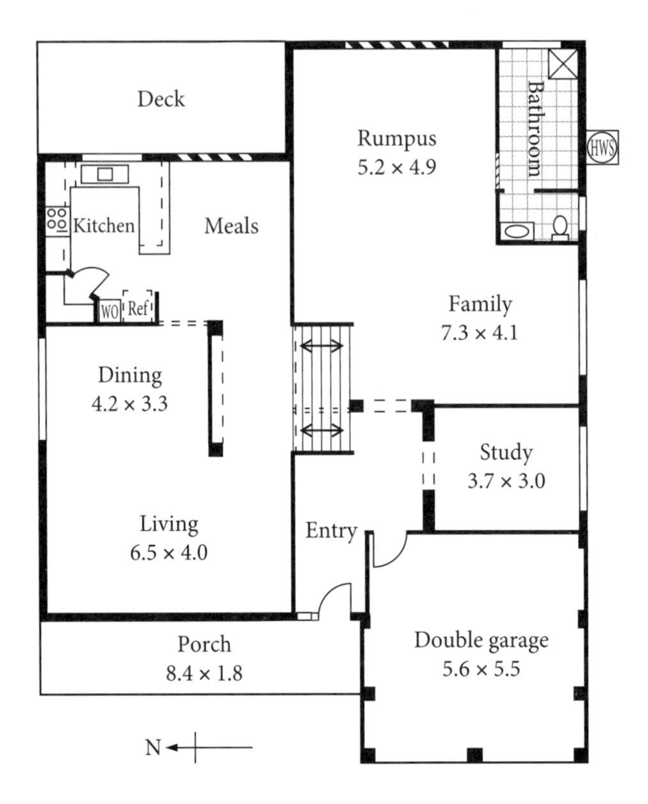

a The porch is to be tiled with terracotta tiles. How many square metres of tiles are needed?

b If each tile measures 30 cm by 30 cm, how many tiles are needed to tile the porch?

c The dining and living areas are to be carpeted. How much (correct to the nearest \$100) will it cost to carpet the rooms if the price of carpet is \$145 per m^2 (including laying)?

Solution

a Area of porch = 8.4×1.8
$= 15.12\,\text{m}^2$

b Area of tile = $30\,\text{cm} \times 30\,\text{cm}$
$= 0.3\,\text{m} \times 0.3\,\text{m}$ (converting 30 cm to m)
$= 0.09\,\text{m}^2$

No. of tiles needed = $15.12 \div 0.09$
$= 168$

c Area of dining room = 4.2×3.3
$= 13.86\,\text{m}^2$

Area of living room = 6.5×4
$= 26\,\text{m}^2$

Total area = $13.86 + 26$
$= 39.86\,\text{m}^2$

Cost of carpet = $39.86 \times \$145$
$= \$5779.70$
$\approx \$5800$ (to the nearest \$100)

The trapezoidal rule for estimating area

The **trapezoidal rule** is a method and formula that uses trapeziums to approximate the area of an irregular shape, often with a curved boundary.

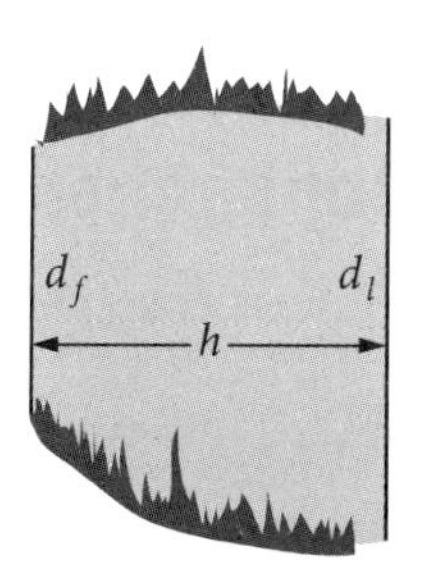

$$\text{Area} = \frac{\text{width of strip}}{2}(\text{first measurement} + \text{last measurement})$$

OR

$$A \approx \frac{h}{2}(d_f + d_l)$$

where h = distance between measurements
d_f = first measurement
d_l = last measurement.

Hint

The formula $A \approx \frac{h}{2}(d_f + d_l)$ is provided on the HSC exam reference sheet (and at the back of this book).

Example 20

Vertical measurements of a dam are taken at 10 m intervals.

a Use 2 applications of the trapezoidal rule to estimate the area of the dam, correct to one decimal place.

b Find the volume of water in the dam if the dam has a constant depth of 2.5 m.

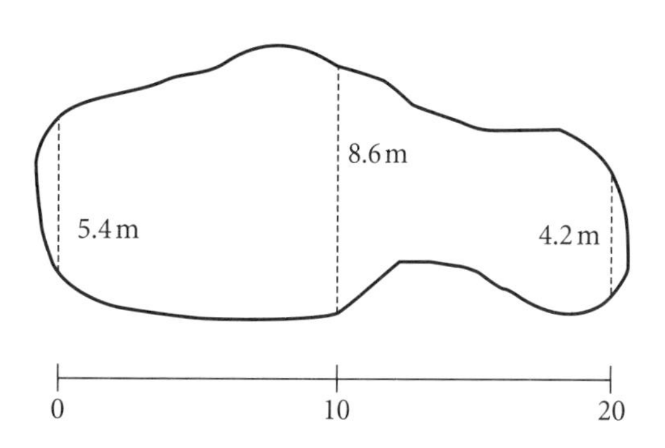

Solution

a Left trapezium:

$d_f = 5.4,\ d_l = 8.6,\ h = 10$

$$A \approx \frac{h}{2}(d_f + d_l)$$
$$\approx \frac{10}{2}(5.4 + 8.6)$$
$$\approx 70$$

Right trapezium:

$d_f = 8.6,\ d_l = 4.2,\ h = 10$

$$A \approx \frac{h}{2}(d_f + d_l)$$
$$\approx \frac{10}{2}(8.6 + 4.2)$$
$$\approx 64$$

The approximate total area of the dam is $70 + 64 = 134\ \text{m}^2$.

b $V = Ah$
$= 134 \times 2.5$
$= 335\ \text{m}^3$

So the volume of the dam is approximately $335\ \text{m}^3$.

TOPIC SUMMARY

Practice set 1

Multiple-choice questions

Solutions start on page 95.

Question 1

Which units express the rate of the cost of bananas in a supermarket?

A \$/h **B** \$/m **C** \$/kg **D** \$/L

Question 2

What is the simplest form of the ratio 3000 : 450?

A 15 : 2 **B** 3 : 2 **C** 3 : 20 **D** 20 : 3

Question 3

The ratio of cows to bulls on a rural property is 33 : 2.

If there are 198 cows, how many bulls are there?

A 3 **B** 6 **C** 12 **D** 17

Question 4

A concrete mix contains cement, sand and gravel in the ratio 1 : 3 : 5.

If a driveway requires 36 m^3 of concrete, how much sand is required?

A 4 m^3 **B** 9 m^3 **C** 12 m^3 **D** 20 m^3

Question 5

What is 1.5 g/mL expressed in g/L?

A 15 g/L **B** 150 g/L **C** 1500 g/L **D** 15 000 g/L

Question 6

A 250 mL bottle of cough mixture contains 21.6 g/40 mL of sorbitol.

How much sorbitol is in the bottle?

A 13.5 g **B** 33 g **C** 135 g **D** 463 g

Question 7

A horse runs at 12 m/s.

At this speed, how far could it run in 2 minutes?

A 24 m **B** 720 m **C** 1.4 km **D** 1.44 km

Question 8

The cost of a 600 mL bottle of juice is \$3.75.

What is the cost per litre?

A \$1.60 **B** \$4.50 **C** \$6.20 **D** \$6.25

Question 9

On the open road, a car's fuel economy is 7.8 L/100 km.

Approximately how many litres of fuel will the car use to travel 425 km?

A 25 L **B** 33 L **C** 54 L **D** 66 L

Question 10

If a plane is travelling at an average cruising speed of 900 km/h, how long should it take to travel a distance of 600 km?

A 40 min **B** 66 min **C** 90 min **D** 1 h 50 min

Question 11

If a water hose is left running at 75 L/min, how long will it take to fill a pool that requires 45 000 L?

A 6 hours **B** 10 hours **C** 25 hours **D** 2 days

Question 12

A truck travels 350 m in 20 seconds.

What is its average speed in km/h?

A 17.5 km/h **B** 35 km/h **C** 63 km/h **D** 70 km/h

Question 13

What is the cost of running an 1800 W dishwasher for a 45-minute cycle, given the electricity charge is $0.326/kWh?

A $0.44 **B** $2.61 **C** $4.40 **D** $26.41

Question 14

A map scale reads 1 : 20 000.

What is the actual distance for every 1 cm on the map?

A 20 m **B** 200 m **C** 2000 m **D** 20 000 m

Question 15

In a recipe, the ratio of flour to sugar is 5 : 3, and the ratio of sugar to butter is 4 : 1.

What is the ratio of flour to butter?

A 5 : 1 **B** 5 : 8 **C** 20 : 1 **D** 20 : 3

Question 16

A greyhound wins a 515 m race in 29.62 seconds.

What is its average speed for the race in km/h?

A 10 km/h **B** 17 km/h **C** 52 km/h **D** 63 km/h

Question 17

A trout farmer took 600 trout from his dam, tagged and returned them. When a sample of 120 was captured a few days later, 35 were tagged.

Approximately how many trout are in the dam?

A 2057 **B** 17 500 **C** 21 000 **D** 25 200

Question 18

In a class, one-third of the students walk to school, one-quarter take the bus and the rest travel by car.

What is the ratio of students who walk to those who travel by car?

A 3:4 **B** 3:5 **C** 4:5 **D** 4:7

Question 19

The ratio of dogs to cats in a pet store is 3:5. If 2 dogs and 1 cat are sold, the ratio of dogs to cats becomes 1:2.

How many dogs remain in the store?

A 1 **B** 4 **C** 7 **D** 9

Question 20 ©NESA 2014 HSC EXAM, QUESTION 17

A child who weighs 14 kg needs to be given 15 mg of paracetamol for every 2 kg of body weight. Every 10 mL of a particular medicine contains 120 mg of paracetamol.

What is the correct dosage of this medicine for the child?

A 5.6 mL **B** 8.75 mL **C** 11.43 mL **D** 17.5 mL

Practice set 2

Short-answer questions

Solutions start on page 97.

Question 1 (2 marks) 

A bullet train travels at an average speed of 240 km/h.

How far does it travel in 20 min? 2 marks

Question 2 (2 marks) 

A water-saving showerhead has a flow rate of 7.5 L/min.

How much water is used in a week where the shower is used daily for a 10-minute shower? 2 marks

Question 3 (2 marks)

Show that the unit price of a 2 L bottle of milk for $2.50 is a better buy than a 750 mL carton of milk for $1.05. 2 marks

Question 4 (2 marks) 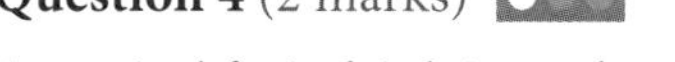

A tropical fruit drink is made up of pineapple juice, orange juice and apple juice in the ratio 5 : 3 : 2.

How many millilitres of pineapple juice are required to make up 2.5 litres of this tropical drink? 2 marks

Question 5 (2 marks)

To travel overseas, Rachel exchanges A$1000 (Australian dollars) for US$1330 (US dollars).
On the same day, Tania wants to exchange US$400 for Australian dollars.

How much will Tania receive in Australian dollars? 2 marks

Question 6 (4 marks)  

Andrew, Brandon and Cosmo are the first three batters in the school cricket team.
In a recent match, Andrew scored 30 runs, Brandon scored 25 runs and Cosmo scored 40 runs.

a What is the ratio of Andrew's to Brandon's to Cosmo's runs scored, in simplest form? 2 marks

b In this match, the ratio of the total number of runs scored by Andrew, Brandon and Cosmo to the total number of runs scored by the whole team is 19 : 36. How many runs were scored by the whole team? 2 marks

Question 7 (2 marks)

How many kilowatt-hours of energy are used when a 4.8 kW air conditioner is operating fully and is uninterrupted between 10 am and 9 pm each day for 125 days of the year? 2 marks

Question 8 (2 marks)

In a cricket match, Suraya's team bats for 20 overs. So far, after 12 overs, Suraya's team has scored 96 runs. Her team needs a total of 182 runs to win.

For the remaining overs, what is the average run rate required per over for her team to win the match? 2 marks

Question 9 (3 marks)

A basketball court has length 28 m and width 15 m.

a What are the dimensions of the court, in centimetres, on a scale drawing if the scale used is 1 : 200? 2 marks

b What scale is used if the length of the basketball court on a scale drawing is 20 cm? Give your answer in simplest ratio form. 1 mark

Question 10 (3 marks)

The floor plan of the ground floor of a split-level home is shown. All measurements are in metres. The diagram is not drawn to scale.

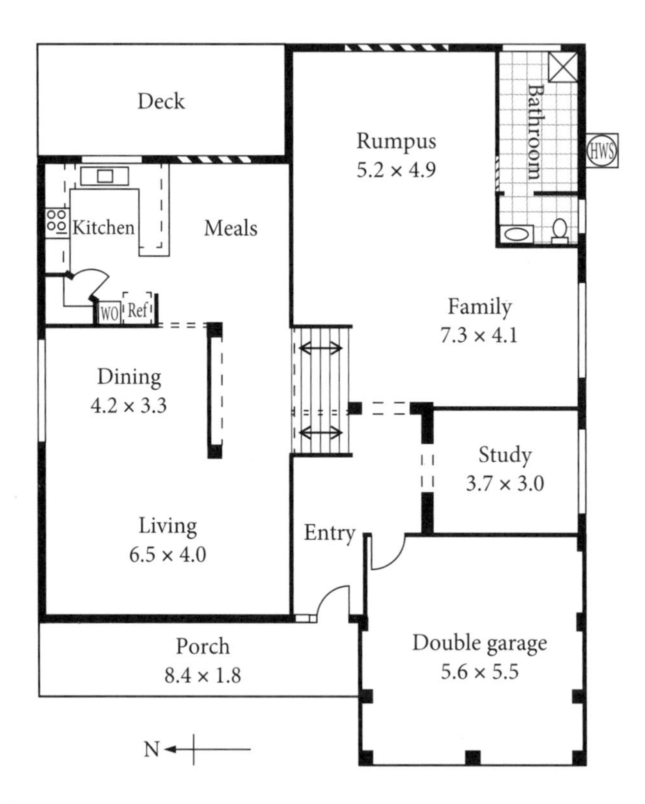

a What is the floor area of the largest room (excluding the garage) on the ground floor? 2 marks

b Which compass direction is the front of the house facing? 1 mark

Question 11 (3 marks)

Joshua owns a van that has a fuel consumption of 10.9 L/100 km in the city and 8.5 L/100 km in the country. Joshua travels 12 000 km per year in the city and 20 000 km per year in the country.

If the cost of petrol is $1.45 per litre in the city and $1.55 per litre in the country, what is the total cost of petrol for Joshua in 1 year? 3 marks

Question 12 (2 marks) ©NESA 2012 HSC EXAM, QUESTION 26(f)

The capture–recapture method was used to estimate a population of seals in 2012.

- 60 seals were caught, tagged and released.
- Later, 120 seals were caught at random.
- 30 of these 120 seals had been tagged.

The estimated population of seals in 2012 was 11% less than the estimated population for 2008.

What was the estimated seal population for 2008? 2 marks

Question 13 (3 marks)

A hospital patient is given an antibiotic fluid intravenously by a drip. The patient needs 3.5 L of fluid per day. Each 1 mL of fluid contains 10 drops.

What is the fluid rate in drops per minute? Answer to the nearest whole number. 3 marks

Question 14 (4 marks)

Charlie runs a 400 m race. He runs the first 300 m at an average speed of 9.08 m/s and the last 100 m at 9.64 m/s.

a Calculate, to the nearest hundredth of a second, the time Charlie takes to complete the race. 2 marks

b What is Charlie's average speed for the entire 400 m race in km/h? Give your answer correct to two decimal places. 2 marks

Question 15 (3 marks)

A 150 W refrigerator is operating all day. The electricity charge is 32.8c/kWh during peak times of the day (6 am to 8 pm), and 21.6c/kWh during off-peak times.

How much will it cost to run this fridge for a year? 3 marks

Question 16 (4 marks)

In a fundraiser raffle, 200 tickets are sold for $5 each. There are 3 cash prizes and the remainder of the money collected is donated to charity.

FUNDRAISER RAFFLE

1st prize is **$200**

2nd prize is **$100**

3rd prize is **$50**

a When all tickets are sold, what is the ratio of prize money to the money donated to charity? Give your answer in simplest ratio form. 2 marks

b The fundraising committee decides to sell twice as many tickets next year. Each of the 400 tickets will be sold for double the price: $10 each. The committee also decides that the 3 money prizes will be doubled.

When all 400 tickets are sold, how much more will be donated to charity next year? 2 marks

Question 17 (4 marks) ©NESA 2013 HSC EXAM, QUESTION 30(c)

Joel mixes petrol and oil in the ratio 40 : 1 to make fuel for his leaf blower.

a Joel pours 5 litres of petrol into an empty container to make fuel for his leaf blower.

How much oil should he add to the petrol to ensure that the fuel is in the correct ratio? 1 mark

b Joel has 4.1 litres of fuel left in his container after filling his leaf blower.

He wishes to use this fuel in his lawnmower. However, his lawnmower requires the petrol and oil to be mixed in the ratio 25 : 1.

How much oil should he add to the container so that the fuel is in the correct ratio for his lawnmower? 3 marks

Question 18 (4 marks)

Holly is an apprentice plumber who is paid \$12.40 per hour when working on weekdays.

A further 60% per hour is paid when she works on Saturdays. If Holly works on a Sunday, she is paid her Saturday hourly rate plus a bonus of \$80.

Last week, Holly worked for 35 hours from Monday to Friday, 6 hours on Saturday and had an emergency callout on Sunday. Her total pay for the 7 days was \$732.24.

How many hours did Holly work on Sunday? 4 marks

Question 19 (4 marks)

When shopping, Mitchell notices two brands of cordial.

Brand A advertises that a drink is made with 1 part cordial to 4 parts water. It is sold in a 3-litre bottle for \$3.15.

Brand B recommends 1 part cordial to 7 parts water. It is sold in a 2.5-litre size bottle for \$3.90.

If the recommended ratio mixtures are followed and all the bottle contents are used, which brand is cheaper per litre of drink? 4 marks

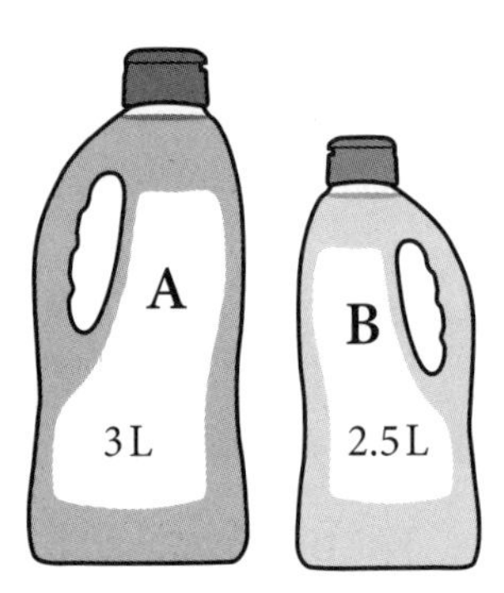

Question 20 (5 marks) ©NESA 2020 HSC EXAM, QUESTION 27

The shaded region on the diagram represents a garden. The scale is 1 cm = 5 m.

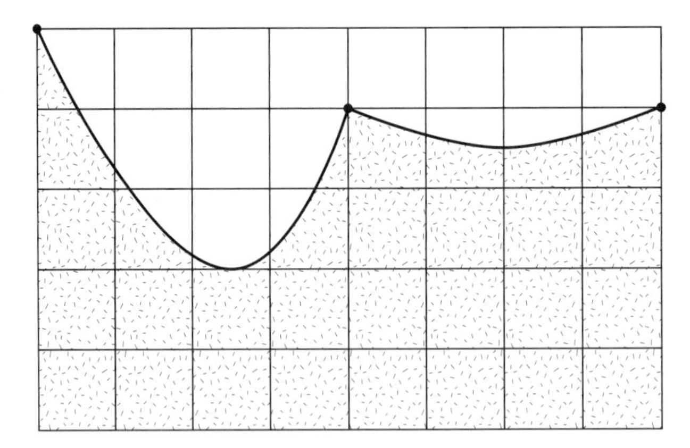

a Use two applications of the trapezoidal rule to calculate the approximate area of the garden. 3 marks

b Should the answer to part **a** be more than, equal to or less than the actual area of the garden? Referring to the diagram above, briefly explain your answer. 2 marks

Practice set 1

Worked solutions

1 C

Bananas are sold by mass, so the appropriate cost rate is dollars per kilogram.

2 D

$3000:450 = \frac{3000}{450} : \frac{20}{3} = 20:3$

3 C

Bulls : cows = 33 : 2 = 198 : ?

198 cows ÷ 33 = 6

Total bulls = 6 × 2 = 12

4 C

Total parts = 1 + 3 + 5 = 9

1 part = 36 ÷ 9 = 4 m^3

Sand = 3 parts = 3 × 4 = 12 m^3

5 C

1 L = 1000 mL

1.5 g/L × 1000 = 1500 g/L

6 C

21.6 g/40 mL = 0.54 g/mL

Amount of sorbitol = 250 × 0.54 = 135 g

7 D

$D = S \times T$

$= 12\,\text{m} \times 60 \times 2$

$= 1440\,\text{m}$

$= (1440 \div 1000)\,\text{km}$

$= 1.44\,\text{km}$

8 D

\$3.75 ÷ 0.6 L = \$6.25/L

9 B

7.8 L ÷ 100 km × 425 km = 33.15 ≈ 33 L

10 A

$T = \frac{D}{S}$

$= \frac{600}{900}\,\text{h}$

$= \frac{2}{3}\,\text{h}$

$= \frac{2}{3} \times 60\,\text{min}$

$= 40\,\text{min}$

11 B

Time = 45 000 ÷ 75

= 600 min

= (600 ÷ 60) h

= 10 h

12 C

350 m ÷ 20 s = 17.5 m/s

17.5 m/s × 3600 = 63 000 m/h

63 000 m/h ÷ 1000 = 63 km/h

13 A

1800 W ÷ 1000 = 1.8 kW

45 min ÷ 60 = 0.75 h

Cost = 1.8 kW × 0.75 h × \$0.326 = \$0.44

14 B

1 cm on the map represents an actual distance of 20 000 cm.

20 000 cm = (20 000 ÷ 100) m = 200 m

15 D

Flour : sugar = 5 : 3

Sugar : butter = 4 : 1

The same value for sugar is needed to compare all ingredients. Use a common multiple of 3 × 4 = 12.

Flour : sugar = 5 : 3 = 20 : 12

(flour = 5 × 4 = 20, sugar = 3 × 4 = 12)

Sugar : butter = 4 : 1 = 12 : 3

(sugar = 4 × 3 = 12, butter = 1 × 3 = 3)

So flour : sugar : butter = 20 : 12 : 3

So flour : butter = 20 : 3.

16 D

$$T = \frac{D}{S} = \frac{515\,\text{m}}{29.62\,\text{s}} = 17.3869\ldots\ \text{m/s}$$

$17.3869\ldots\ \text{m/s} \times 3600 \div 1000$
$= 62.592\ldots\ \text{km/h}$
$\approx 63\,\text{km/h}$

17 A

Let P be the estimated population of trout.

By comparing the proportion of tagged trout in the population and the recaptured sample:

Population: tagged = 600, total = P
Sample: tagged = 35, total = 120

$$\frac{600}{P} = \frac{35}{120}$$

$$\frac{P}{600} = \frac{120}{35}$$

$$P = \frac{120}{35} \times 600$$

$= 2057.143\ldots$
≈ 2057 trout

OR

$$P = \frac{\text{number tagged in population}}{\text{number tagged in sample}} \times \text{total sample}$$

$$P = \frac{600}{35} \times 120$$

$= 2057.143\ldots$
≈ 2057 trout

18 C

Fraction of students travelling by car

$$= 1 - \frac{1}{3} - \frac{1}{4} = \frac{5}{12}$$

$$\text{Walk} : \text{car} = \frac{1}{3} : \frac{5}{12} = \frac{1}{3} \times 12 : \frac{5}{12} \times 12 = 4:5$$

19 C

Dogs : cats = 3 : 5

Using guess and check, possible ratios are:

3 dogs : 5 cats
6 dogs : 10 cats
9 dogs : 15 cats
12 dogs : 20 cats

For 3 dogs, 5 cats:
After selling, there are $3 - 2 = 1$ dog, $5 - 1 = 4$ cats, but $1:4 \neq 1:2$.

For 6 dogs, 10 cats:
After selling, there are $6 - 2 = 4$ dogs, $10 - 1 = 9$ cats, but $4:9 \neq 1:2$.

For 9 dogs, 15 cats:
After selling, there are $9 - 2 = 7$ dogs, $15 - 1 = 14$ cats, and $7:14 = 1:2$.

So 7 dogs left.

20 B

Mass required = $(14 \div 2) \times 15\,\text{mg} = 105\,\text{mg}$

Each 10 mL contains 120 mg, so 1 mL contains 12 mg.

Dosage = $(105 \div 12)\,\text{mL} = 8.75\,\text{mL}$

Practice set 2

Worked solutions

Question 1

$$D = S \times T$$
$$= 240\,\text{km/h} \times \frac{20}{60}\,\text{h}$$
$$= 80\,\text{km}$$

OR

$$240\,\text{km/h} = (240 \div 60)\,\text{km/min}$$
$$= 4\,\text{km/min}$$

$$D = S \times T$$
$$= 4\,\text{km/min} \times 20\,\text{min}$$
$$= 80\,\text{km}$$

Question 2

Amount of water = 7.5 L × 10 min × 7 days
= 525 L

Question 3

Unit price of bottle = $\frac{\$2.50}{2\text{L}}$ = \$1.25/L

750 mL ÷ 1000 = 0.75 L

Unit price of carton = $\frac{\$1.05}{0.75\text{L}}$ = \$1.40/L

So the 2 L bottle is the better buy because \$1.25/L is less than \$1.40/L.

Question 4

2.5 L = 2500 mL
Total parts = 5 + 3 + 2 = 10
1 part = 2500 mL ÷ 10 = 250 mL

Pineapple juice = 5 parts
= 5 × 250
= 1250 mL

Question 5

A\$1000 = US\$1330

So US\$1 = A\$(1000 ÷ 1330)
= A\$0.7518…

So US\$400 = A\$(400 × 0.7518…)
= 300.7518…
≈ A\$300.75

OR

A\$1000 = US\$1330

So A\$1 = US\$(1330 ÷ 1000)
= US\$1.33

So to convert US\$400 to A\$:

US\$400 = A\$(400 ÷ 1.33)
= 300.7518…
≈ A\$300.75

Question 6

a $30:25:40 = \frac{30}{5} : \frac{25}{5} : \frac{40}{5} = 6:5:8$

b Total number of runs scored by Andrew, Brandon and Cosmo = 30 + 25 + 40 = 95

Let t be the total number of runs scored by the whole team.

$95:t = 19:36$

So $\frac{95}{t} = \frac{19}{36}$

$\frac{t}{95} = \frac{36}{19}$ OR $\frac{t}{95} = \frac{36}{19}$

$19 \times 5 = 95$

So $t = 36 \times 5$
$= 180$

OR

$t = \frac{36}{19} \times 95$
$= 180$

The team scored 180 runs.

Question 7

Number of hours from 10 am to 9 pm = 11

Energy used = 4.8 kW × 11 h × 125 days
= 6600 kWh

Question 8

Number of runs needed = 182 – 96 = 86

Number of overs left = 13 to 20 overs
= 8 more overs

Average run rate needed = 86 runs ÷ 8 overs
= 10.75 runs/over

Question 9

a Convert dimensions to cm:

28 m = 2800 cm, 15 m = 1500 cm

Scaled length = 2800 cm ÷ 200
= 14 cm

Scaled width = 1500 cm ÷ 200
= 7.5 cm

WORKED SOLUTIONS

b Scale = 20 cm : 28 m
= 20 cm : 2800 cm
= 1 : 140

Question 10

a The family room is the largest room.
Its area is $7.3 \times 4.1 = 29.93\,\text{m}^2$.

b The compass rose shows that the left side of the house is facing north. The front of the house is the side that has the porch and double garage, at the bottom of the plan, so it is facing west.

Question 11

Petrol used in city $= 12\,000\,\text{km} \times \dfrac{10.9\,\text{L}}{100\,\text{km}}$
$= 1308\,\text{L}$

Petrol used in country $= 20\,000\,\text{km} \times \dfrac{8.5\,\text{L}}{100\,\text{km}}$
$= 1700\,\text{L}$

Cost for year = 1308 × \$1.45 + 1700 × \$1.55
= \$4531.60

Question 12

Let P be the estimated population of seals.

By comparing the proportion of tagged seals in the population and the recaptured sample:

Population: tagged = 60, total = P
Sample: tagged = 30, total = 120

$$\frac{60}{P} = \frac{30}{120}$$
$$\frac{P}{60} = \frac{120}{30}$$
$$\frac{P}{60} = 4$$
$$P = 4 \times 60 = 240 \text{ seals}$$

OR

$$P = \frac{\text{number tagged in population}}{\text{number tagged in sample}} \times \text{total sample}$$
$$= \frac{60}{30} \times 120$$
$$= 240 \text{ seals}$$

This is 11% less than the estimated population for 2008.

Let x = estimated population for 2008.
100% – 11% = 89%

89% of x = 240
$0.89x = 240$
$$x = \frac{240}{0.89} = 269.6629\ldots$$

Estimated population for 2008 is 270 seals.

Question 13

3.5 × 1000 = 3500 mL
= 3500 × 10 drops
= 35 000 drops

Fluid rate = 35 000 drops/day
= 35 000 ÷ 24 ÷ 60 drops/min
= 24.3055…
≈ 24 drops/min

Question 14

a $T = \dfrac{D}{S}$

$$T = \frac{300}{9.08} + \frac{100}{9.64}$$
$= 43.4130\ldots$
$\approx 43.41\,\text{s}$

b $S = \dfrac{D}{T}$

$$= \frac{400\,\text{m}}{43.41\,\text{s}}$$
= 9.214 466… m/s
= 9.214 466… × 3600 ÷ 1000
= 33.1720…
≈ 33.17 km/h

Question 15

150 ÷ 1000 = 0.15 kW

Peak hours (6 am to 8 pm) = 14

Off-peak hours = 24 – 14 = 10

Peak energy cost for a year
= 0.15 kW × 14 h × 365 days × \$0.328
≈ \$251.41

Off-peak energy cost for a year
= 0.15 kW × 10 h × 365 × 0.216
= \$118.26

Total cost = \$251.41 + \$118.26
= \$369.67

Question 16

a Prize money = \$200 + \$100 + \$50 = \$350

Total ticket sales = 200 × \$5 = \$1000

Charity donation = \$1000 – \$350 = \$650

Prize money : charity donation = 350 : 650
$$= \frac{350}{50} : \frac{650}{50}$$
= 7 : 13

b Prize money = 2 × \$350 = \$700

Total ticket sales = 400 × \$10 = \$4000

Charity donation = \$4000 − \$700 = \$3300

Extra charity donation = \$3300 − \$650 = \$2650

Question 17

a Let x be the amount of oil required.

Petrol : oil = 40 : 1 = 5 : x

$40 \div 8 = 5$

So $x = 1 \div 8$
$= 0.125$ L or 125 mL

OR

$$\frac{40}{1} = \frac{5}{x}$$
$$\frac{1}{40} = \frac{x}{5}$$
$$\frac{x}{5} = \frac{1}{40}$$
$$x = \frac{1}{40} \times 5$$
$$= \frac{5}{40}$$
$= 0.125$ L or 125 mL

b Petrol : oil = 25 : 1

But remaining 4.1 L has petrol : oil = 40 : 1

Total parts = 40 + 1 = 41

1 part = 4.1 ÷ 41 = 0.1 L

So petrol = 40 × 0.1 = 4 L
oil = 1 × 0.1 = 0.1 L

For new lawnmower mixture, let y be the amount of oil to mix with 4 L of petrol.

Petrol : oil = 25 : 1 = 4 : y

$$\frac{25}{1} = \frac{4}{y}$$
$$\frac{1}{25} = \frac{y}{4}$$
$$\frac{y}{4} = \frac{1}{25}$$
$$y = \frac{1}{25} \times 4$$
$$y = \frac{4}{25}$$
$= 0.16$ L

So oil to be added (to current mixture)
= 0.16 − 0.1 = 0.06 L or 60 mL

Question 18

Saturday hourly rate = \$12.40 × 160% = \$19.84

Monday to Friday earnings = \$12.40 × 35 = \$434

Saturday earnings = \$19.84 × 6 = \$119.04

Let x be the number of hours Holly worked on Sunday.

Sunday earnings = \$19.84 × x + \$80

Total earnings = \$434 + \$119.04 + \$19.84x + \$80
= \$732.24

\$633.04 + \$19.84x = \$732.24
\$19.84x = \$732.24 − \$633.04
= \$99.20
x = \$99.20 ÷ \$19.84
= 5

Holly worked for 5 hours on Sunday.

Question 19

Brand A:

C : W = 1 : 4 = 3 : ?

1 × 3 = 3 L cordial
4 × 3 = 12 L water

Total drink volume = 3 + 12 = 15 L

Unit price of drink = \$3.15 ÷ 15 L = \$0.21/L

Brand B:

C : W = 1 : 7 = 2.5 : ?

1 × 2.5 = 2.5 L cordial
7 × 2.5 = 17.5 L water

Total drink volume = 2.5 + 17. 5 = 20 L

Unit price of drink = \$3.90 ÷ 20 L = \$0.195/L

Hence, brand B is cheaper per litre of drink.

WORKED SOLUTIONS

Question 20

a $A = \frac{h}{2}(d_f + d_l)$

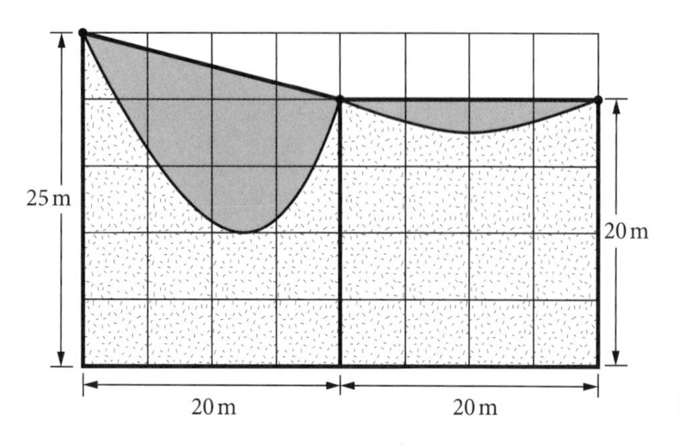

Left side: $h = 4 \times 5 = 20\,\text{m}$

$d_f = 5 \times 5 = 25\,\text{m}$

$d_l = 4 \times 5 = 20\,\text{m}$

$A = \frac{20}{2}(25 + 20) = 450\,\text{m}^2$

Right side: $h = 20\,\text{m}$

$d_f = 4 \times 5 = 20\,\text{m}$

$d_l = 4 \times 5 = 20\,\text{m}$

$A = \frac{20}{2}(20 + 20) = 400\,\text{m}^2$

Total area $= 450 + 400$

$= 850\,\text{m}^2$

b The answer in part **a** should be more than the actual area of the garden because the approximation involves 2 trapeziums (see diagram) that cover more area than that which is shaded.

HSC exam topic grid (2011–2020)

This grid shows the coverage of this topic in past HSC exams by question number. The past exam papers can be downloaded from the NESA website (www.educationstandards.nsw.edu.au) by selecting 'Year 11 – Year 12', 'HSC exam papers'. NESA marking feedback and guidelines can also be found there.

Before 2019, 'Mathematics Standard 2' was called 'Mathematics General 2' and, before 2014, 'General Mathematics'. For these exams, select 'Year 11 – Year 12', 'Resources archive', 'HSC exam papers archive'.

	Rates	Ratios	Scale and scale drawings	Trapezoidal rule (introduced 2019)
2011	21		24(a)	13†
2012	26(g)	**26(f)***	27(c)	
2013	26(d)	**30(c)**		
2014	**17**, 18, 20, 22, 27(a)(i), 27(b)		28(d)	28(d)(iii)†
2015	26(b), 26(g), 27(b), 30(a), 30(d)	26(a)*	29(c)	
2016	9, 11, 15, 26(c), 28(b)	28(a)*		
2017	2, 6, 14, 26(a), 26(b), 27(d)(iii)	26(c)*		
2018	5, 8, 26(g), 27(a), 27(d)(i), 28(c)	10*	26(g)	28(a)†
2019 new course	2, 24, 33(a), 41(a)	**18**	41(a)	41(b)
2020	3	23	**27**	**27**

Questions in **bold** can be found in this chapter.
* Capture–recapture method.
† Uses Simpson's rule, which is no longer in the course, but can be solved by the trapezoidal rule, giving a similar answer.

CHAPTER 4 INVESTMENTS, LOANS AND ANNUITIES

MS-F4 Investments and loans 104

F4.1 Investments 104

F4.2 Depreciation and loans 107

MS-F5 Annuities 110

4

INVESTMENTS, LOANS AND ANNUITIES

Denoted content is common with the Mathematics Advanced course.

Investments

- Simple interest:

$$I = Prn$$

- Compound interest:
  future value (FV), present value (PV)

$$FV = PV(1 + r)^n$$
$$I = FV - PV$$

- Comparing simple interest and compound interest
- Inflation and appreciation
- Shares: dividend and dividend yield

$$\text{Dividend yield} = \frac{\text{dividend per share}}{\text{market value per share}} \times 100\%$$

Loans and credit cards

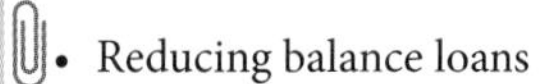

- Reducing balance loans
- Credit cards: compound interest

Depreciation

- Straight-line method of depreciation:

$$S = V_0 - Dn$$

- Declining-balance method of depreciation:

$$S = V_0(1 - r)^n$$

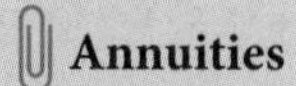

Annuities

- Annuities: modelling as a recurrence relation
- Present and future value
- Present and future value tables

Glossary

annuity
A compound interest investment from which payments are made or received regularly for a fixed period of time.

appreciated value
The value an asset has increased to over time.

appreciation
An increase in the value of an asset over time (for example, artwork, antique jewellery, prestige car, land or a house).

asset
Item or resource with monetary value, such as property, machinery or equipment, used by a business to earn income.

brokerage
The commission that a stockbroker charges for buying and selling shares for a client.

compound interest
Interest on a loan or investment calculated on the initial principal and the accumulated interest from previous periods. Differs from **simple interest**.

credit card
A card linked to an account that allows consumers to purchase goods and services without cash and pay for them later. Compound interest is charged daily if the balance owing is not paid by the due date.

depreciation
A loss in the value of an asset over time, as a result of wear and tear, age or being replaced by a newer item. There are 2 methods for depreciating an asset.

In the **straight-line method** of depreciation, the value of the depreciating asset decreases by the same amount during each time period:

$$S = V_0 - Dn.$$

In the **declining-balance method** of depreciation, the value of the depreciating asset decreases by the same percentage (depreciation rate) during each time period:

$$S = V_0(1 - r)^n.$$

dividend
A part of the profits earned by a company and distributed to a shareholder. It is a fixed amount of money offered per share to the shareholder.

dividend yield
The **dividend** expressed as a percentage of the current market price of the share (similar to an interest rate).

$$\text{Dividend yield} = \frac{\text{dividend per share}}{\text{market value per share}} \times 100\%$$

A+ DIGITAL FLASHCARDS
Revise this topic's key terms and concepts by scanning the QR code or typing the URL into your browser.

https://get.ga/a-hsc-maths-standard-2

future value
The final amount of an investment or annuity after compound interest has been added. In finance, the compound interest formula $A = P(1 + r)^n$ is known as the future value formula and is written as

$$FV = PV(1 + r)^n.$$

inflation
An increase in the prices of all goods and services in a nation's economy. The yearly percentage increase is called the annual inflation rate. The compound interest formula is used to calculate the new prices.

interest-free period
A period of time when interest is not charged on an amount borrowed, for example, 55 days for a credit card account.

present value
The amount of money (or principal) that could be invested now under compound interest to produce a future value over a given period of time. The future value formula can be rearranged as

$$PV = \frac{FV}{(1 + r)^n}.$$

principal
The original amount of money invested or borrowed.

recurrence relation
A relation or pattern where each new value depends on the same formula being applied to the previous value. Each new principal is equal to the previous principal increased by a percentage that is the interest rate. Compound interest, loans and annuities are examples.

reducing balance loan
A loan where the interest charged is calculated on the amount still owing (the reducing balance of the loan) after each payment is made rather than on the original principal.

salvage value
The current value of an asset after depreciation.

share
One part of a company's capital that can be bought, entitling the shareholder to a part of the company's profits.

simple interest
Also known as flat-rate interest. Interest on an investment or loan earned or charged on the original principal only. Differs from **compound interest**.

GLOSSARY

Topic summary

Investments and loans (MS-F4)

F4.1 Investments

Review of simple interest (Year 11)

Simple interest (or flat-rate interest) is interest calculated as a percentage of the **principal** (the amount of money invested or borrowed).

$$I = Prn$$

where I = interest
P = principal or initial amount
r = interest rate per period, expressed as a decimal
n = number of periods

Denoted content for this topic is common with the Mathematics Advanced course.

Compound interest and future value

Compound interest is interest that is added to the principal and reinvested (compounded). In the financial world, the principal is called the **present value** (PV) and the final amount is called the **future value** (FV).

$$FV = PV(1 + r)^n \quad \text{and} \quad I = FV - PV$$

where FV = future value (final amount)
PV = present value (principal, initial value)
n = number of compounding periods
r = interest rate per compounding period, expressed as a decimal
I = compound interest earned.

Hint
Only the FV formula is provided on the HSC exam reference sheet.

Example 1

Calculate the future value when \$5000 is invested for 3 years at 4% p.a. compounded yearly.

Solution

Using $FV = PV(1 + r)^n$

PV = \$5000, r = 4% p.a. = 0.04, n = 3 years

$FV = 5000\,(1 + 0.04)^3$
$= \$5624.32$

So the future value of \$5000 at 4% p.a. for 3 years is \$5624.32.

Hint
This question only asks for the future value of this investment, which is the final balance in the account. If the question asks for the compound interest earned, you need to calculate the difference between the future value and the present value.
Interest $= FV - PV$
$= 5624.32 - 5000$
$= \$624.32$

TOPIC SUMMARY

Example 2

Calculate the future value and the interest earned when \$5000 is invested for 3 years at 4% p.a. compounded monthly.

Hint
The interest rate and the time period must have the same units, so both should be changed to months because interest is compounded monthly.

Solution

$PV = \$5000$, $r = 4\%$ p.a. $= \dfrac{0.04}{12}$ per month (decimal rate), $n = 3 \times 12 = 36$ months

$$
\begin{aligned}
FV &= PV(1 + r)^n \\
&= 5000\left(1 + \frac{0.04}{12}\right)^{36} \\
&\approx \$5636.36
\end{aligned}
$$

Note that, compared to Example 1 where the same present value has interest compounded yearly, this investment compounded monthly earns more interest. The difference in the future values, \$5636.36 – \$5624.32 = \$12.04, indicates extra interest.

Length of compounding period

If the present value, the term and the interest rate remain the same, but the compounding period is more frequent, such as compounding monthly compared to yearly, then more interest is earned with a higher future value. This is because in compound interest investments (and loans) the interest is added to the principal, so interest is earned on the added interest. The smaller the compounding period, the more often interest is added to the principal.

Example 3

What is the difference in interest between the following investments over 20 years?

Investment A: \$10 000 earning simple interest at 4% p.a.
Investment B: \$10 000 earning compound interest at 4% p.a. compounded yearly

Solution

Investment A:

$PV = \$10\,000$, $r = 4\%$ p.a. $= 0.04$, $n = 20$ years

$$
\begin{aligned}
I &= Prn \\
&= 10\,000 \times 0.04 \times 20 \\
&= \$8000
\end{aligned}
$$

Investment B:

$PV = \$10\,000$, $r = 4\%$ p.a. $= 0.04$, $n = 20$ years

$$
\begin{aligned}
FV &= PV(1 + r)^n \\
&= 10\,000(1 + 0.04)^{20} \\
&= \$21\,911.23
\end{aligned}
$$

$$
\begin{aligned}
I &= 21\,911.23 - 10\,000 \\
&= \$11\,911.23
\end{aligned}
$$

Difference = \$11 911.23 – \$8000 = \$3911.23

So the compound interest investment earns \$3911.23 more interest over 20 years than the simple investment.

Present value

The present value of an investment is the amount of money that, if invested now, would equal a given value in the future. The future value formula can be rearranged to change its subject to PV.

$$
\begin{aligned}
FV &= PV(1 + r)^n \\
PV(1 + r)^n &= FV \\
PV &= \frac{FV}{(1 + r)^n}
\end{aligned}
$$

Inflation and appreciation

Inflation is an increase in the prices of all goods and services. The yearly percentage increase is called the annual inflation rate. This is usually accompanied by increases in workers' wages and salaries to 'keep up with inflation'. Calculating the new price of goods and services after inflation is an application of increasing a quantity by a percentage repeatedly. This means we can use the future value formula.

$$FV = PV(1 + r)^n$$

Appreciation is the increase in value of an item (such as artwork, gold, a prestige car, land or a house) over time. It is the opposite of **depreciation**, which is a decrease in value over time of items such as cars, furniture and electrical appliances. Calculating the value of items after appreciation is another application of repeated percentage increase, so we can use the future value formula again.

Example 4

A loaf of bread is currently priced at \$3.20. If the average annual inflation rate is 2.2%, what is the predicted price of a loaf of bread in 4 years' time?

Solution

$PV = \$3.20$, $r = 2.2\%$ per year $= 0.022$, $n = 4$ years

$$\begin{aligned} FV &= PV(1 + r)^n \\ &= 3.20(1 + 0.022)^4 \\ &= 3.4910\ldots \\ &\approx \$3.49 \end{aligned}$$

Shares

A **share** represents a unit of ownership in a company. The stock exchange is where the buying and selling of shares happens. The Australian market is known as the Australian Securities Exchange (ASX).

When you own shares in a company, you are a shareholder and earn a **dividend** (a type of interest) on your investment. The dividend can be collected by the investor (you) or can be reinvested in more shares in the hope of earning a higher dividend next time. The **dividend yield** is the dividend as a percentage of the market price of the share (similar to an interest rate).

$$\text{Dividend yield} = \frac{\text{dividend per share}}{\text{market price per share}} \times 100\%$$

Example 5

Rosalie purchases 2000 shares. The market price is \$4.21 when the company offers a dividend of 23 cents per share. How much does Rosalie receive in dividends and what is the dividend yield, correct to two decimal places?

Solution

$$\begin{aligned} \text{Dividend} &= 2000 \times \$0.23 \\ &= \$460 \end{aligned}$$

$$\begin{aligned} \text{Dividend yield} &= \frac{0.23}{4.21} \times 100\% \\ &= 5.4631\ldots \\ &\approx 5.46\% \end{aligned}$$

F4.2 Depreciation and loans

Depreciation

Depreciation is a decrease in the value of an **asset** over time as a result of wear and tear, age or being out of date. There are 2 methods of depreciation.

For the straight-line method of depreciation, the value of the depreciating asset decreases by the same amount during each time period.

$$S = V_0 - Dn$$

where S = **salvage value** of the asset after n periods
V_0 = initial value of the asset
D = amount of depreciation per period
n = number of periods.

For the declining-balance method of depreciation, the value of the depreciating asset decreases by the same percentage during each time period.

$$S = V_0(1 - r)^n,$$

where r is the depreciation rate per period, expressed as a decimal.

Hint
Both formulas are provided on the HSC exam reference sheet.

Example 6

An asset purchased for \$25 000 depreciates by 15% p.a. using the declining-balance method.

a Calculate the salvage value after 10 years.

b After how many years will the salvage value be closest to \$2500?

Solution

a $V_0 = 25\,000$, $r = 15\%$ per year $= 0.15$, $n = 10$ years

$S = 25\,000(1 - 0.15)^{10}$
$\approx \$4921.86$

b Using $S = V_0(1 - r)^n$ and substitution of values:

$2500 = 25\,000(1 - 0.15)^n$ then we need to find n:

As the calculation in part **a** shows a salvage value of approximately \$5000, continue to guess and check for values of n above 10 to get closest to \$2500.

$2500 = 25\,000(0.85)^n$

Guess and check: $n = 12$ years

$S = 25\,000(0.85)^{12}$
$= \$3556.04$ (too big)

Guess and check: $n = 15$ years

$S = 25\,000(0.85)^{15}$
$= \$2183.86$ (too small)

Guess and check: $n = 14$ years

$S = 25\,000(0.85)^{14}$
$= \$2569.24$ (closest to \$2500)

Hence, after 14 years the salvage value reaches closest to \$2500.

Hint
For guess-and-check questions, show your working to display values that are both bigger and smaller than your final answer.

Reducing balance loans

A **reducing balance loan** is a compound interest loan that is repaid by making regular payments. The interest paid is calculated on the amount still owing (the reducing balance of the loan) after each payment is made. As the balance decreases, so does the amount of interest paid.

You can save interest and pay off a reducing balance loan more quickly by:

- making repayments more frequently, for example, weekly or fortnightly rather than monthly
- making extra payments of any size from time to time, for example, if you receive a bonus or pay rise at work
- increasing the size of each repayment (even by just a few dollars).

Example 7

Carla borrows \$300 000 at a reducible interest rate of 12% p.a. She makes repayments of \$5000 at the end of each month.

The table shows the progress of the loan, and the balance owing is shown on the graph.

Note: $r = 12\%\text{ p.a.} = \dfrac{0.12}{12} = 0.01$ (1% per month)

No. of months (n)	Principal (P)	Interest (I)	Repayment (R)	Balance owing ($P + I - R$)
1	\$300 000	$300\,000 \times \dfrac{0.12}{12}$ = \$3000	\$5000	300 000 + 3000 − 5000 = \$298 000
2	\$298 000	A	\$5000	B

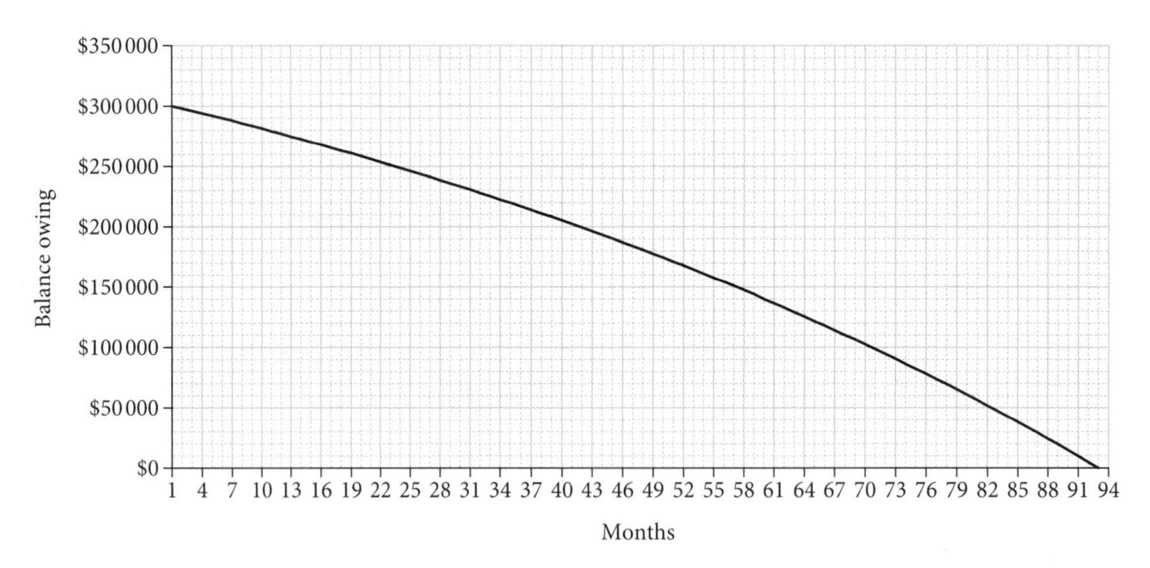

a Calculate the missing values in the table, A and B, for the second month.

b What is the approximate balance owing after 2 years?

c If Carla continues to make all her \$5000 monthly repayments, she will repay this loan and its interest in 93 months. How much interest will she have paid over the 93 months?

Solution

a $A = \text{interest} = 298\,000 \times \frac{0.12}{12} = \$2980$

$B = \text{balance owing} = 298\,000 + 2980 - 5000 = \$295\,980$

b 2 years = 24 months. From the graph, balance owing is about $250 000.

c Total repayments = 93 × 5000 = $465 000

Interest = total repayments – principal borrowed
= 465 000 – 300 000
= $165 000

> **Hint**
> Make sure you read the axes titles and the scales of the graph.

Credit cards

Financial institutions can issue **credit cards** to consumers, which can be used to buy goods and services without cash and then pay for them later. Using a credit card is like having a short-term loan.

Many cards have an **interest-free period** of 55 days to repay the full amount owing. If the full amount is not paid in full by the due date, then compound interest is charged on the outstanding amount each day from the date of purchase.

Interest-free periods do not apply to cash advances. These have a relatively high interest rate applied for each day until the withdrawal is paid back to the financial institution.

A credit limit is the maximum amount set by a financial institution (for example $5000) that the credit-card holder is able to spend.

Example 8

Consider the credit card statement shown.

a Calculate the closing balance for the statement on the right.

b Based on the closing balance, what is the available credit if the credit limit for this card is $2000?

c The credit card charge is 0.052% compound interest per day on any outstanding balances. How much interest is charged on the closing balance, which is overdue on the credit card for 14 days?

ACCOUNT SUMMARY	
Opening balance	$122.16 DR
Payments	$150.00 CR
New purchases	$1022.38 DR
Charges	$1.19 DR
Closing balance	$

Solution

a DR: debit (–) and CR: credit (+)

Closing balance
= –122.16 + 150.00 – 1022.38 – 1.19
= –$995.73

$995.73 is owing.

b Available credit = 2000 – 995.73
= $1004.27

c $FV = PV(1 + r)^n$

$PV = 955.73$, $r = 0.052\%$ per day $= 0.000\,52$, $n = 14$ days

$FV = 995.73(1 + 0.000\,52)^{14}$
$= 1003.0034\ldots$
$\approx \$1003.00$

$I = FV - PV$
$= \$1003.00 - \$995.73$
$= \$7.27$

Interest charged is $7.27.

> **Hint**
> The time period and interest rate must be in the same unit.

Annuities (MS-F5)

Annuities

An **annuity** is a compound interest investment from which payments are made or received regularly for a fixed period of time, for example, superannuation. Each contribution is made at the end of each compounding period. The future value of an annuity is the total value of all contributions and the interest that has compounded each period.

Example 9

A contribution of \$5000 is invested at the end of each year for 3 years at 4% p.a. This table shows the progress of the annuity.

Year	Balance at start of year (P)	Interest earned (I)	Contribution at end of year (a)	Balance at end of year, $P + I + a$
1	\$0	\$0	\$5000	\$5000
2	\$5000	5000 × 4% = \$200	\$5000	5000 + 200 + 5000 = \$10 200
3	\$10 200	10 200 × 4% = \$408	\$5000	

a What is the future value?

b How much interest has been earned?

Hint
The regular contributions of \$5000 keep increasing the balance and hence the interest increases.
This is different to one calculation for compound interest for 3 years: $5000(1 + 0.04)^3 = \$5624.32$

Solution

a Future value after 3 years is the final balance = 10 200 + 408 + 5000
= \$15 608

b Add values in the interest column

Interest earned = 0 + 200 + 408
= \$608

OR

FV – total contributions made

Interest earned = 15 608 – 3 × 5000
= \$608

Using a spreadsheet can extend annuity tables for many years.

This annuity can also be modelled by a **recurrence relation**. A recurrence relation is a formula or method that uses the previous amount to calculate the next amount.

End of year 1: $V_1 = \$5000$

End of year 2:
$$V_2 = V_1(1 + 0.04)^1 + 5000$$
$$= 5000(1.04) + 5000$$
$$= \$10\,200$$

End of year 3:
$$V_3 = V_2(1 + 0.04)^1 + 5000$$
$$= 10\,200(1.04) + 5000$$
$$= \$15\,608$$

Example 10 ©NESA 2020 HSC EXAM, QUESTION 34

Tina inherits \$60 000 and invests it in an account earning interest at a rate of 0.5% per month. Each month, immediately after the interest has been paid, Tina withdraws \$800.

The amount in the account immediately after the nth withdrawal can be determined using the recurrence relation

$$A_n = A_{n-1}(1.005) - 800,$$

where $n = 1, 2, 3, \ldots$ and $A_0 = 60\,000$.

a Use the recurrence relation to find the amount of money in the account immediately after the third withdrawal.

b Calculate the amount of interest earned in the first 3 months.

Solution

a $r = 0.5\%$ per month $= 0.005$, $A_0 = 60\,000$ (initial amount)

After 1st withdrawal:

$$\begin{aligned} A_1 &= A_0(1.005) - 800 \\ &= 60\,000(1.005) - 800 \\ &= \$59\,500 \end{aligned}$$

After 2nd withdrawal:

$$\begin{aligned} A_2 &= A_1(1.005) - 800 \\ &= 59\,500(1.005) - 800 \\ &= \$58\,997.50 \end{aligned}$$

After 3rd withdrawal:

$$\begin{aligned} A_3 &= A_2(1.005) - 800 \\ &= 58\,997.50(1.005) - 800 \\ &= \$58\,492.49 \end{aligned}$$

b Total withdrawals after 3 months = 800×3 = \$2400

Difference in balance = \$60 000 − \$58 492.49 = \$1507.51

$$\begin{aligned} \text{Interest} &= 2400 - 1507.51 \\ &= \$892.49 \end{aligned}$$

Present value of an annuity

The present value of an annuity is the single amount of money that, if invested now, would give the same future value as the regular contributions of the annuity.

Example 11

What is the single sum of money that could be invested to give the same future value as a contribution of \$5000 invested at the end of each year for 3 years at 4% p.a., as shown in Example 9?

Solution

As calculated in Example 9, the future value of these contributions is \$15 608.

The single amount is the present value:

$FV = \$15\,608$, $r = 4\%$ per year, $n = 3$ years

$$\begin{aligned} PV &= \frac{FV}{(1 + r)^n} \\ &= \frac{15608}{(1 + 0.04)^3} \\ &= 13\,875.455\,17\ldots \\ &\approx \$13\,875.46 \end{aligned}$$

Tables for present and future value of an annuity

Calculations for annuities can be tedious. An alternative is to use a spreadsheet or a table of interest factors and a calculator. Annuity tables show the future value or present value of an ordinary annuity when $1 is invested at the end of the period at the given interest rate for the given number of periods. These tables are provided in questions when needed.

When using a future value or present value table:

1. Find the time period, n, and the rate of interest, r, for the compounding period.
2. Look at the intersection of n and r in the table.
3. Multiply the contribution per period by the table value to find the future value or present value.
4. To calculate the interest earned:

Interest earned = future value – total contributions

Example 12

This table gives the future value of an annuity with a contribution of $1 per period. The values are rounded to four decimal places.

A contribution of $5000 is invested at the end of each year for 3 years at 4% p.a.

a What is the future value?

b How much interest is earned?

Future value of ordinary annuity with contribution of $1 per period

Period,	Interest rate per period			
n	1%	2%	3%	4%
1	1.0000	1.0000	1.0000	1.0000
2	2.0100	2.0200	2.0300	2.0400
3	3.0301	3.0604	3.0909	3.1216
4	4.0604	4.1216	4.1836	4.2465

Solution

a $n = 3$ years, $r = 4\%$ p.a.

Future value of ordinary annuity with contribution of $1 per period

Period,	Interest rate per period			
n	1%	2%	3%	4%
1	1.0000	1.0000	1.0000	1.0000
2	2.0100	2.0200	2.0300	2.0400
3	3.0301	3.0604	3.0909	3.1216
4	4.0604	4.1216	4.1836	4.2465

From the future value table:

Future value for $1 contribution per year = $3.1216

Future value for $5000 contribution per year = 5000×3.1216
= $15 608

b Interest = FV – total contributions
= $15\,608 - 3 \times 5000$
= $608

Example 13

A contribution of \$5000 is invested at the end of each quarter for 3 years at 4% p.a.

a What is the future value?

b How much interest is earned?

c What is the single amount that could be invested today to be equal to the future value of these contributions?

Hint
The time period and the interest rate must both be changed to suit quarterly contributions and compounding periods.

Solution

a $r = 4\%$ p.a. $\div 4 = 1\%$ per quarter

$n = 3$ years $\times 4 = 12$ quarters

From the future value table:

Future value for \$1 contribution per quarter
= \$12.6825

Future value for \$5000 contribution quarter
= \$5000 × 12.6825
= \$63 412.50

Future value of ordinary annuity with contribution of \$1 per period

Period, n	Interest rate per period			
	1%	2%	3%	4%
1	1.0000	1.0000	1.0000	1.0000
2	2.0100	2.0200	2.0300	2.0400
3	3.0301	3.0604	3.0909	3.1216
4	4.0604	4.1216	4.1836	4.2465
5	5.1010	5.2040	5.3091	5.4163
6	6.1520	6.3081	6.4684	6.6330
7	7.2135	7.4343	7.6625	7.8983
8	8.2857	8.5830	8.8923	9.2142
9	9.3685	9.7546	10.1591	10.5828
10	10.4622	10.9497	11.4639	12.0061
11	11.5668	12.1687	12.8078	13.4864
12	12.6825	13.4121	14.1920	15.0258

b Interest = FV – total contributions
= \$63 412.50 – 12 × \$5000
= \$3412.50

c $PV = \frac{FV}{(1+r)^n}$ $\quad r = 1\% = 0.01$

$= \frac{63412.50}{(1+0.01)^{12}}$

$\approx$ \$56 275.37

Example 14

Using the present value table, what is the single sum of money that could be invested to give the same future value of a contribution of \$5000 that is invested at the end of each quarter for 3 years at 4% p.a.?

Note: This is the same question as part **c** of Example 13, except using a table rather than a formula.

Present value of \$1 contribution per period

Period,	Interest rate per period					
n	1%	2%	3%	4%	5%	6%
1	0.9901	0.9804	0.9709	0.9615	0.9524	0.9434
2	1.9704	1.9416	1.9135	1.8861	1.8594	1.8334
3	2.9410	2.8839	2.8286	2.7751	2.7232	2.6730
4	3.9020	3.8077	3.7171	3.6299	3.5460	3.4651
5	4.8534	4.7135	4.5797	4.4518	4.3295	4.2124
6	5.7955	5.6014	5.4172	5.2421	5.0757	4.9173
7	6.7282	6.4720	6.2303	6.0021	5.7864	5.5824
8	7.6517	7.3255	7.0197	6.7327	6.4632	6.2098
9	8.5660	8.1622	7.7861	7.4353	7.1078	6.8017
10	9.4713	8.9826	8.5302	8.1109	7.7217	7.3601
11	10.3676	9.7868	9.2526	8.7605	8.3064	7.8869
12	11.2551	10.5753	9.9540	9.3851	8.8633	8.3838

Solution

$n = 3 \text{ years} \times 4 = 12 \text{ quarters}$

$r = 4\% \text{ p.a.} \div 4 = 1\% \text{ per quarter}$

$PV = 5000 \times 11.2551$
$= \$56\,275.50$

Note: This answer is 13 cents more than the \$56 275.37 that was calculated using the PV formula in part **c** of Example 13 because the table has values that are rounded to only four decimal places.

Practice set 1

Multiple-choice questions

Solutions start on page 125.

Question 1

Which of the following formulas is correct for compound interest problems?

A $I = Prn$ **B** $I = Pr^n$ **C** $I = PV(1 + r)^n$ **D** $FV = PV(1 + r)^n$

Question 2

An interest rate of 4.8% p.a. is equivalent to which percentage per month?

A 0.4% **B** 0.18% **C** 1.2% **D** 2.4%

Question 3

What is the future value when $5000 is invested for 6 years at 3% p.a. compounded yearly?

A $900 **B** $5900 **C** $5970.26 **D** $30 900

Question 4

Which of the following provides the best financial return if $25 000 is invested for 10 years?

A simple interest at 6% p.a.
B interest compounded annually at 6% p.a.
C interest compounded quarterly at 6% p.a.
D interest compounded monthly at 6% p.a.

Question 5

What is the interest paid on a loan of $300 000 with fortnightly repayments of $741.33 over 25 years?

A $18 533.25 **B** $181 864.50 **C** $318 533.25 **D** $481 864.50

Question 6

Michelle owns 2700 shares in an IT company. When the market value of each share was $4.55, the company declared a dividend of 6.2%.

What income does Michelle receive?

A $761.67 **B** $1981.45 **C** $3679.12 **D** $9571.07

Question 7

An apartment is purchased for $436 000.

What is the appreciated value of this apartment 3 years later if the average annual inflation rate is 2.8%?

A $448 208 **B** $473 659 **C** $1 220 800 **D** $1 343 624

Question 8

A four-wheel-drive utility van is bought for $76 000. Based on the declining-balance method of depreciation, each year the value of the utility van depreciates by 10%. The table shows the van's first 3 years of depreciation.

Year	Salvage value
1	$68 400
2	$61 560
3	$55 404

How much does the van depreciate by after 3 years?

A $12 996 **B** $14 440 **C** $20 596 **D** $22 800

Question 9 ©NESA 2019 HSC EXAM, QUESTION 13

The graphs show the future values over time of $\$P$, invested at three different rates of compound interest.

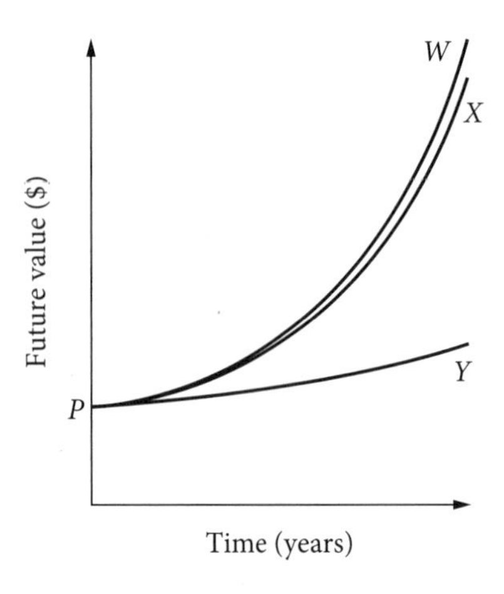

Which of the following correctly identifies each graph?

A

W	5% p.a. compounding annually
X	10% p.a. compounding annually
Y	10% p.a. compounding quarterly

B

W	5% p.a. compounding annually
X	10% p.a. compounding quarterly
Y	10% p.a. compounding annually

C

W	10% p.a. compounding quarterly
X	10% p.a. compounding annually
Y	5% p.a. compounding annually

D

W	10% p.a. compounding annually
X	10% p.a. compounding quarterly
Y	5% p.a. compounding annually

Question 10

Joanna borrows \$600 000 from the bank to pay for a house. The loan is a reducing balance loan with fixed equal monthly repayments over 29 years. The graph shows the balance owing on the loan over time.

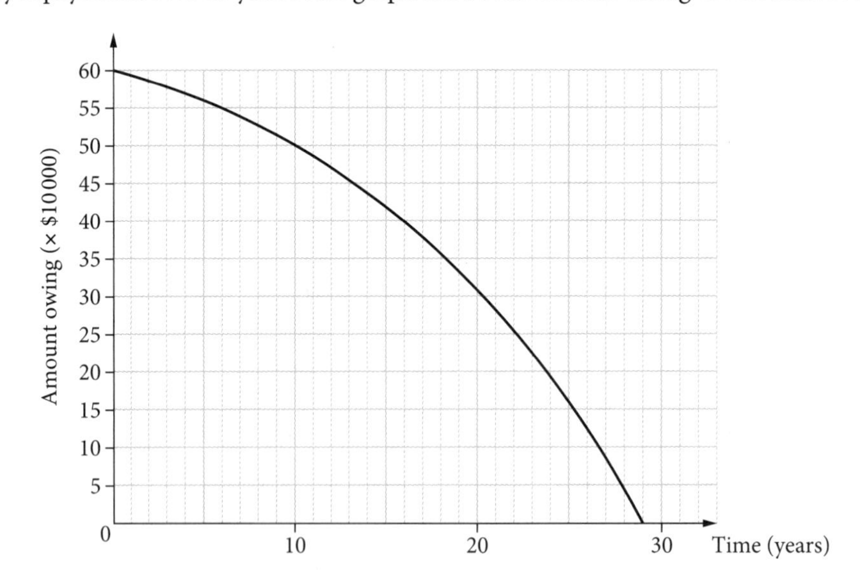

How much has been paid off the loan after 16 years?

A \$20 000 **B** \$40 000 **C** \$200 000 **D** \$400 000

Question 11

A company pays its shareholders a dividend of 68 cents when the market share price is \$12.34.

What is the dividend yield paid?

A 5.5% **B** 6.8% **C** 8.4% **D** 18.1%

Question 12

A single amount of \$15 000 is invested for 3 years, earning interest at the rate of 2.4% p.a. compounded monthly.

Which expression will give the future value of the investment?

A $FV = 15\,000(1 + 0.024)^3$

B $FV = 15\,000(1 + 0.024)^{36}$

C $FV = 15\,000\left(1 + \frac{0.024}{12}\right)^3$

D $FV = 15\,000\left(1 + \frac{0.024}{12}\right)^{36}$

Question 13 ©NESA 2015 HSC EXAM, QUESTION 10

A piece of machinery, initially worth \$56 000, depreciates at 8% p.a.

Which graph best shows the salvage value of this piece of machinery over time?

A

Value (\$'000)
Time in years

B

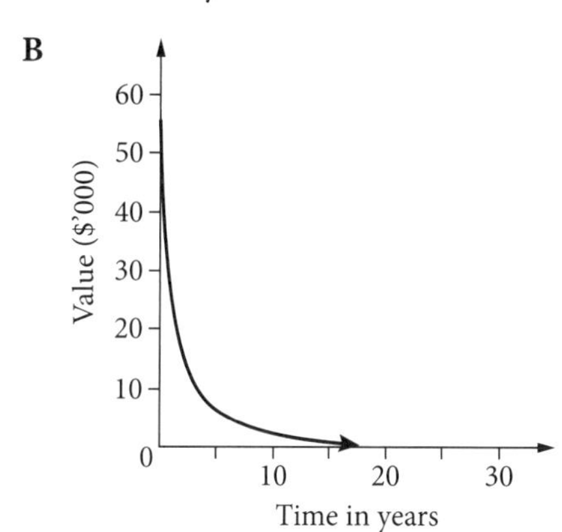

C

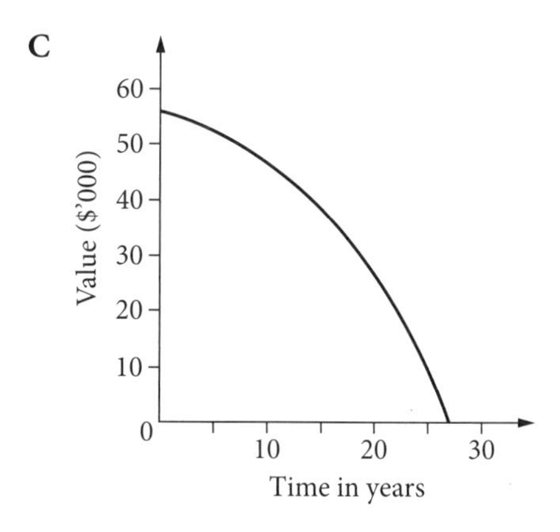

D

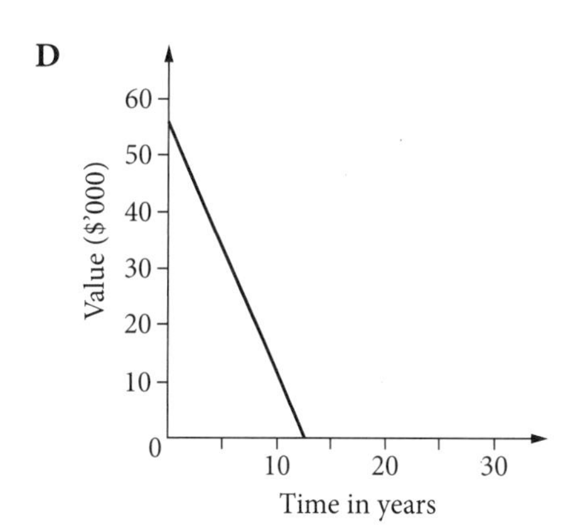

Question 14

A credit card has an outstanding balance of \$400, which is charged 22 days interest at a rate of 17% p.a. compounded daily.

What is the interest charged on this balance?

A \$4.10 **B** \$4.12 **C** \$404.12 **D** \$1496

9780170459204

PRACTICE SET 1

Question 15

A tennis racquet costs $320 today.

What was the value of the racquet, to the nearest dollar, 5 years ago if the annual inflation rate is constant at 2.7% p.a.?

A $255 **B** $280 **C** $311 **D** $366

Question 16

The table shows the present value interest factors, for an annuity of $1 per period, for various interest rates, r, and periods, n.

Present value interest factors

Period, n	Interest rate per period					
	1%	**2%**	**3%**	**4%**	**5%**	**6%**
1	0.9901	0.9804	0.9709	0.9615	0.9524	0.9434
2	1.9704	1.9416	1.9135	1.8861	1.8594	1.8334
3	2.9410	2.8839	2.8286	2.7751	2.7232	2.6730
4	3.9020	3.8077	3.7171	3.6299	3.5460	3.4651
5	4.8534	4.7135	4.5797	4.4518	4.3295	4.2124
6	5.7955	5.6014	5.4172	5.2421	5.0757	4.9173
7	6.7282	6.4720	6.2303	6.0021	5.7864	5.5824
8	7.6517	7.3255	7.0197	6.7327	6.4632	6.2098
9	8.5660	8.1622	7.7861	7.4353	7.1078	6.8017
10	9.4713	8.9826	8.5302	8.1109	7.7217	7.3601

Giovanna plans to invest $10 000 every 6 months for 5 years. Her investment will earn interest at a rate of 4% p.a. compounded half-yearly.

What is the present value of this annuity?

A $44 518 **B** $47 135 **C** $81 109 **D** $89 826

Question 17

Peter borrows $650 000 to buy an apartment. Interest is charged at 4.2% p.a. compounded fortnightly.

How much does he owe at the end of the first fortnight, after he has made a repayment of $1800?

A $647 150 **B** $648 200 **C** $649 250 **D** $650 475

Question 18 ©NESA 2016 HSC EXAM, QUESTION 8

The table shows the future value of an investment of $1000, compounding yearly, at varying interest rates for different periods of time.

Future values of an investment of $1000

Number of years	Interest rate per annum				
	1%	2%	3%	4%	5%
1	1010.00	1020.00	1030.00	1040.00	1050.00
2	1020.10	1040.40	1060.90	1081.60	1102.50
3	1030.30	1061.21	1092.73	1124.86	1157.63
4	1040.60	1082.43	1125.51	1169.86	1215.51
5	1051.01	1104.08	1159.27	1216.65	1276.28

Based on the information provided, what is the future value of an investment of $2500 over 3 years at 4% p.a.?

A $1124.86 **B** $2812.15 **C** $3624.86 **D** $5312.15

Question 19

What single amount must be invested now, at 4% p.a. compounded quarterly, so that in 5 years it will have grown to $30 000?

A $13 691.61 **B** $24 586.33 **C** $24 657.81 **D** $28 543.97

Question 20

Christina invested $45 000 into an account, which earns $3112 in interest after 2 years.

If the interest is compounded yearly, what is the interest rate per annum, correct to one decimal place?

A 3.4% **B** 3.5% **C** 6.9% **D** 7.2%

Practice set 2

Short-answer questions

Solutions start on page 127.

Question 1 (2 marks) 

If a woodwork company's \$500 000 machinery depreciates by \$10 000 in the first year, what is the annual rate of depreciation? 2 marks

Question 2 (2 marks) 

Patricia's share portfolio is shown in the table.

Number of shares	Company	Market value	Dividend yield
1200	Eastpac Bank	\$36.77	4.1%
5500	PHB Mining	\$1.56	2.8%

Calculate the total dividend Patricia earned from these shares. 2 marks

Question 3 (2 marks) 

In successive years, the inflation rates are 2.8% and 3.2%. If a jar of coffee was \$16.95 at the beginning of this time, what is the expected cost of the jar at the end of these 2 years? 2 marks

Question 4 (4 marks)

Jenny begins an annuity fund to which she contributes \$10 000 regularly at the end of each year. She earns compound interest on the balance at the start of the year. At the end of 3 years, Jenny has \$31 216 in this annuity fund.

Year	Balance at start of year	Interest	Contribution at end of year	Balance at end of year
1	\$0	\$0	\$10 000	\$10 000
2	\$10 000	\$400	\$10 000	\$20 400
3	\$20 400	\$816	\$10 000	\$31 216

a How much interest has this account earned in 3 years? 1 mark

b What is the annual interest rate for this account? 1 mark

c Complete the annuity for the fourth year to find the balance at the end of 4 years. 2 marks

Year	Balance at start of year	Interest	Contribution at end of year	Balance at end of year
4				

Question 5 (3 marks)

The table below shows monthly repayments for loans over terms from 15 to 20 years.

Monthly loan repayments

Principal	Term of loan (years)					
	15	16	17	18	19	20
$50 000	$492.37	$477.25	$464.15	$452.73	$442.73	$433.92
$60 000	$590.85	$572.70	$556.98	$543.28	$531.27	$520.70
$70 000	$689.32	$668.15	$649.81	$633.83	$619.82	$607.48
$80 000	$787.80	$763.60	$742.64	$724.37	$708.36	$694.26
$90 000	$886.27	$859.05	$835.47	$814.92	$796.91	$781.05
$100 000	$984.74	$954.50	$928.30	$905.46	$885.45	$867.83

a If Belinda borrows $90 000, how much does she repay all together if the loan is for 18 years? 1 mark

b What is the interest saved if a loan of $60 000 is paid over 15 years rather than over 20 years? 2 marks

Question 6 (4 marks)

a What is the future value when $25 000 is invested for 6 years earning 4.8% p.a. compounded yearly? 1 mark

b How much extra interest is earned if $25 000 is invested for 6 years at 4.8% p.a. compounded monthly compared to compounded yearly? 3 marks

Question 7 (2 marks)

Part of a credit card statement is shown. This credit card company charges 0.052% compound interest per day on any outstanding balances.

How much interest is charged on the closing balance, which is 21 days overdue?

ACCOUNT SUMMARY	
Opening balance	$106.10 DR
Payments	$110.00 CR
New purchases	$612.29 DR
Charges	$2.04 DR
Closing balance	$

Question 8 (2 marks)

The table gives the future value in dollars of an annuity with a contribution of $1 per period. The values are rounded to four decimal places.

Future value of ordinary annuity with contribution of $1 per period

Period, n	Interest rate per period, I						
	1%	2%	3%	4%	5%	6%	8%
1	1.0000	1.0000	1.0000	1.0000	1.0000	1.0000	1.0000
2	2.0100	2.0200	2.0300	2.0400	2.0500	2.0600	2.0800
3	3.0301	3.0604	3.0909	3.1216	3.1525	3.1836	3.2464
4	4.0604	4.1216	4.1836	4.2465	4.3101	4.3746	4.5061
5	5.1010	5.2040	5.3091	5.4163	5.5256	5.6371	5.8666
6	6.1520	6.3081	6.4684	6.6330	6.8019	6.9753	7.3359

Julia opens an account that earns 6% p.a. compounded yearly. She contributes $20 000 at the end of every year. How much interest will the annuity have earned after 5 years? 2 marks

Question 9 (2 marks)

Anh says that during the first year, an investment with a simple interest rate of 6% p.a. will earn the same interest as if the investment was compounded yearly at 6% p.a. Is he correct? Justify your answer with any appropriate calculations. 2 marks

Question 10 (4 marks)

This graph shows the declining value of a machine at different annual rates of depreciation.

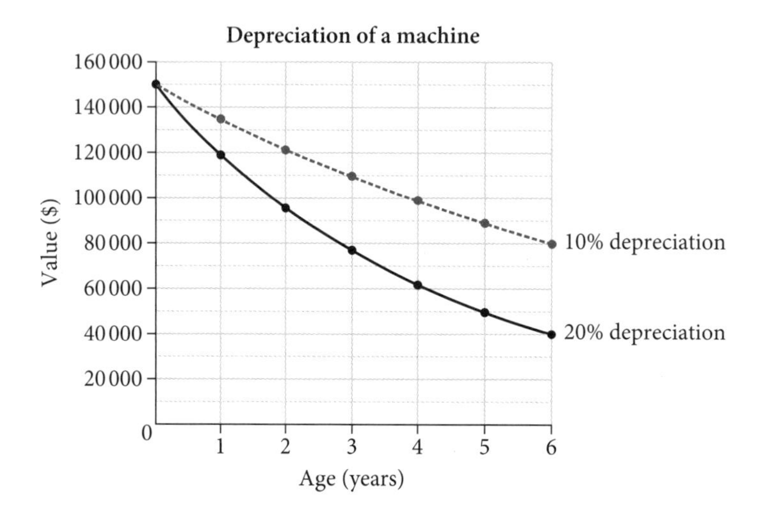

a From the graph, what is the difference in the salvage value of the machine between the 10% and the 20% depreciation rates after 4 years? 2 marks

b Using the declining-balance formula, find the salvage value of the machine after 7 years using the rate of 20% depreciation. 2 marks

Question 11 (2 marks)

Ilonka borrows $90 000 to buy a caravan and is charged reducible interest at 8% p.a.
The repayment, R, is set by the bank at $2000 per month.

The loan balance sheet shows the interest charged and the balance owing at the end of the first month. The interest charged and the repayments continue each month.

Month	Principal at start of month, P	Monthly interest, I	Balance at end of month, $P+I-R$
1	$90 000	$600	$88 600
2	$88 600	A	B

Calculate the values of A and B for the second month. 2 marks

Question 12 (3 marks)

A cafe business purchases a coffee machine for $5400 and each year the machine depreciates by a fixed percentage rate.

The table displays the salvage value of the coffee machine for the first 3 years.

Age (years)	Salvage value
1	$4050
2	$3037.50
3	$2278.13

a What is the annual depreciation rate of the coffee machine? 2 marks

b What will be the salvage value of the coffee machine after 10 years? 1 mark

Question 13 (4 marks)

a Farrokh invests $3500 at 8% p.a. compounded quarterly for 3 years. What is the final balance of his investment? 2 marks

b Would Farrokh have earned a better financial result if he had chosen an annuity of $360 per quarter for 3 years at the same interest rate? Use the future value annuity table to justify your answer. 2 marks

Future value of ordinary annuity with contribution of $1 per period

Period,	Interest rate per period, *I*%								
n	1%	2%	3%	4%	5%	6%	8%	10%	12%
1	1.0000	1.0000	1.0000	1.0000	1.0000	1.0000	1.0000	1.0000	1.0000
2	2.0100	2.0200	2.0300	2.0400	2.0500	2.0600	2.0800	2.1000	2.1200
3	3.0301	3.0604	3.0909	3.1216	3.1525	3.1836	3.2464	3.3100	3.3744
4	4.0604	4.1216	4.1836	4.2465	4.3101	4.3746	4.5061	4.6410	4.7793
5	5.1010	5.2040	5.3091	5.4163	5.5256	5.6371	5.8666	6.1051	6.3528
6	6.1520	6.3081	6.4684	6.6330	6.8019	6.9753	7.3359	7.7156	8.1152
7	7.2135	7.4343	7.6625	7.8983	8.1420	8.3938	8.9228	9.4872	10.0890
8	8.2857	8.5830	8.8923	9.2142	9.5491	9.8975	10.6366	11.4359	12.2997
9	9.3685	9.7546	10.1591	10.5828	11.0266	11.4913	12.4876	13.5795	14.7757
10	10.4622	10.9497	11.4639	12.0061	12.5779	13.1808	14.4866	15.9374	17.5487
11	11.5668	12.1687	12.8078	13.4864	14.2068	14.9716	16.6455	18.5312	20.6546
12	12.6825	13.4121	14.1920	15.0258	15.9171	16.8699	18.9771	21.3843	24.1331

Question 14 (2 marks)

An excavator bought for $45 900 is depreciated using the declining-balance method, with a rate of depreciation of 4% per half-year. What is the salvage value of the excavator after 3 years? 2 marks

Question 15 (4 marks)

In June, Maria bought 3000 shares in a mining company at $1.88 per share. She paid a brokerage fee of 2.2% of the purchase price. In September, Maria was paid a dividend of 35 cents per share. In December, Maria sold all 3000 shares for $2.35 per share. Again, her brokerage fee was 2.2% of the transaction of these sales. 4 marks

Calculate Maria's net profit from her sale of the shares.

Question 16 (5 marks)

A farmer bought a tractor for $185 000. Each year when the tractor insurance is renewed, the value of the tractor depreciates at a rate of 7.5% of the previous year's value.

a By how much does the tractor depreciate in the first 5 years? 2 marks

b By guessing and checking, work out the year in which the depreciated value of the tractor will first fall below half of its original value. 3 marks

Question 17 (4 marks)

For 2 years, Josh will deposit $5000 at the end of each 6-month period into an annuity fund earning interest at 4.6% p.a. compounded half-yearly.

a What is the final balance of this account at the end of 2 years? 2 marks

b What single amount of money could Josh have invested today into an account that will give the same future value when offered the same interest rate for 2 years? 2 marks

Question 18 ©NESA 2015 HSC EXAM, QUESTION 29(b) (2 marks)

Jamal borrowed \$350 000 to be repaid over 30 years, with monthly repayments of \$1880. However, after 10 years, he made a lump sum payment of \$80 000. The monthly repayment remained unchanged. The graph shows the balances owing over the period of the loan. 2 marks

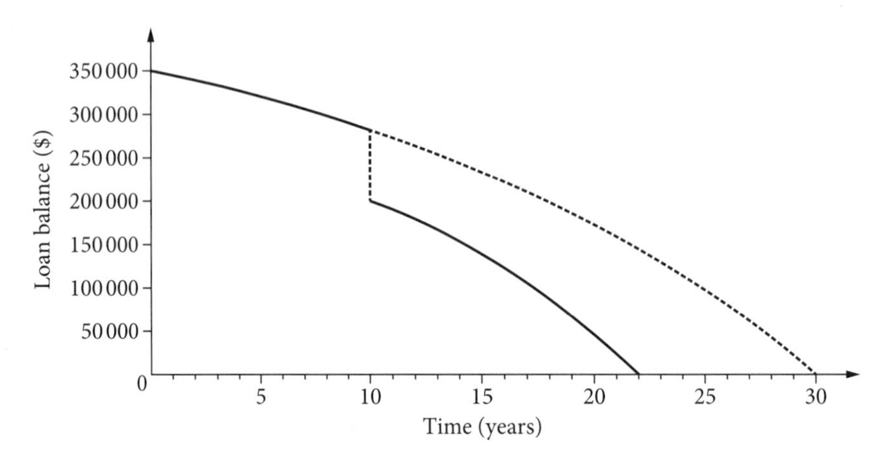

Over the period of the loan, how much less will Jamal pay by making the lump sum payment?

Question 19 (4 marks)

The declining-balance formula, $S = V_0 \times 0.85^n$, gives the salvage value after n years of an item that was initially worth V_0 dollars.

a What is the annual rate of depreciation for this item? 1 mark

b As the item continues to depreciate, what percentage of the initial value remains after 5 years? Give your answer correct to two decimal places. 1 mark

c The graph shows the salvage value of the item over n years.

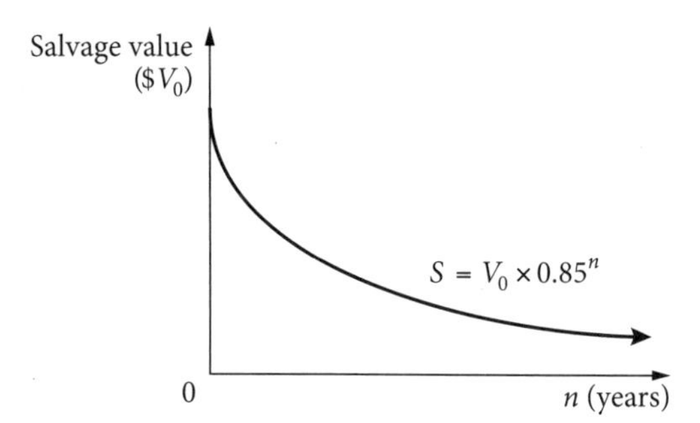

On the same axis above, draw a graph indicating the depreciation of the same item if it is depreciating at a rate of 10% p.a. 2 marks

Question 20 (4 marks)

Kathrine inherits \$50 000 and invests it into an account earning interest at a rate of 0.25% per month. Each month, immediately after the interest is paid, Kathrine withdraws \$1000.

The amount in the account immediately after the nth withdrawal can be determined using the recurrence relation:

$$A_n = A_{n-1}(1.0025) - 1000,$$

where $n = 1, 2, 3, \ldots$ and $A_0 = 50\,000$ (initial amount invested)

Hence, after the first month:

$$\begin{aligned} A_1 &= A_0(1.0025) - 1000 \\ &= 50\,000(1.0025) - 1000 \\ &= \$49\,125 \end{aligned}$$

a Use the recurrence relation to find the amount of money in the account immediately after the third withdrawal. 2 marks

b Calculate the amount of interest earned in the first 3 months. 2 marks

Practice set 1

Worked solutions

1 D

The compound interest formula is also known as the future value formula.

2 A

4.8% p.a. ÷ 12 months = 0.4% per month

3 C

3% = 0.03 (decimal)

$FV = 5000(1 + 0.03)^6$
$= \$5970.26$

4 D

If the interest rate is the same, then compound interest always earns more than simple interest, and interest that is compounded more often earns more interest, so interest compounded monthly gives the highest future value.

5 B

Interest = total repayments – amount borrowed
= 741.33 × 26 × 25 – 300 000
= $181 864.50

6 A

Dividend earned = 6.2% × 2700 × 4.55
= $761.67

7 B

2.8% = 0.028 (decimal)

$FV - 436\,000(1 + 0.028)^3$
$= \$473\,659$

8 C

Depreciation = 76 000 – 55 404
= $20 596

9 C

Line *W* has the highest future value, so it matches with 10% compounded quarterly (higher interest rate and frequency of compounding). Line *Y* has the lowest future value.

Line *X* is only a little below line *W*, so it is 10% annually (compounding less often).

10 C

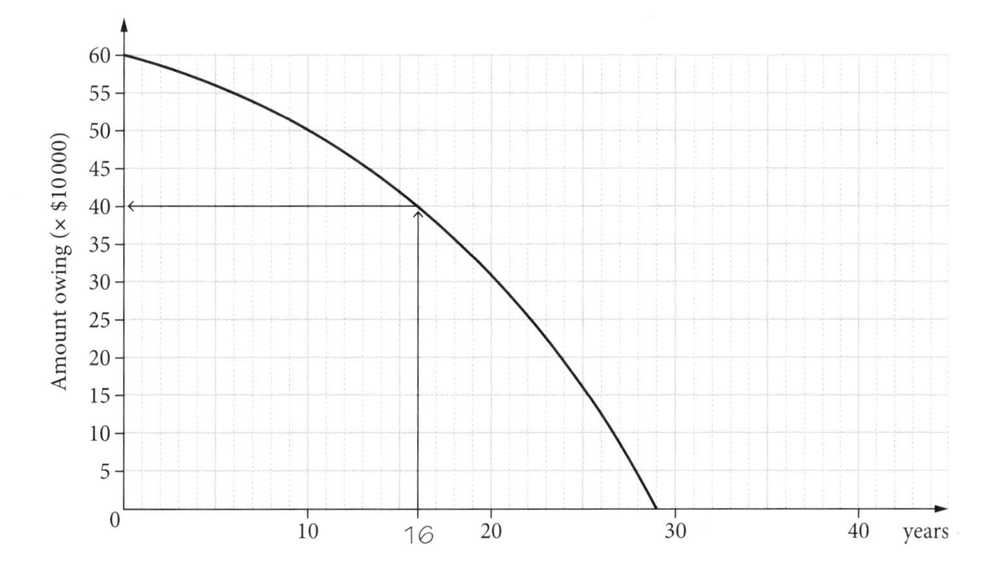

40 × 10 000 = $400 000 balance owing, so paid-off loan at this stage = 600 000 – 400 000 = $200 000

11 A

68 cents = \$0.68

Dividend yield $= \dfrac{0.68}{12.34} \times 100 = 5.5\%$

12 D

$r = 2.4\,\%$ p.a. $= \dfrac{0.024}{12}$ (decimal rate per month)

$n = 3$ years $\times 12 = 36$ months

Hence, $FV = 15000\left(1 + \dfrac{0.024}{12}\right)^{36}$

13 A

Declining-balance depreciation never reaches a salvage value of \$0.

14 B

$r = 17\%$ p.a. $= \dfrac{0.017}{365}$ per day, $n = 22$ days

$$FV = 400\left(1 + \frac{0.17}{365}\right)^{22} = \$404.12$$

$$I = FV - PV = 404.12 - 400 = \$4.12$$

15 B

$$PV = \frac{FV}{(1+r)^n} = \frac{320}{(1+0.027)^5} = \$280.09 \approx \$280$$

16 D

Compounding every 6 months:

$r = 4\% \div 2 = 2\%$

$n = 5 \times 2 = 10$ periods

From table, $PV = 8.9826$ for each \$1

Hence, $10\,000 \times 8.9826 = \$89\,826$

17 C

$$FV = 650\,000\left(1 + \frac{0.042}{26}\right)^1 = \$651\,050$$

Balance owing $= 651\,050 - 1800 = \$649\,250$

18 B

Future values of an investment of \$1000

Number of years	Interest rate per annum				
	1%	2%	3%	(4%)	5%
1	1010.00	1020.00	1030.00	1040.00	1050.00
2	1020.10	1040.40	1060.90	1081.60	1102.50
(3)	1030.30	1061.21	1092.73	(1124.86)	1157.63
4	1040.60	1082.43	1125.51	1169.86	1215.51
5	1051.01	1104.08	1159.27	1216.65	1276.28

$\dfrac{2500}{1000} \times 1124.86 = \$2812.15$

19 B

$r = 4\%$ p.a. $\div 4 = 1\%$ per quarter

$n = 5$ years $\times 4 = 20$ quarters

$$PV = \frac{30\,000}{(1+0.01)^{20}} = \$24\,586.33$$

20 A

$$FV = PV + I = 45\,000 + 3112 = \$48\,112$$

$$48\,112 = 45\,000(1+r)^2$$

$$\frac{48\,112}{45\,000} = (1+r)^2$$

$$\sqrt{\frac{48\,112}{45\,000}} = 1 + r$$

$$r = \sqrt{\frac{48\,112}{45\,000}} - 1 = 0.033\,999\ldots \times 100\% \approx 3.4\%$$

Practice set 2

Worked solutions

Question 1

$\frac{10\,000}{500\,000} \times 100 = 2\%$

Question 2

Total share dividends
= Eastpac Bank + PHB mining
$= 1200 \times 36.77 \times 4.1\% + 5500 \times 1.56 \times 2.8\%$
$= 1809.08 + 240.24$
= $2049.32

Question 3

Increase $16.95 by 2.8% and 3.2%.

In 2 years, coffee cost $= 16.95 \times 102.8\% \times 103.2\%$
$= 16.95 \times 1.028 \times 1.032$
= $17.98 (nearest cent)

Question 4

a Interest = 400 + 816 = $1216
(or $31\,216 - 3 \times 10\,000$ = $1216)

b $\frac{400}{10\,000} \times 100 = 4\%$

c

Year	Balance at start of year	Interest	Contribution at end of year	Balance at end of year
4	$31 216	31 216 × 4% = $1248.64	$10 000	31 216 + 1248.64 + 10 000 = $42 464.64

Question 5

a $814.92/month × 12 months × 18 years
= $176 022.72

b $520.70 \times 12 \times 20 - 590.85 \times 12 \times 15$ = $18 615

Question 6

a $FV = 25\,000(1 + 0.048)^6$
= $33 121.33

b PV = $25 000

r = 4.8% p.a. ÷ 12 = 0.004 per month

n = 6 years × 12 = 72 months

$FV = 25\,000(1 + 0.004)^{72}$
≈ $33 324.78

Extra interest earned is difference in FV
$= 33\,324.78 - 33\,121.33$
= $203.45

Question 7

Closing balance $= -106.10 + 110.00 - 612.29 - 2.04$
= –$610.43

So $610.43 is owing.

$FV = PV(1 + r)^n$

PV = $610.43
$r = 0.052\% = 0.000\,52$ per day
n = 21 days

$FV = 610.43(1 + 0.000\,52)^{21}$
= $617.13

So interest $= FV - PV$
$= 617.13 - 610.43$
= $6.70

Question 8

Period, n = 5 years, interest rate each year, r = 6%

Future value for $1 contribution per year = $5.6371

Future value for $20 000 contribution per year
= 20 000 × 5.6371 = $112 742

Interest = FV – total contributions
$= 112\,742 - 5 \times 20\,000$
$= 112\,742 - 100\,000$
= $12 742

WORKED SOLUTIONS

Question 9

Yes, Anh is correct.

For example, let $PV = \$100$, $r = 6\%$ p.a., $n = 1$ year

Simple interest, $I = Prn$
$= 100 \times 0.06 \times 1$
$= \$6$

Compound interest, $FV = 100(1 + 0.06)^1$
$= \$106$

Hence, $I = FV - PV = 106 - 100 = \$6$

This would be the case for any value of PV.

Question 10

a Using the graph:
from 10% curve, $S = \$100\,000$
from 20% curve, $S = \$60\,000$

Difference $= 100\,000 - 60\,000$
$= \$40\,000$

b $S = 150\,000(1 - 0.2)^7$
$= \$31\,457.28$

Question 11

$A = 88\,600 \times 0.08 \div 12 = \$590.67$

$B = 88\,600 + 590.67 - 2000 = \$87\,190.67$

Question 12

a Amount of depreciation in the first year
$= 5400 - 4050 = \$1350$

Rate of depreciation p.a. $= \dfrac{1350}{5400} \times 100 = 25\%$

b $I = 5400(1 - 0.25)^{10}$
$= \$304.09$

Question 13

a $PV = \$3500$
$r = 8\%$ p.a. $\div 4 = 0.02$ per quarter
$n = 3$ years $\times 4 = 12$ quarters

$FV = 3500(1 + 0.02)^{12}$
$\approx \$4438.85$

b From table: $r = 2\%$ and $n = 12$.

$FV = 360 \times 13.4121$
$= \$4828.36$

Yes, Farrokh would be best with this quarterly annuity because the future value is $389.51 more.

Question 14

$V_0 = 45\,900$, $r = 4\% = 0.04$ per half-year,
$n = 3$ years $= 6$ half-years

$S = 45\,900(1 - 0.04)^6$
$= \$35\,928.58$

Question 15

Cost of buying shares $= 3000 \times 1.88 = \$5640$

Brokerage fee on buying shares $= 2.2\% \times 5640$
$= \$124.08$

Dividend earnings $= 0.35 \times 3000 = \$1050$

Income of shares sold $= 3000 \times 2.35 = \$7050$

Brokerage fee on selling shares $= 2.2\% \times 7050$
$= \$155.10$

Profit $= -5640 - 124.08 + 1050 + 7050 - 155.10$
$= \$2180.82$

Question 16

a $7.5\% = 0.075$

$S = 185\,000(1 - 0.075)^5$
$= \$125\,279.61$

In 5 years, the tractor depreciated by
$185\,000 - 125\,279.61 = \$59\,720.39$

b Half original value $= 185\,000 \div 2 = \$92\,500$

Guess and check: $n = 10$ years

$S = 185\,000(1 - 0.075)^{10}$
$= \$84\,837.73$ (too small)

Guess and check: $n = 7$ years

$S = 185\,000(1 - 0.075)^7$
$= \$107\,192.37$ (too big)

Guess and check: $n = 8$ years

$S = 185\,000(1 - 0.075)^8$
$= \$99\,152.94$ (too big)

Guess and check: $n = 9$ years

$S = 185\,000(1 - 0.075)^9$
$= \$91\,716.47$ (closest below \$92 500)

Hence, after 9 years the salvage value first falls below half its original value.

Question 17

a Contributions compounded every 6 months

$n = 4$ half-years, $r = 0.046$ p.a. $\div$ 2 = 0.023 per half-year

n (half-year)	Balance at start of period	Interest	Contribution at end of period	Balance at end of period
1	\$0	\$0	\$5000	\$5000
2	\$5000	5000×0.023 = \$115	\$5000	5000 + 115 + 5000 = \$10 115
3	\$10 115	$10\,115 \times 0.023$ = \$232.65	\$5000	10 115 + 232.65 + 5000 = \$15 347.65
4	\$15 347.65	$15\,347.65 \times 0.023$ = \$353.00	\$5000	15 347.65 + 353 + 5000 = \$20 700.65

Hence, the final balance at the end of 2 years is \$20 700.65.

b $PV = \dfrac{20700.65}{(1 + 0.023)^4} = \$18\,900.85$

Question 18

First 10 years of repayments
= 10 years × 12 months × \$1880/month
= \$225 600

Lump sum payment + 80 000

Loan is paid by 22 years.

Remaining 12 years of repayments
= 12 years × 12 months × \$1880/month
= \$270 720

Total payments
= 225 600 + 80 000 + 270 720
= \$576 320

Compared to original repayments to be over 30 years
= 30 years × 12 months × 1880
= \$676 800

Hence Jamal's savings
= 676 800 − 576 320
= \$100 480

Question 19

a Annual depreciation rate: $1 - r = 0.85$

$r = 1 - 0.85$
$= 0.15$
$= 15\%$

b $S = V_0 \times 0.85^5$

$0.85^5 = 0.4437\ldots$
$= 44.37\%$

c 10% p.a. depreciation would have the salvage value $S = V_0 \times (1 - 0.1)^n$,

$S = V_0 \times 0.9^n$

The salvage value will decrease more slowly than 15% depreciation, so its graph should be a little higher.

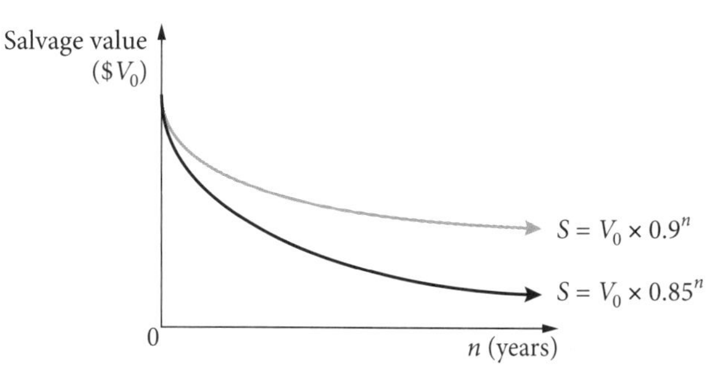

Question 20

a $r = 0.25\% = 0.0025$

$A_0 = 50\,000$ (initial amount)

$$A_1 = A_0(1.0025) - 1000$$
$$= 50\,000(1.0025) - 1000$$
$$= \$49\,125$$

$$A_2 = A_1(1.0025) - 1000$$
$$= 49\,125(1.0025) - 1000$$
$$= \$48\,247.81$$

$$A_3 = A_2(1.0025) - 1000$$
$$= 48\,247.81(1.0025) - 1000$$
$$= \$47\,368.43$$

b Total withdrawals after 3 months
$= 1000 \times 3 = \$3000$

Difference in balance:
$50\,000 - 47\,368.43 = \$2631.57$ reduction

Interest $= 3000 - 2631.57$
$= \$368.43$

HSC exam topic grid (2011–2020)

This grid shows the coverage of this topic in past HSC exams by question number. The past exam papers can be downloaded from the NESA website (www.educationstandards.nsw.edu.au) by selecting 'Year 11 – Year 12', 'HSC exam papers'. NESA marking feedback and guidelines can also be found there.

Before 2019, 'Mathematics Standard 2' was called 'Mathematics General 2' and, before 2014, 'General Mathematics'. For these exams, select 'Year 11 – Year 12', 'Resources archive', 'HSC exam papers archive'.

	Investments and shares	Depreciation	Reducing balance loans	Annuities
2011	23(c)	28b	10*, 22	
2012	9	16, 26(b)	24, 26(c)*	
2013	9, 26(e), 28(d)	28(e)		
2014	30(a)	9		21
2015	17, 26(d)	**10**	29(a)*, **29(b)**	30(c)
2016	**8**		17*, 27(d)	28(d)
2017	10, 26(e)	11	28(c)	27(c)
2018	19	26(h)	28(d)*, 29(e)	26(c)
2019 new course	3, 9, **13**, 21	37	27*	42
2020	4, 21, 29	11	22*	14, **34**, 37

Questions in **bold** can be found in this chapter.
* Credit cards.

CHAPTER 5
BIVARIATE DATA AND THE NORMAL DISTRIBUTION

MS-S4 Bivariate data analysis 134
MS-S5 The normal distribution 139

All content for this topic is common with the Mathematics Advanced course.

BIVARIATE DATA AND THE NORMAL DISTRIBUTION

Scatterplots

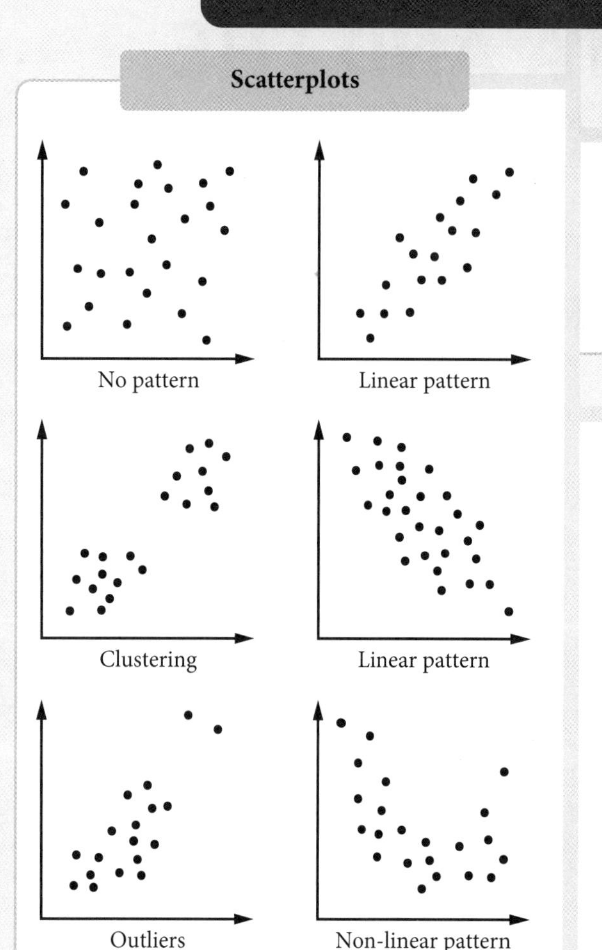

Relationship/correlation

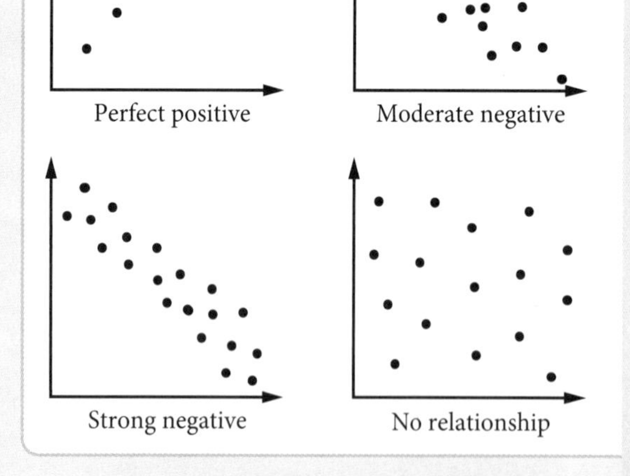

Line of best fit

- Line of best fit by eye
- Least-squares regression line by calculator
- Interpolation: data prediction within the set
- Extrapolation: data prediction outside the set

The normal distribution

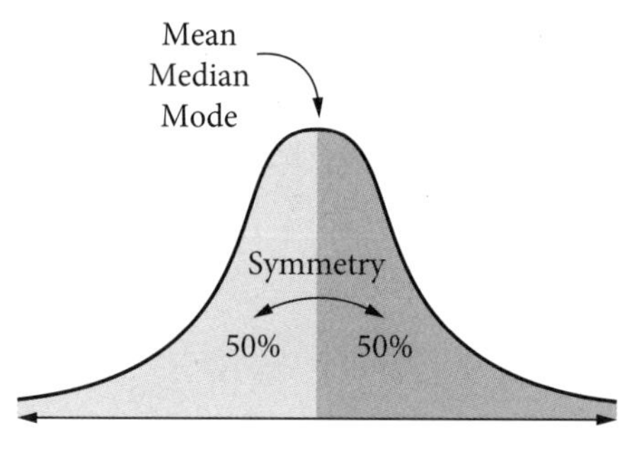

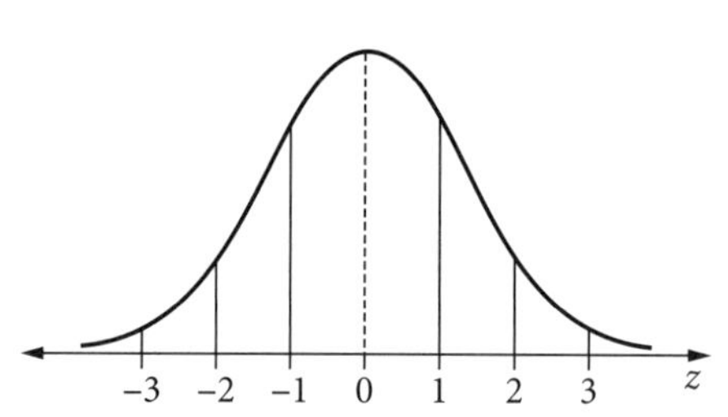

- Approximately 68% of scores have z-scores between −1 and 1.
- Approximately 95% of scores have z-scores between −2 and 2.
- Approximately 99.7% of scores have z-scores between −3 and 3.

z-scores

$$z\text{-score} = \frac{\text{score} - \text{mean}}{\text{standard deviation}}$$

$$z = \frac{x - \mu}{\sigma}$$

Pearson's correlation coefficient (r)

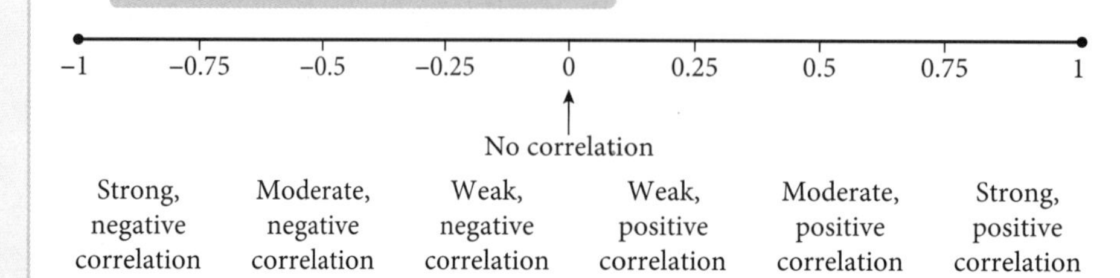

All formulas for this topic appear on the HSC exam reference sheet (and at the back of this book).

Glossary

bivariate data
Data for 2 variables, such as height and weight, for the same group that can be used to determine whether a relationship exists.

correlation
The strength of a relationship between 2 variables.

dependent variable
A variable with an outcome that depends on another variable, usually y, and plotted as y-values on a scatterplot.

empirical rule
The 68%–95%–99.7% rule that relates to the distribution of scores about the mean on a normal distribution bell curve.

extrapolation
Estimation of values outside the range of a data set.

independent variable
A variable with an outcome that does not depend on another variable, usually x, and plotted as x-values on a scatterplot.

interpolation
Estimating values inside the range of a data set.

least-squares regression line
The line of best fit, a straight line that is as close as possible to points plotted on a graph.

linear
Relating to a straight line.

line of best fit
A line that best represents all the points on a scatterplot.

mean
The 'average'; a single statistical value that represents all values in a set of data.

median
The middle value in a set of ordered data.

https://get.ga/a-hsc-maths-standard-2

A+ DIGITAL FLASHCARDS
Revise this topic's key terms and concepts by scanning the QR code or typing the URL into your browser.

mode
The most common score in a set of scores.

normal curve/distribution
A data distribution in the shape of a bell, where the data is symmetrical about the centre.

outlier
A very low or very high data value that lies well outside the other values in a data set.

Pearson's correlation coefficient (r)
A statistical value that represents the strength of the linear relationship between 2 variables.

raw score
The original score obtained before it is converted to a z-score.

scatterplot
A graph that displays bivariate data (2 variables); the points plotted on the number plane are used to determine whether a relationship exists between 2 variables.

standard deviation
A statistical value that represents the spread of a set of data from the mean.

symmetry
A balance of data about the centre.

z-score
A value that represents how far away a raw score is from the mean, how many standard deviations above (if positive) or below (if negative) the mean.

Topic summary

Bivariate data analysis (MS-S4)

Scatterplots

Scatterplots represent **bivariate data**, which means the data is made up of 2 different variables. Points can be plotted on a number plane to determine whether a relationship exists between the 2 variables.

- *Linear* means 'straight line'. A pattern of points in the shape of a line is a linear pattern.

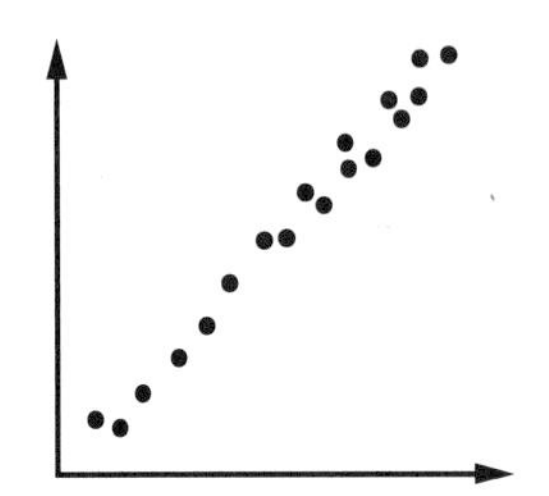

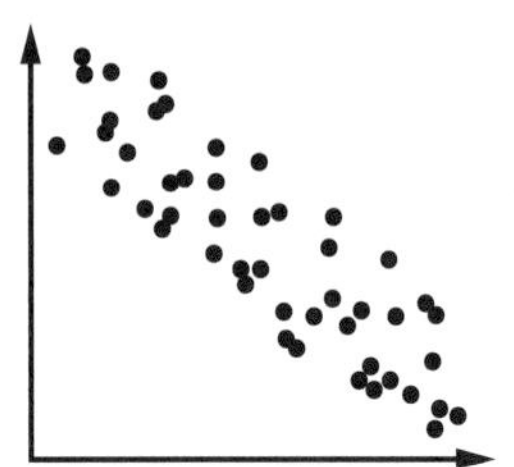

- *Non-linear* means 'curved line'. A pattern of points in the shape of a curve is non-linear.

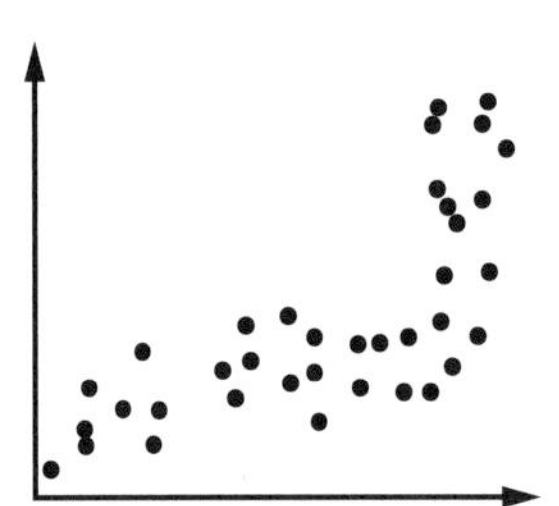

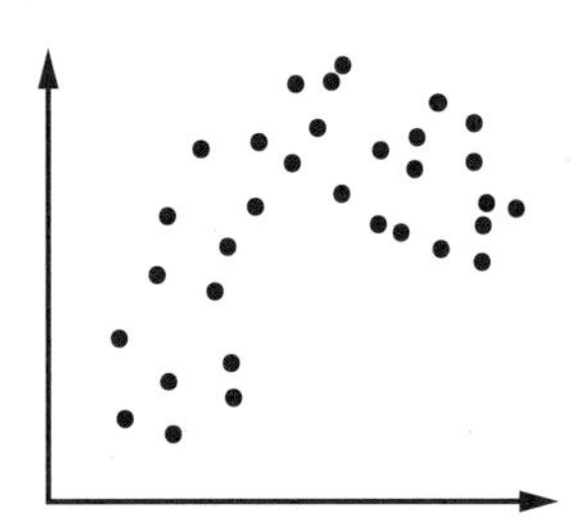

- *No pattern* means points are randomly dispersed on a number plane, with no identifiable shape or form.

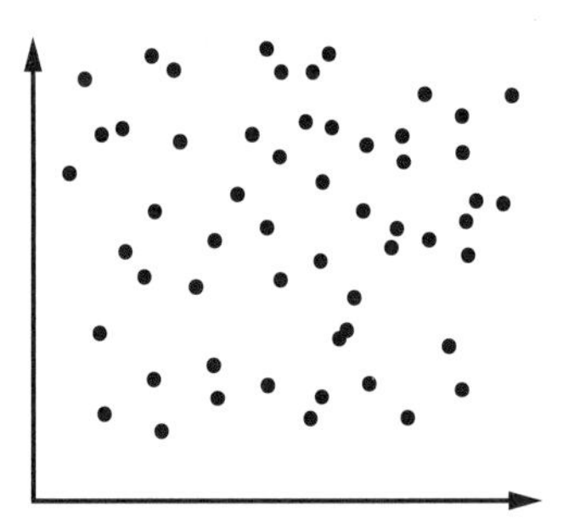

If a linear pattern is formed between 2 variables, then they have a relationship called a **correlation**, which can be defined by its direction and strength.

Direction

Positive: As x increases, y also increases, like a line going 'uphill'.

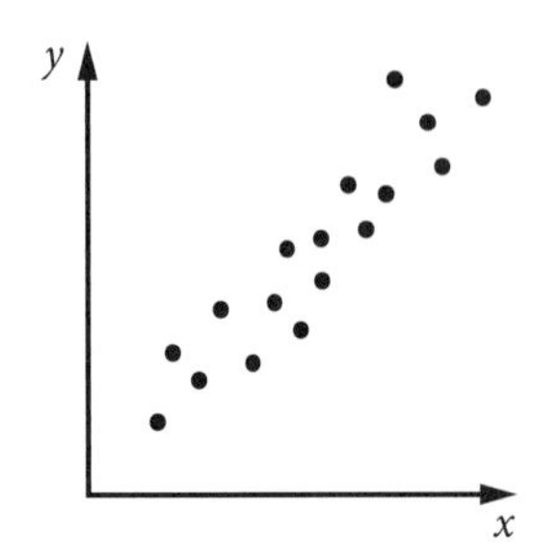

Negative: As x increases, y decreases, like a line going 'downhill'.

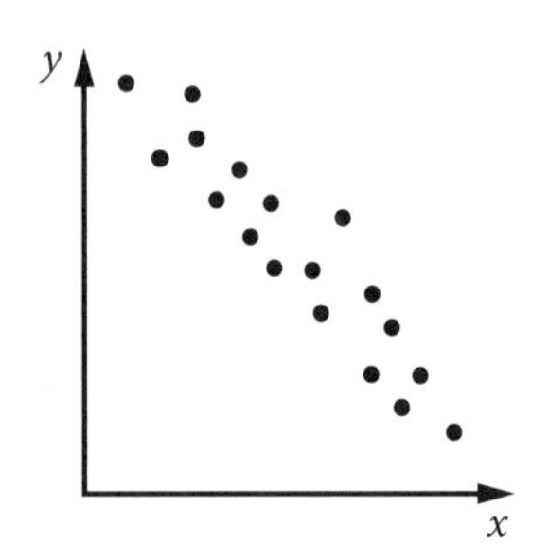

Strength

Perfect: an exact straight line

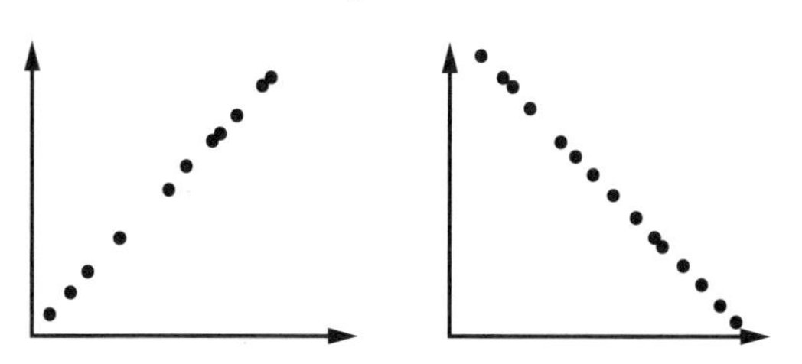

Strong: a mostly thin, straight line

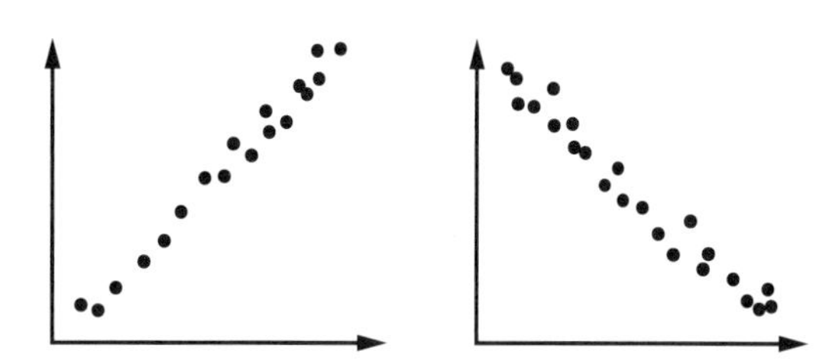

Moderate: a thick/wider straight line

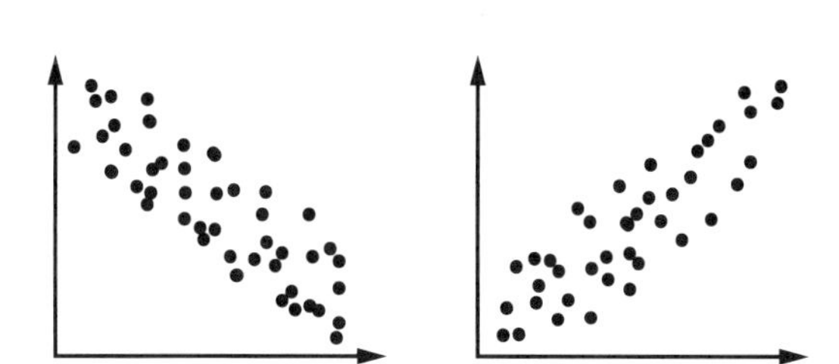

Weak: not in a line but with the general direction of a line

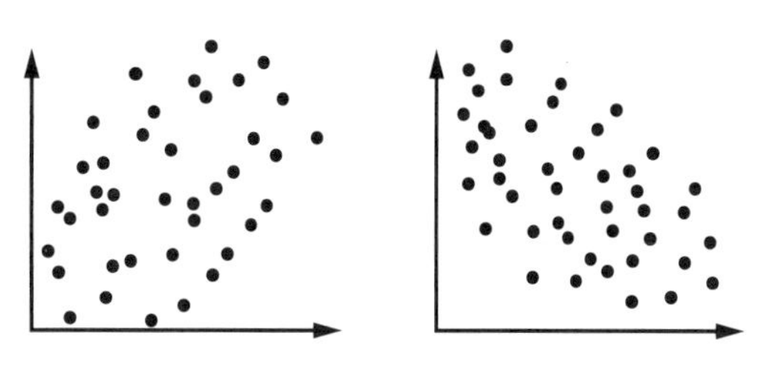

Example 1

Describe the correlation of this scatterplot.

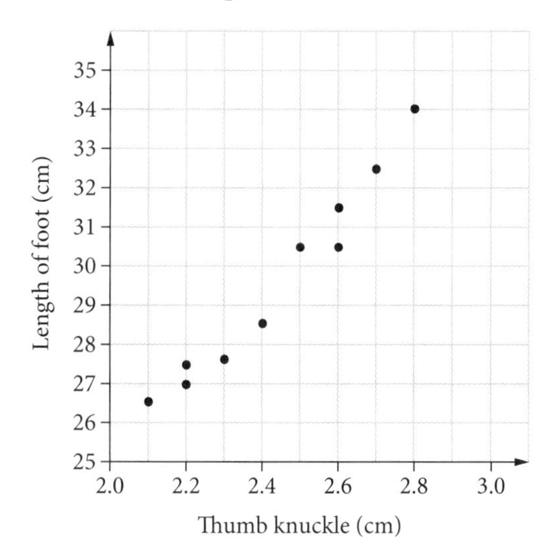

> **Hint**
> Direction and strength *must* be used to define the **correlation** between 2 variables.

> **Hint**
> Correlation is simply an association.
> The correlation is strong but this does not mean that the 2 variables have a causal relationship.

Solution

The correlation is strong, positive: the bigger the thumb knuckle, the longer the foot length.

Pearson's correlation coefficient

Pearson's correlation coefficient is the measure of the strength of the **linear** relationship between 2 variables. It is a value, r, on a scale between 1 and -1, where -1 is a perfect negative correlation, 0 is no correlation and 1 is a perfect positive correlation.

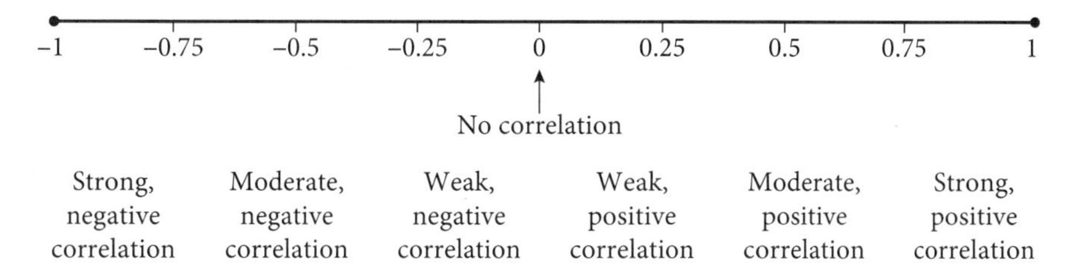

Example 2

Describe the correlation of each pair of variables.

a number of kilometres travelled and amount of fuel in a petrol tank

b temperature of a day and number of layers of clothing worn

c head circumference and number of hamburgers eaten in a day

Solution

a strong, positive – the more fuel, the more kilometres travelled

b strong, negative – the hotter the temperature, the fewer layers of clothing worn

c no correlation

Pearson's correlation coefficient is calculated using your calculator (and other technology). One of the outcomes of this course is to use technology. Just as with standard deviation, you are expected to be able to use your calculator to calculate Pearson's correlation coefficient during an exam.

Example 3

What is Pearson's correlation coefficient for the bivariate data for these body measurements of a group of teenage friends?

Weight (kg)	51	58	63	65	62	58
Height (cm)	162	159	162	178	170	167

Solution

In the calculator's STAT mode, the data entered gives the correlation coefficient.

Operation	Casio scientific	Sharp scientific
Start statistics mode.	MODE STAT A + BX	MODE STAT LINE
Clear the statistical memory.	SHIFT 1 Edit, Del-A	2ndF DEL
Enter data.	SHIFT 1 Data to get table Enter in X column 51 = 58 =, etc. Enter in Y column 162 = 159 =, etc. AC to leave table	51 2ndF STO 162 M+ 58 2ndF STO 159 M+ etc.
Calculate the correlation coefficient.	SHIFT 1 Reg r =	ALPHA r =

$r = 0.6033$ (four decimal places)

Hence, this is a moderate, positive correlation – the taller a person is, the heavier they are likely to be.

Line of best fit

A **line of best fit** is drawn by eye on a scatterplot to model a strong linear relationship.

A line of best fit:

- represents most or all data points as closely as possible
- passes through as many points as possible
- has roughly the same number of points above and below it
- is drawn so that the distances between each point and the line are as small as possible.

The equation of a straight line is

$$y = mx + c,$$

where $m(\text{gradient}) = \frac{\text{rise}}{\text{run}} = \frac{\text{vertical change in position}}{\text{horizontal change in position}}$

x = **independent variable**
y = **dependent variable**
c = y-intercept (vertical intercept).

Hint
To calculate the gradient, choose 2 points that the line of best fit goes through *exactly*.

Example 4

Find the equation of the line of best fit that was drawn by eye on this scatterplot.

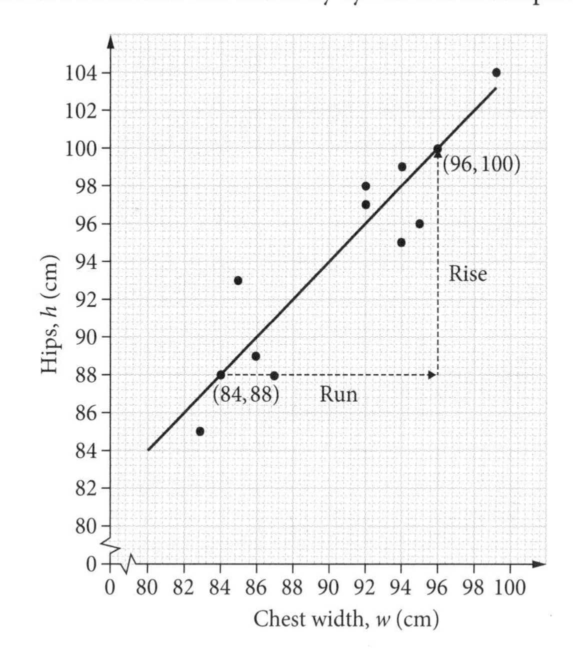

Solution

Select 2 points on the line: (84, 88) and (96, 100).

Use the 2 points to calculate the gradient (m) of the line:

$$m = \frac{\text{rise}}{\text{run}} = \frac{100 - 88}{96 - 84} = \frac{12}{12} = 1$$

The vertical intercept (c) is not shown on this graph because the axes do not start at 0. To find the value of the y-intercept, substitute a point such as (84, 88) and $m = 1$ into the general equation $y = mx + c$:

$$\begin{aligned} y &= 1x + c \\ 88 &= 1(84) + c \\ c &= 88 - 84 \\ &= 4 \end{aligned}$$

Using the variables from the axes labels, h and w, the equation of the line is $h = w + 4$.

So the line of best fit that models the data indicates that hip measurements are 4 cm longer than chest measurements for this group.

Interpolation and extrapolation

A line of best fit can be used to predict what might happen:

- between the points on a scatterplot, within the range of data (called **interpolation**)
- beyond the points on the scatterplot, outside the range of data (called **extrapolation**).

Example 5

From the previous example, calculate the hip measurement for a chest width of 90 cm using the equation $h = w + 4$.

Solution

$$h = 90 + 4$$
$$= 94$$

So a person in this group with a chest measurement of 90 cm would be expected to have a hip measurement of 94 cm.

Least-squares regression line

The **least-squares regression line** is made using Pearson's correlation coefficient, the means and standard deviations of x- and y-values. This line of best fit best is the line that represents all of the points on a scatterplot. As was done for the correlation coefficient, the gradient and y-intercept of this line can be found on a scientific calculator using the statistics STAT mode (or other technology). Again, you may be asked to do this during an exam. On a calculator, the equation is written as $y = a + bx$ (or $y = bx + a$), where b = gradient and a = vertical intercept.

Example 6

a Use your calculator to find the gradient and vertical intercept to write the equation of the least-squares regression line for this set of variables.

Weight, w (kg)	51	58	63	65	62	58
Height, H (cm)	162	159	162	178	170	167

b From this equation, what is the predicted weight, to the nearest kilogram, for a height of 175 cm?

Solution

Operation	Casio scientific	Sharp scientific
Start statistics mode.	MODE STAT A + BX	MODE STAT LINE
Clear the statistical memory.	SHIFT 1 Edit, Del-A	2ndF DEL
Enter data.	SHIFT 1 Data to get table Enter in X column 51 = 58 =, etc. Enter in Y column 162 = 159 =, etc. AC to leave table	51 2ndF STO 162 M+ 58 2ndF STO 159 M+ etc.
On these calculators, the gradient is b and the y-intercept is a ($b = 0.83665\ldots$, $a = -116.552\ldots$)	SHIFT 1 Reg b = SHIFT 1 Reg a =	ALPHA b = ALPHA a =

a In STAT mode, gradient, b, is 0.836 65 and vertical intercept, a, is 116.55. The equation of the line is $y = 0.83665x + 116.55$. Using the variables for weight, w, and height, H, the equation is $H = 0.83665w + 116.55$.

b

$$175 = 0.83665w + 116.55$$
$$175 - 116.55 = 0.83665w$$
$$\frac{58.45}{0.83665} = w$$
$$w = 69.86$$
$$\approx 70 \text{ kg}$$

So, based on this data, a person with a height of 175 cm would be expected to have a weight of 70 kg.

9780170459204

The normal distribution (MS-S5)

The **normal distribution**, normal curve or bell curve is a statistical distribution in the shape of a bell, where the data has **symmetry** about the centre. In a perfectly symmetrical bell curve, all 3 measures of central tendency (**mean, median** and **mode**) are equal and are at the centre of the distribution.

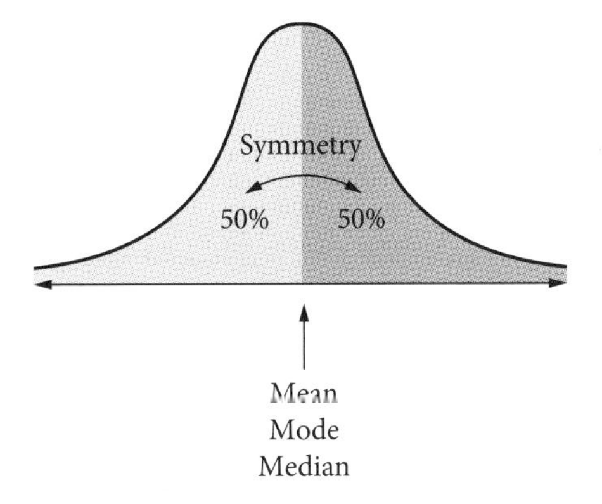

The empirical rule

The following percentages apply to normal distributions only.

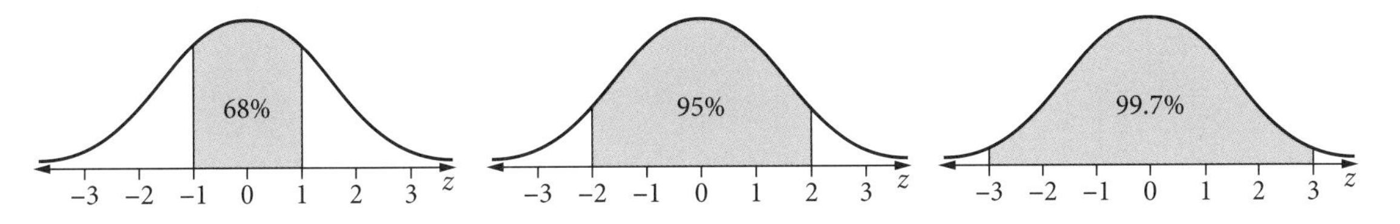

- Approximately 68% of scores lie within 1 **standard deviation** of the mean.
- Approximately 95% of scores lie within 2 standard deviations of the mean.
- Approximately 99.7% of scores lie within 3 standard deviations of the mean.

The **empirical rule** is used to find the percentage of data within each standard deviation of the bell curve.

For example, the percentage of data between 1 and 2 standard deviations above the mean is:

$$\frac{95\% - 68\%}{2} = 13.5\%$$

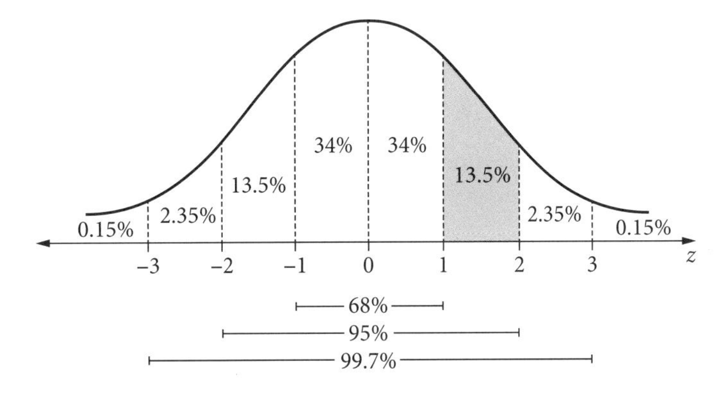

Hint

The empirical rule is on the reference sheet. Some students like to memorise the areas of the 4 individual regions instead, to make calculations easier in problems: 34% ($\frac{1}{2}$ of 68%), 13.5%, 2.35% and 0.15%, as shown in the diagram.

Comparing normal curves

The position, height and width of a normal curve are determined by the mean and the standard deviation of the distribution.

The following normal curves have the same standard deviation but different means.

The data set for graph X has a smaller mean than the data set for graph Y.

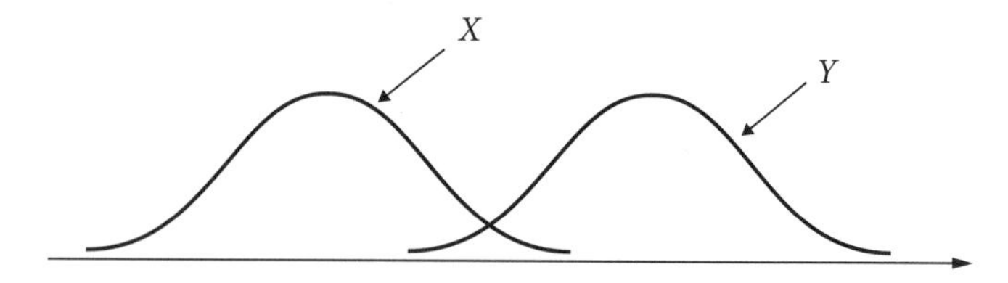

The following normal curves have the same mean but different standard deviations.

The data set for graph W has a smaller standard deviation (less spread) than the data set for graph Z.

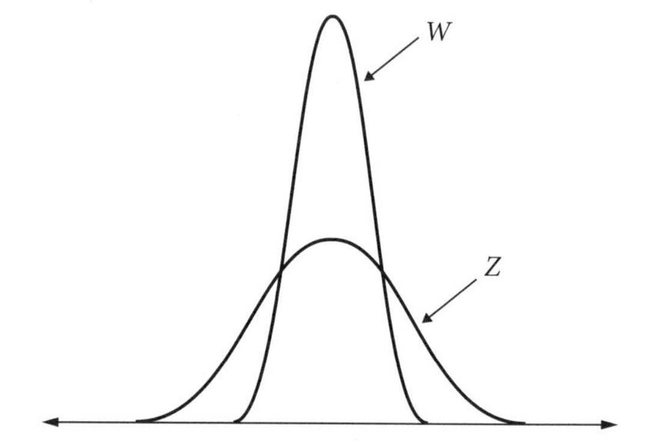

z-scores

z-scores are used to represent how far away a **raw score** is from the mean. A positive z-score means the raw score is above the mean and a negative z-score means the raw score is below the mean.

$$z\text{-score} = \frac{\text{raw score} - \text{mean}}{\text{standard deviation}}$$

This formula is on your HSC exam reference sheet as:

$$z = \frac{x - \mu}{\sigma}$$

where x = score
μ = mean
σ = standard deviation.

Note:

- $z = 0$ indicates that the score, x, is exactly the mean of the distribution.
- $z = 1$ indicates that the score, x, is 1 standard deviation above the mean of the distribution.
- $z = -1$ indicates that the score, x, is 1 standard deviation below the mean of the distribution.

The HSC exam reference sheet shows z-scores of 0, ±1, ±2, ±3 on the normal distribution curve:

- Approximately 68% of scores have z-scores between −1 and 1.
- Approximately 95% of scores have z-scores between −2 and 2.
- Approximately 99.7% of scores have z-scores between −3 and 3.

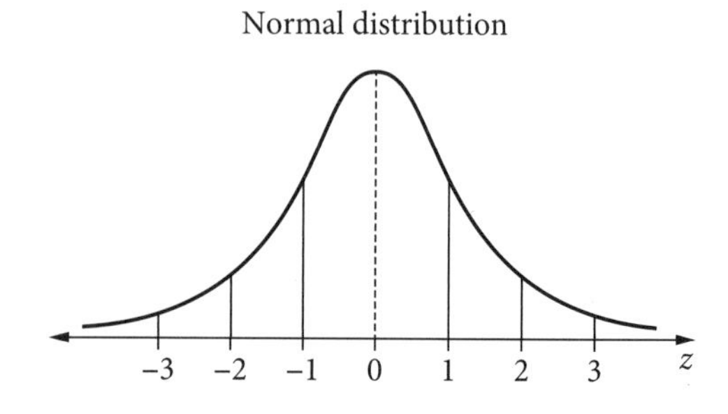

Example 7

The mean of a set of scores is 70 and the standard deviation is 10.

Calculate the z-score for each of the following raw scores.

a 80 **b** 55

Solution

a $z = \frac{x - \mu}{\sigma} = \frac{80 - 70}{10} = 1$

b $z = \frac{x - \mu}{\sigma} = \frac{55 - 70}{10} = -1.5$

So a score of 80 is 1 standard deviation above the mean in this data set.

Example 8

In a normally distributed set of scores with a mean of 18 and a standard deviation of 4, what mark corresponds to a z-score of –2.5?

Solution

$$z = \frac{x - \mu}{\sigma}$$
$$-2.5 = \frac{x - 18}{4}$$
$$-2.5 \times 4 = x - 18$$
$$-10 = x - 18$$
$$-10 + 18 = x$$
$$x = 8$$

Applying z-scores

Probability

z-scores are sometimes used to determine the likelihood of an event. The higher or lower the z-score, the less likely the event is to happen.

For example, a z-score greater than 3 indicates only 0.15% of the values in the data set are above this. Hence, $P(z > 3) = 0.15\% = 0.0015$.

For the normal distribution:

- $P(-1 < z < 1) = 68\% = 0.68$
- $P(-2 < z < 2) = 95\% = 0.95$
- $P(-3 < z < 3) = 99.7\% = 0.997$

Comparing data sets

z-scores can be used to compare data. The higher the z-score, the further away the score is from the mean. We can use z-scores to compare different normal distributions.

Example 9

A factory produces 220-gram packets of mixed lollies. On a particular day, the factory produces packets with a mean weight of 225 grams and a standard deviation of 2 grams. Quality control requires any packet with a z-score of less than −2 to be rejected.

What percentage of packets are rejected?

Solution

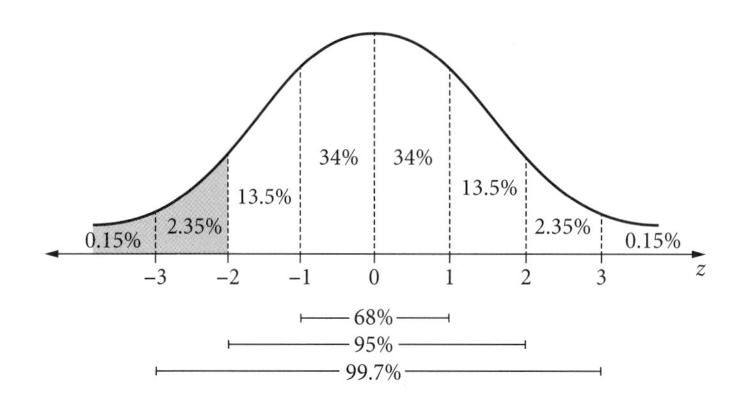

$P(z < -2) = 0.15 + 2.35 = 2.5\%$

OR

95% of packets are within 2 standard deviations, so $P(z < -2) = \dfrac{100\% - 95\%}{2} = 2.5\%$.

So 2.5% of packets are rejected.

Example 10

Susie's marks for exams in Maths, Biology and Drama, as well as the mean and standard deviation for her class, are shown the table.

Subject	Susie's mark	Class mean	Class standard deviation
Maths	75	70	10
Biology	77	70	4
Drama	78	66	12

Relative to the rest of the class, in which subject did Susie achieve her best mark? Justify your answer.

Solution

Calculate the z-score for each exam.

Maths:

$$z = \frac{x - \mu}{\sigma} = \frac{75 - 70}{10} = 0.5$$

Biology:

$$z = \frac{x - \mu}{\sigma} = \frac{77 - 70}{4} = 1.75$$

Drama:

$$z = \frac{x - \mu}{\sigma} = \frac{78 - 66}{12} = 1$$

Biology was Susie's best result relative to the class because it had the highest z-score.

Review of the shape of a distribution (Year 11)

Symmetrical: The values in the data set are evenly spread about the centre.

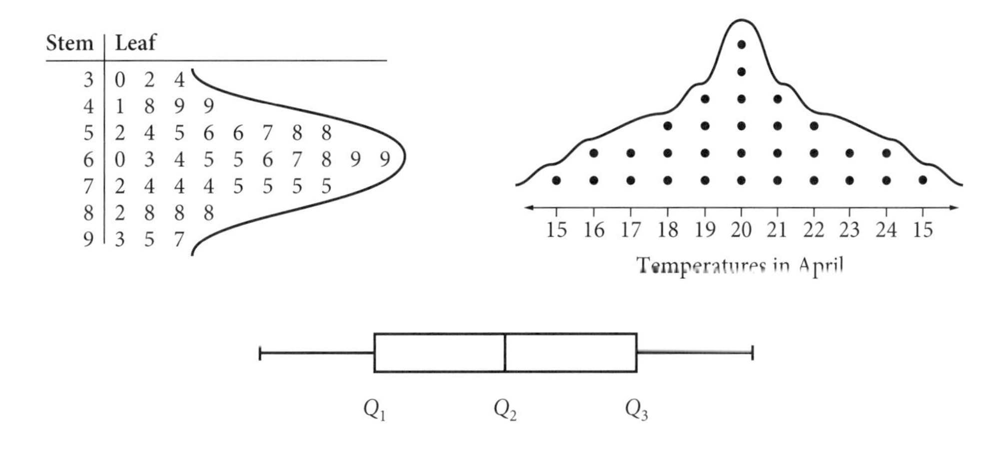

Positively skewed: Data 'skews' (twists) to the left when the mean is greater than the mode and median. The tail of the curve is in the positive direction of the number line.

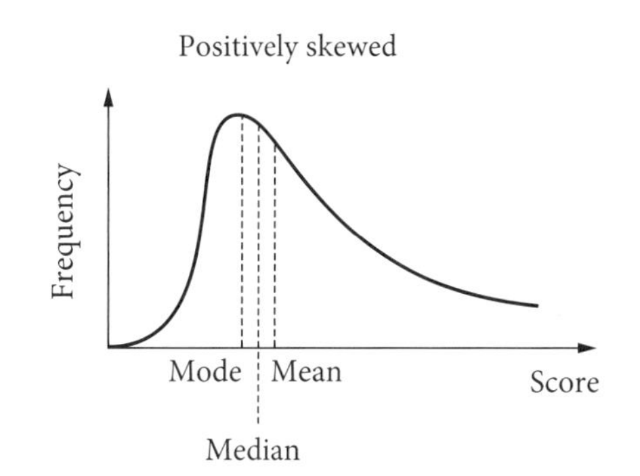

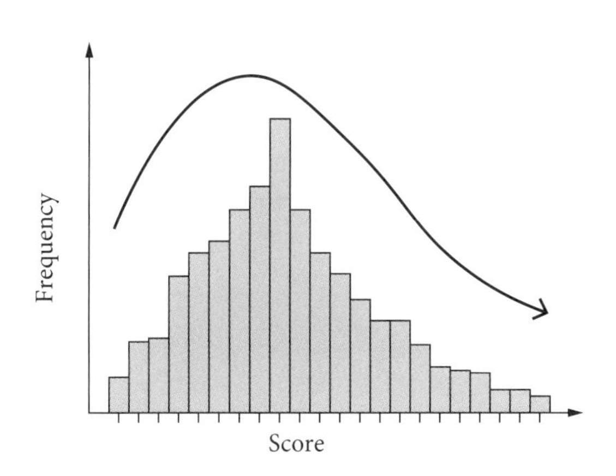

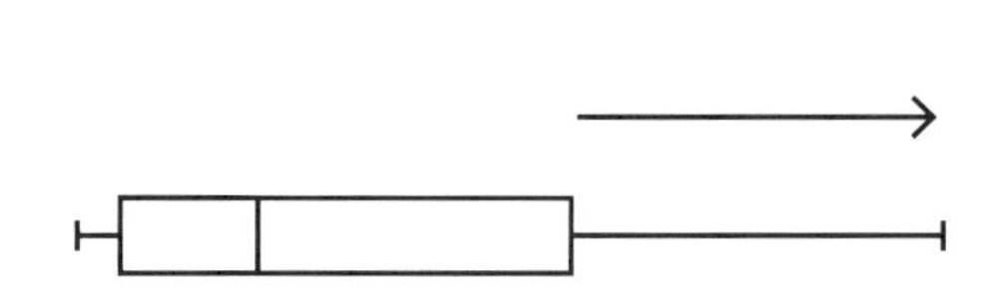

Negatively skewed: Data 'skews' to the right when the mean is less than the mode and median. The tail of the curve is in the negative direction of the number line.

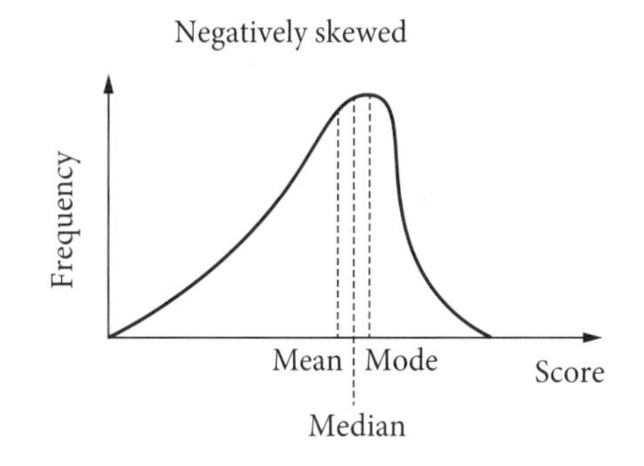

> **Hint**
> Positive and negative correlation are *not* the same as positive and negative skewness.

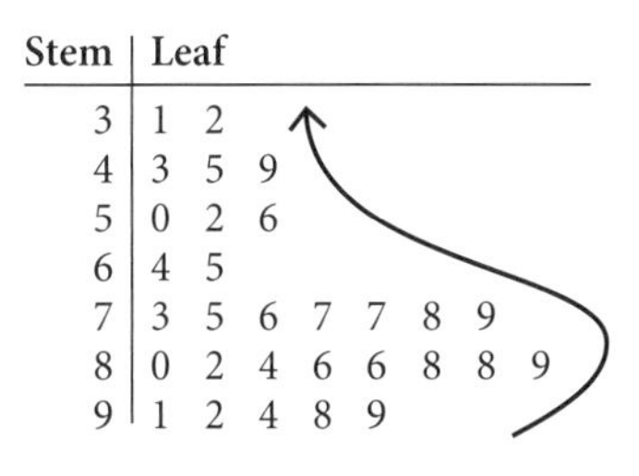

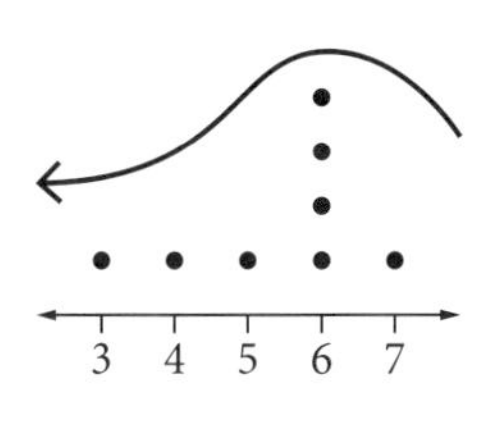

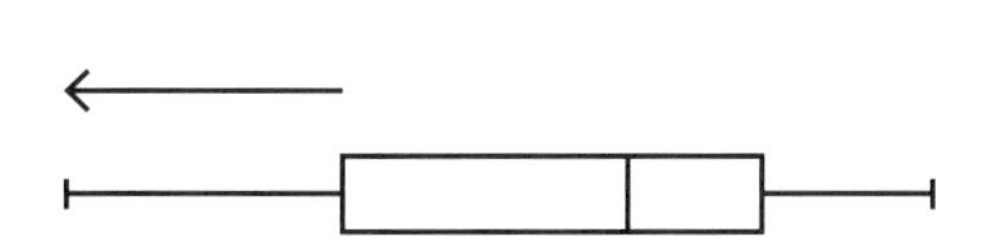

Practice set 1

Multiple-choice questions

Solutions start on page 155.

Question 1

Which scatterplot shows no correlation?

A

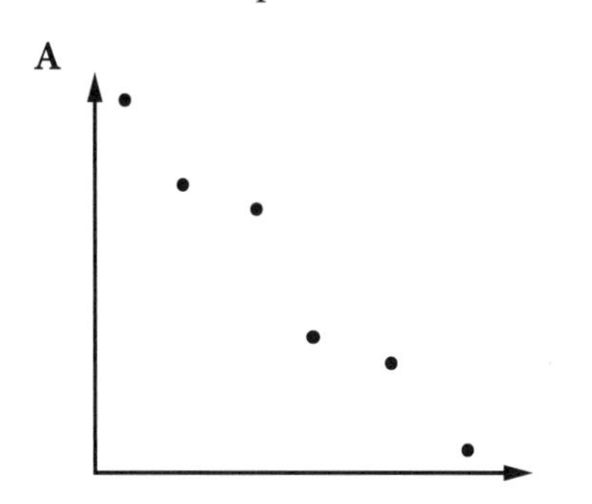

B

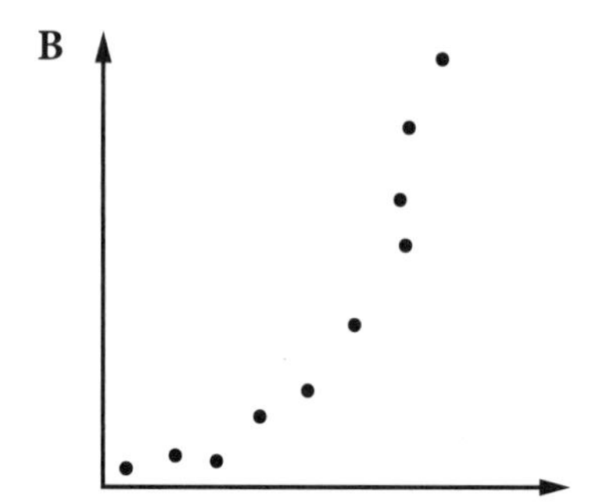

C

D

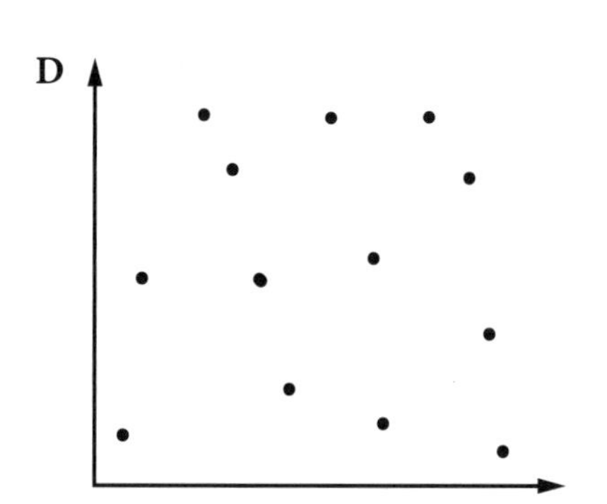

Question 2

Which graph represents a non-linear relationship?

A

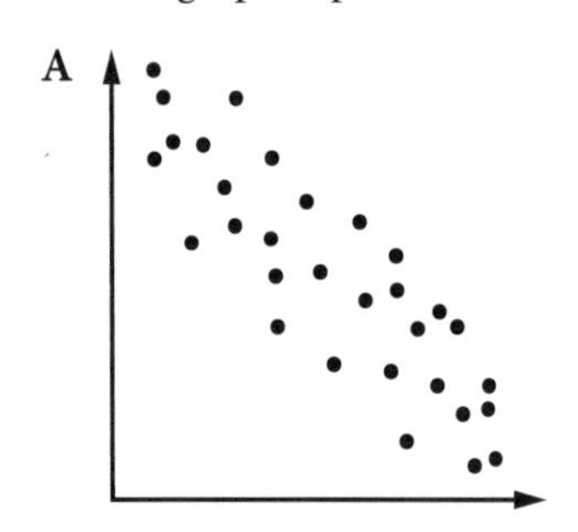

B

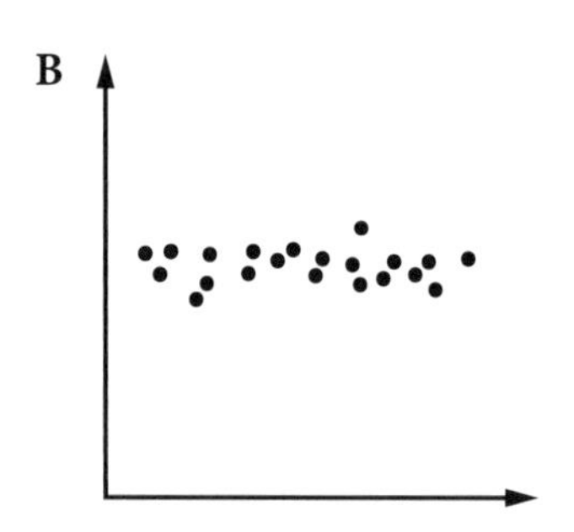

C

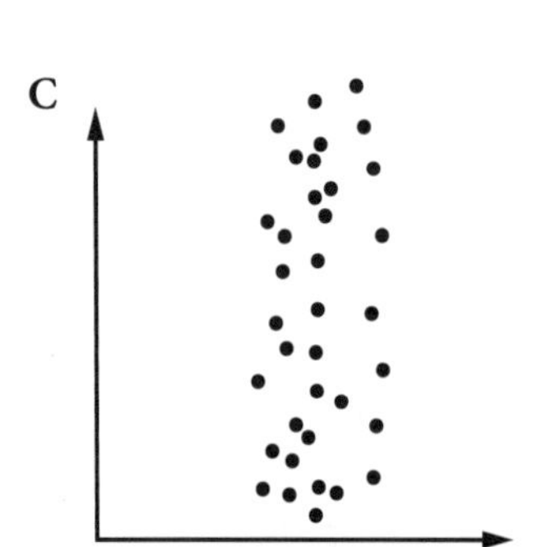

D

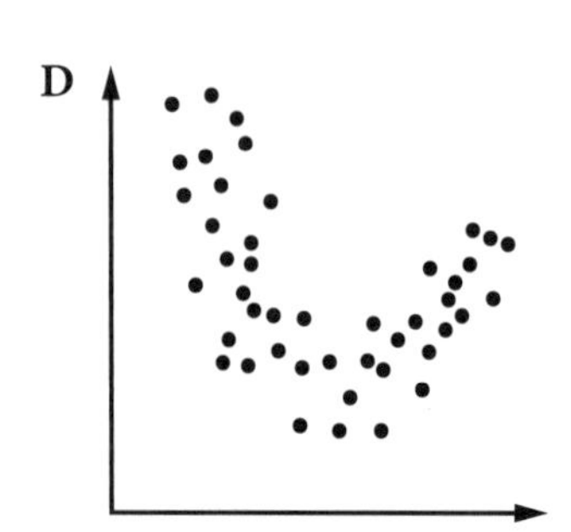

Question 3 ©NESA 2008 HSC EXAM, QUESTION 12

A scatterplot is shown.

Which of the following best describes the correlation between R and T?

A Positive

B Negative

C Positively skewed

D Negatively skewed

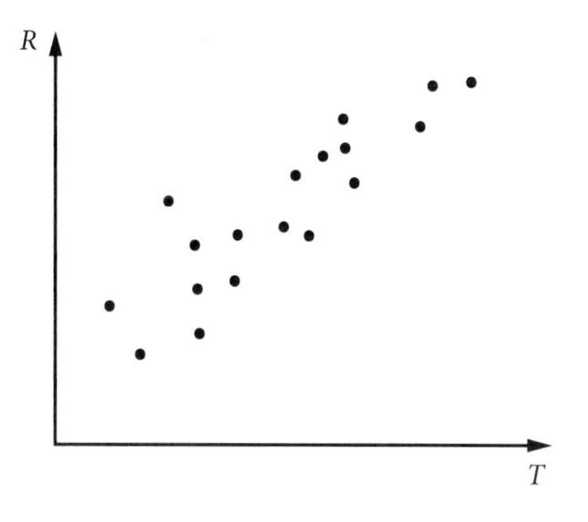

Question 4 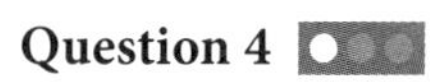

What will Pearson's correlation coefficient, r, be for this scatterplot?

A −2

B −1

C 0

D 1

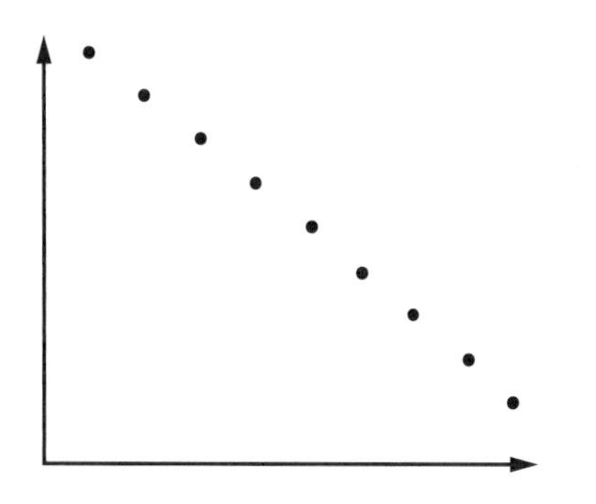

Question 5 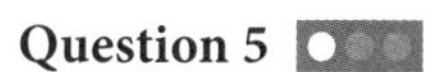

Which statement best describes the relationship between new 2 variables on this scatterplot?

A Strong, positive correlation

B Strong, negative correlation

C Weak, positive correlation

D Weak, negative correlation

Question 6

What does a correlation coefficient of $r = 0.6$ indicate?

A Strong, positive correlation

B Moderate, positive correlation

C Moderate, negative correlation

D Strong, negative correlation

9780170459204

Question 7 ©NESA 2017 HSC EXAM, QUESTION 12

Which of the data sets graphed below has the largest positive coefficient value?

A

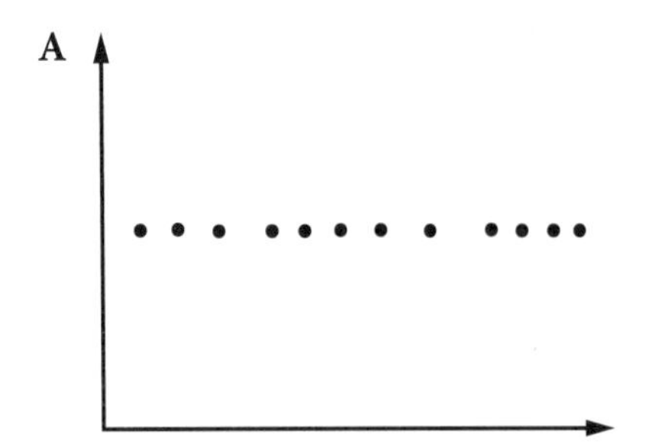

B

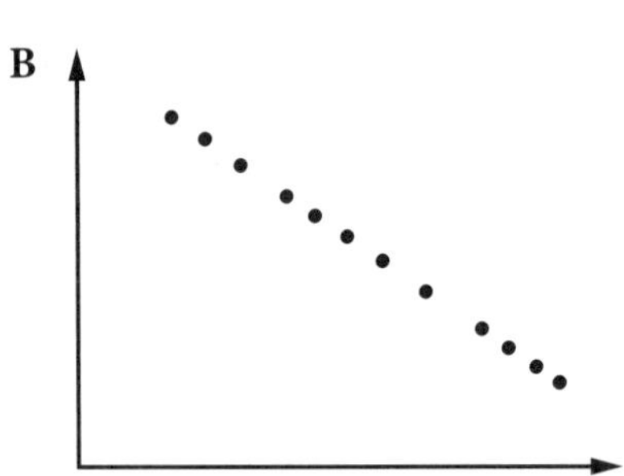

C

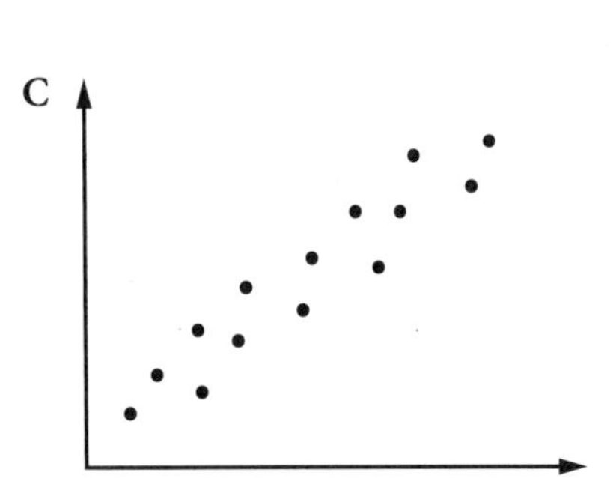

D

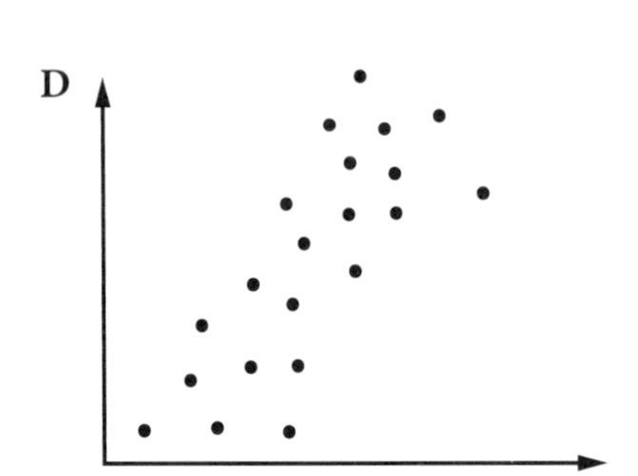

Question 8

Which pair of variables has a strong, negative correlation?

A Amount of road traffic and time to travel to work

B Population of a town and number of schools

C Number of sharks at a beach and number of people swimming

D Number of ice creams sold in a day and the day's maximum temperature

Question 9

Which box plot best represents a data set that is symmetrical?

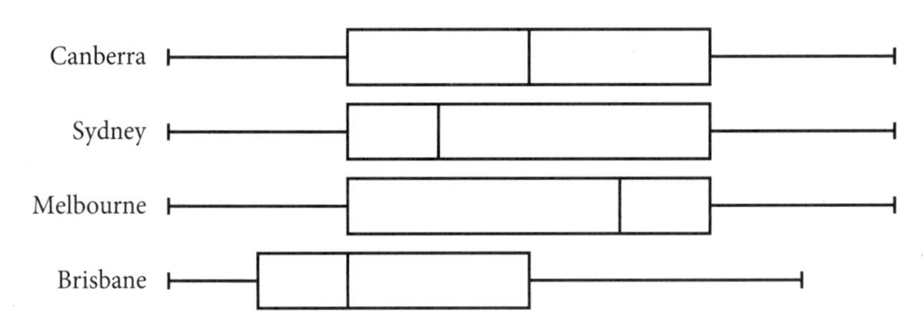

A Canberra

B Sydney

C Melbourne

D Brisbane

Question 10

The mean of a set of scores is 75 and the standard deviation is 10.

Which score has a z-score of -2?

A 45

B 55

C 60

D 65

Question 11 ©NESA 2020 HSC EXAM, QUESTION 8

John recently completed a class test for three of his school subjects. The class scores on each test were normally distributed.

The table shows the subjects and John's scores as well as the mean and standard deviation of the class scores on each test.

Subject	John's score	Mean	Standard deviation
French	82	70	8
Commerce	80	65	5
Music	74	50	12

Relative to the rest of the class, what are John's highest and lowest test results?

	Highest result	Lowest result
A	Commerce	French
B	French	Music
C	Music	French
D	Commerce	Music

Question 12

Effie achieves a z-score of 0.75 in her Music examination.

If the class mean is 65 and standard deviation is 12, what is her raw mark?

A 56 **B** 61 **C** 74 **D** 79

Question 13

The heights (h) and arm spans (s) of 10 male swimmers are graphed and the line of best fit has the equation $s = 1.14h - 24.86$.

According to this equation, what is the height of a male swimmer with an arm span of 195 cm?

A 149 cm **B** 193 cm **C** 197 cm **D** 222 cm

Question 14

A machine produces chocolate bars with a mean length of 80 mm and standard deviation of 6 mm.

If all chocolates less than or more than 3 standard deviations from the mean are rejected, which of the following lengths would be unacceptable?

A 62 mm **B** 86 mm **C** 94 mm **D** 99 mm

Question 15

In a supercar motor race, the mean number of pit stops per car in a 24-hour race is 10 and the standard deviation is 2.

If there are 125 cars in the race, approximately how many cars make between 8 and 12 pit stops?

A 85 **B** 102 **C** 108 **D** 119

Question 16

What percentage of scores lie between z-scores of -1 and 3?

A 65.7% **B** 83.85% **C** 95% **D** 97.35%

Question 17

What is the gradient of the line of best fit for the following graph?

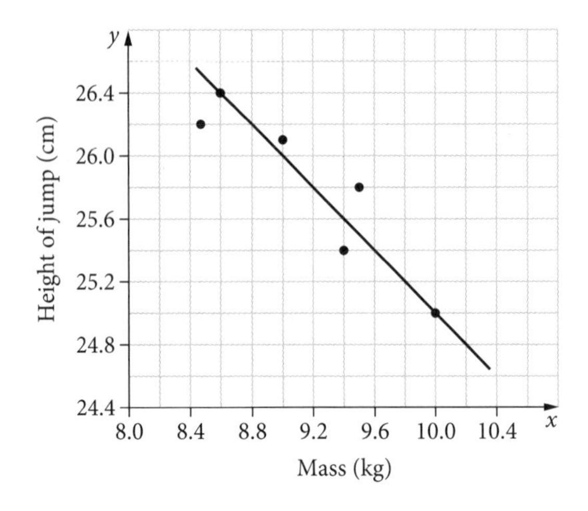

A −1 **B** −0.8 **C** 1 **D** 27

Question 18

What is the equation of the least-squares regression line for the following data?

x	156	167	163	174	164	179	183
y	55	64	68	78	61	86	89

A $y = 1.31x + 150.36$ **B** $y = 1.31x - 150.36$

C $y = -150.36x + 1.31$ **D** $y = 150.36x + 1.31$

Question 19

The arm span (in cm) of a population of girls is normally distributed with a mean of 135 cm and a standard deviation of 4 cm.

If 600 of these girls are randomly selected to have their arm span measured, approximately how many of the girls have an arm span between 139 cm and 147 cm?

A 95 **B** 96 **C** 299 **D** 300

Question 20 ©NESA 2019 HSC EXAM, QUESTION 15

The scores of an examination are normally distributed with a mean of 70 and a standard deviation of 6. Michael received a score on the examination between the lower quartile and the upper quartile of the scores.

Which shaded region most accurately represents where Michael's score lies?

A

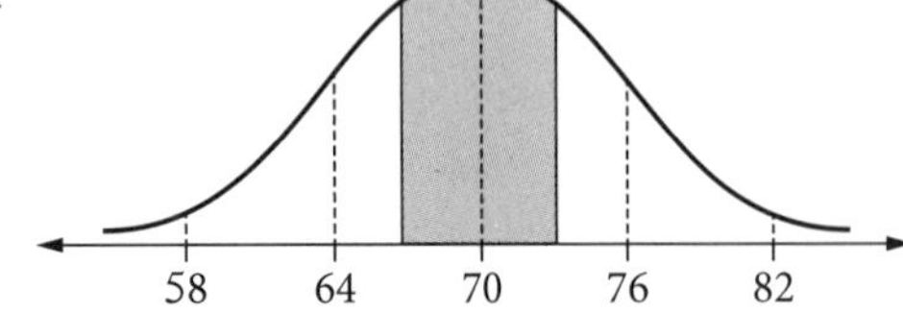

B

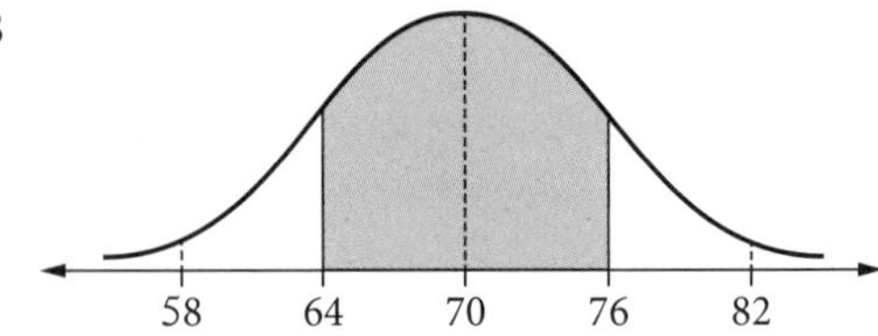

C

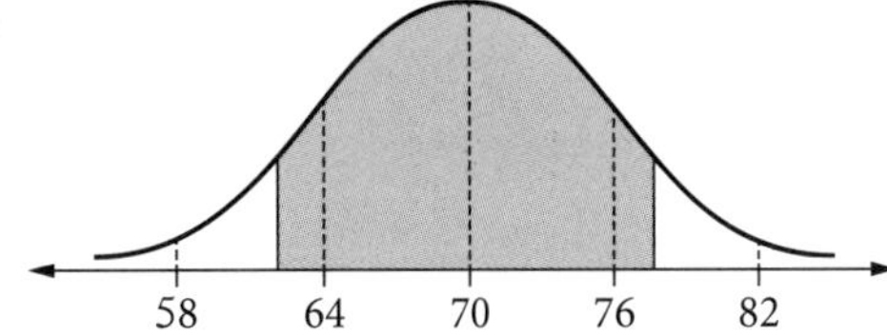

D

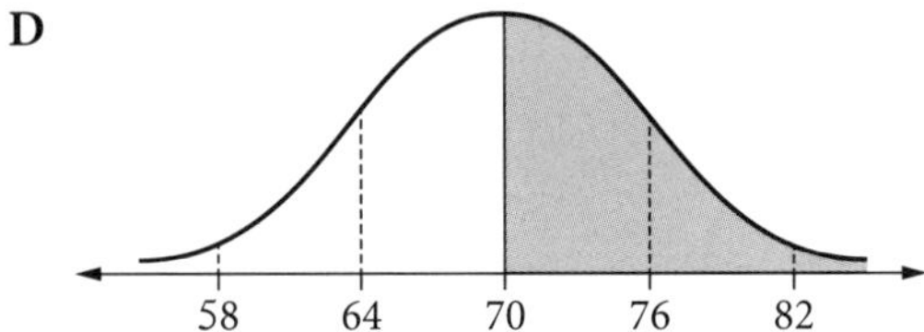

Practice set 2

Short-answer questions

Solutions start on page 157.

Question 1 (3 marks)

The correlation coefficients of the following scatterplots are $r = -0.8$, $r = -0.4$ and $r = 0.85$.

Match each scatterplot to its correlation coefficient. 3 marks

a

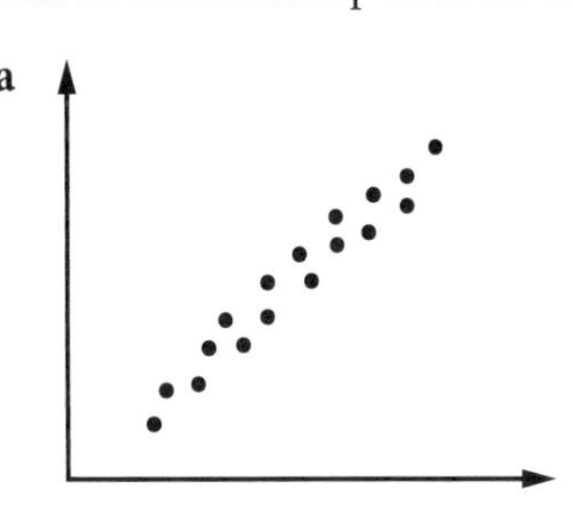

$r =$ ____________

b

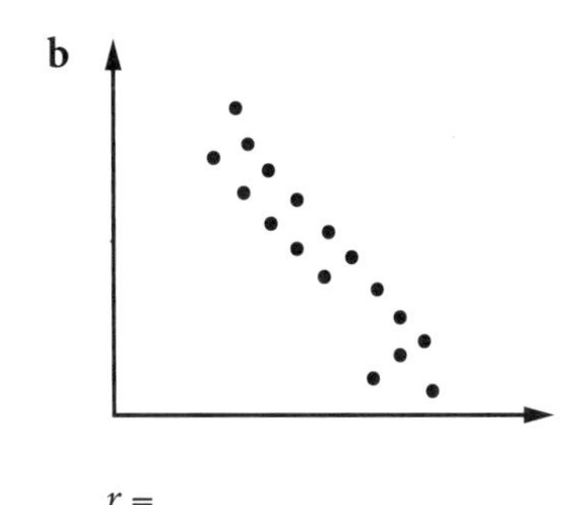

$r =$ ____________

c

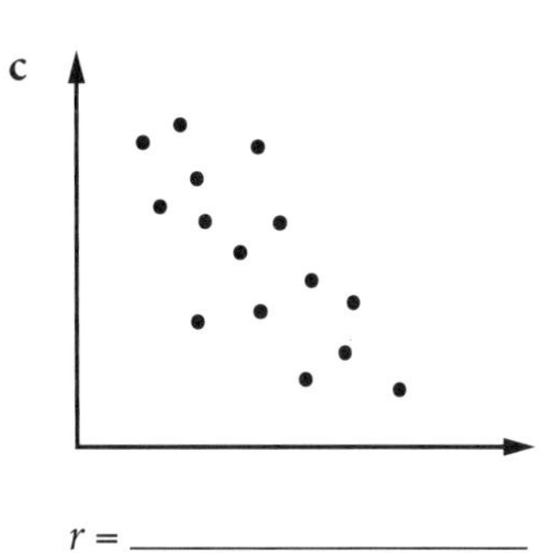

$r =$ ____________

Question 2 (4 marks)

Do the following variables have a positive, negative or zero correlation?

a Number of cups of hot soup sold in a school canteen and daily maximum temperature 1 mark

b Amount of time practising batting in the cricket nets and number of runs scored in a game 1 mark

c Number of umbrellas sold and amount of rain fallen in the week 1 mark

d Amount of water drunk in a day and a person's height 1 mark

Question 3 (4 marks)

a What is the mode for this data set? 1 mark

b What is the median for this data set? 1 mark

c What is the mean for this data set? 1 mark

d Describe the shape of this data set. 1 mark

Stem	Leaf
5	4
6	7 9
7	0 3 4 5 5 9
8	1 2 2 4 7 8 9
9	3 5
10	2 7

Question 4 (2 marks)

What percentage is shaded for each of these normal distributions?

a more than 1 standard deviation from the mean 1 mark

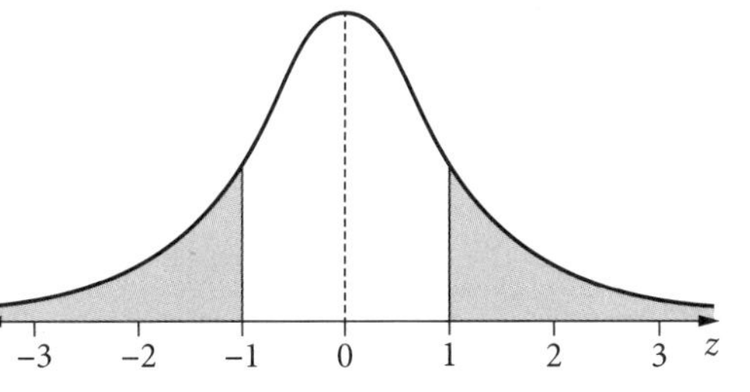

b more than 2 standard deviations from the mean 1 mark

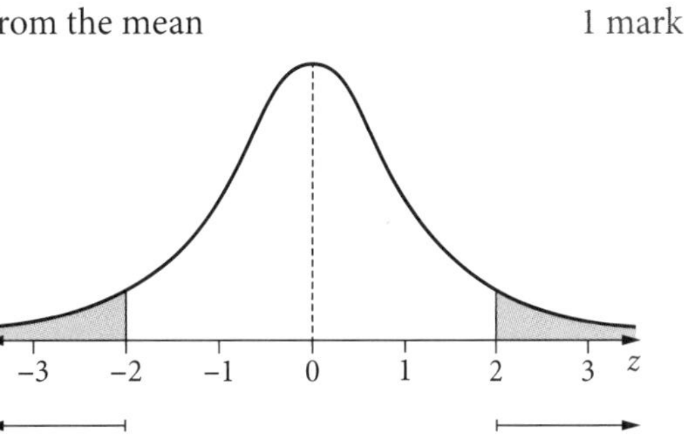

Question 5 (2 marks) 

For each normal distribution curve, shade in:

a 34% of data that is below the mean. 1 mark

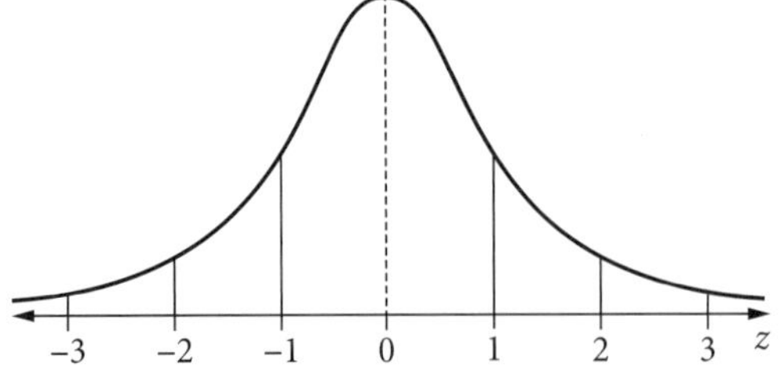

b 16% of data that is above the mean. 1 mark

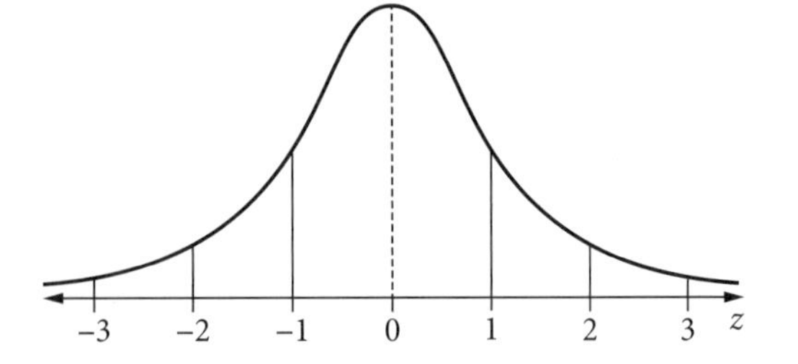

Question 6 (3 marks) 

For the normal distribution shown, the mean is 20 and the standard deviation is 3.

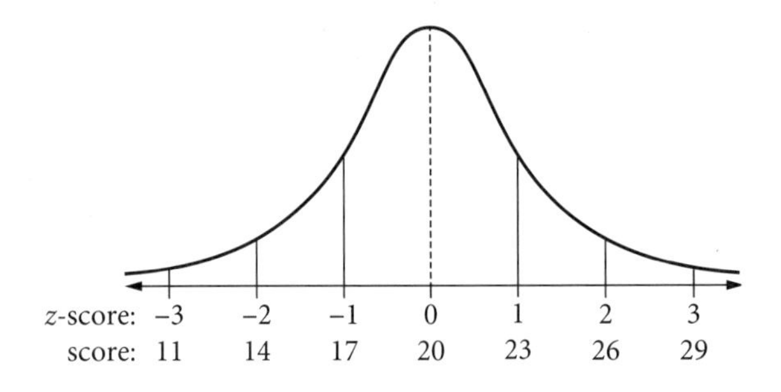

What percentage of scores lie between:

a 17 and 23? 1 mark

b 14 and 26? 1 mark

c 11 and 20? 1 mark

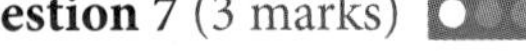

Question 7 (3 marks)

A particular normally distributed set of scores has a mean of 18 and a standard deviation of 4.

Complete the missing raw scores for each z-score in the table. 3 marks

***z*-score**	−3	−2	−1.5	−1	0	1	2	2.75	3
Raw score				14	18				30

Question 8 (2 marks)

The heights of a group of men are normally distributed with a mean of 176 cm and a standard deviation of 10 cm.

a What is the z-score of a man who is 188 cm tall? 1 mark

b What z-score would you calculate for a man who is 176 cm tall? 1 mark

Question 9 (2 marks)

There are 18 000 students sitting for a state-wide science exam. If the results form a normal distribution, how many students would score a result that is more than 1 standard deviation below the mean? 2 marks

Question 10 (3 marks)

Newcastle's mean maximum temperature during winter is 17.3°C with a standard deviation of 2.6°C. (Assume that these maximum temperatures are normally distributed.)

a What temperature has a z-score of –2? 1 mark

b What percentage of winter days in Newcastle have a maximum temperature between 14.7°C and 22.5°C? 1 mark

c How many days in winter would you assume have a maximum temperature between 14.7°C and 22.5°C? 1 mark

Question 11 (3 marks)

Lola sat a Maths test and achieved a score of 72. The class mean mark was 60 with a standard deviation of 8. She also sat a Chemistry test where the mean mark was 66 with a standard deviation of 10.

Lola had the same z-score for both Maths and Chemistry. What is her mark for the Chemistry test? 3 marks

Question 12 (2 marks)

In the table shown, Tran's original marks in a Physics test have a mean of 55 and a standard deviation of 10.

a Complete the table if the marks are to be scaled with a mean of 60 and a standard deviation of 12. 1 mark

Original mark			55		75
***z*-score**	−2	−1	0	1	2
Scaled mark			60		

b If Tran's z-score in this Physics test is 1.8, what was his: 1 mark

i original mark? **ii** scaled mark?

Question 13 (3 marks)

The results of a spelling test for a class are normally distributed. The mean is 25 correct words.

a Faduma's raw score of 14 correct words gives her a z-score of –2.75. What is the standard deviation of the test results? 2 marks

b How many possible words has Sebastian spelt correctly if his z-score lies between 2 and 3 standard deviations above the mean? 1 mark

Question 14 (3 marks)

Given the equation of a least-squares regression line is $y = 0.36x + 41.6$, find:

a y when $x = 17.8$. 1 mark

b x when $y = 209.54$. 2 marks

Question 15 (3 marks)

The machine in a factory that fills bottles with orange juice is set as a normal distribution. It is known that the volume of 95% of the bottles filled is between 496 mL and 524 mL.

What is the machine's setting of its mean capacity and its standard deviation? 3 marks

PRACTICE SET 2

Question 16 (3 marks) ©NESA 2019 HSC EXAM, QUESTION 23

A set of bivariate data is collected by measuring the height and arm span of seven children. The graph is a scatterplot of these measurements.

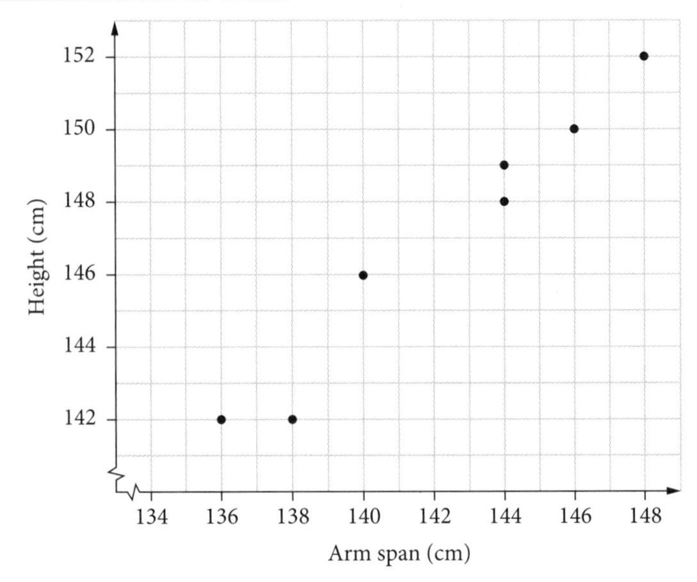

a Calculate Pearson's correlation coefficient, correct to two decimal places. 1 mark

b Identify the strength and direction of the linear association between height and arm span. 1 mark

c The equation of the least-squares regression line is shown.

$$\text{height} = 0.866 \times (\text{arm span}) + 23.7$$

A child has an arm span of 143 cm.

Calculate the predicted height of this child using the equation of the least-squares regression line. 1 mark

Question 17 (6 marks)

The table shows data for a sample of 10 people who each stated their age and the amount of time they used social media per week.

Age, a (years)	12	15	21	21	24	35	40	55	65	70
Social media usage (hours per week)	18	36	30	25	40	15	25	21	10	5

a Plot the points (65, 10) and (70, 5) on the grid below. 1 mark

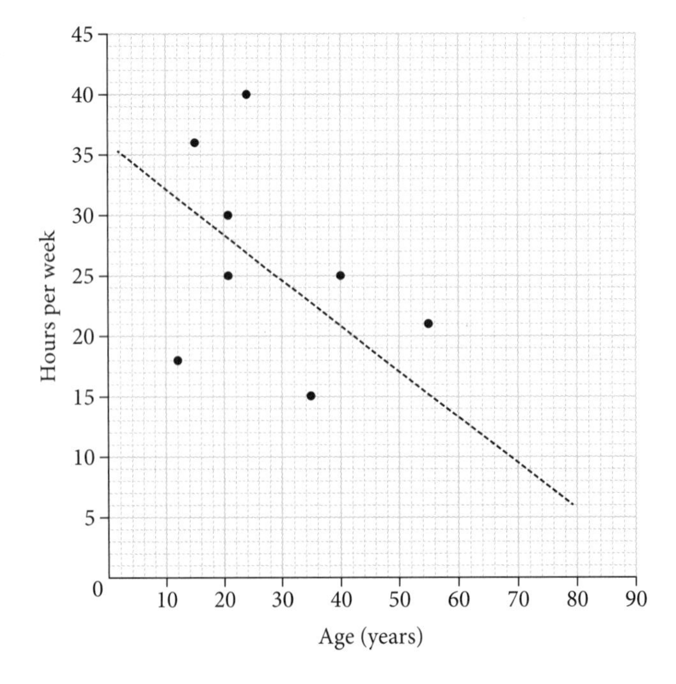

b Calculate Pearson's correlation coefficient to two decimal places. 1 mark

c Describe the strength and association between age and hours of social media usage for this sample. 1 mark

d Using technology, construct the least-squares regression line. Give answers to two decimal places. 1 mark

e Using your equation, predict the time spent on social media per week for an 18-year-old person, correct to the nearest hour. 1 mark

f Explain a limitation to this model. 1 mark

Question 18 (9 marks) ©NESA 2008 HSC EXAM, QUESTION 28(a)

The following graph indicates z-scores of 'height-for-age' for girls aged 5–19 years.

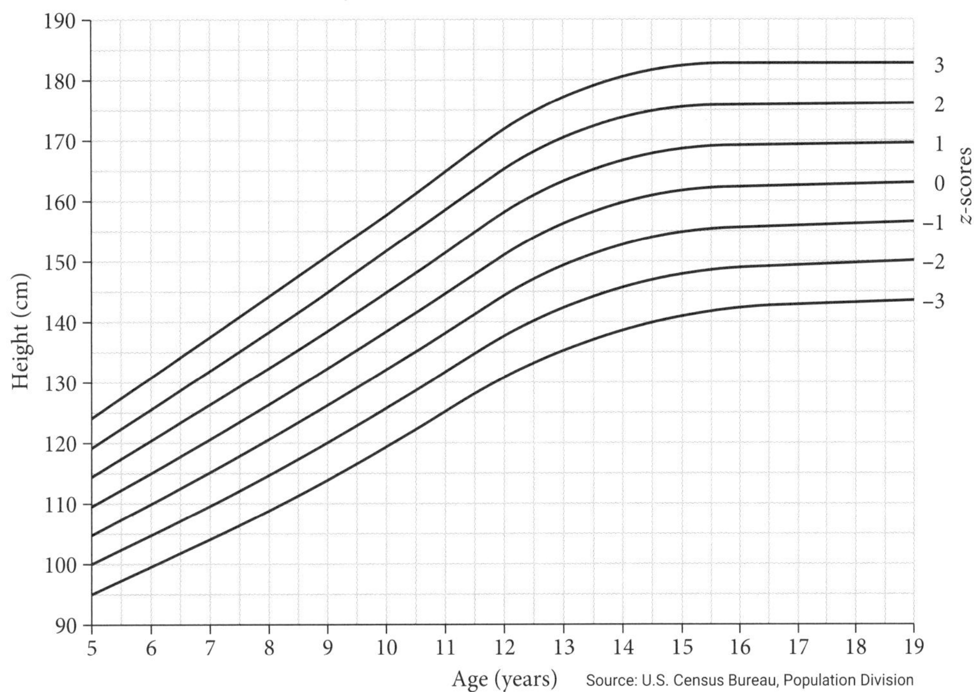

a What is the z-score for a 6-year-old girl of height 120 cm? 1 mark

b Rachel's age is $10\frac{1}{2}$ years.

i If 2.5% of girls the same age are taller than Rachel, how tall is she? 1 mark

ii If Zoe is aged $15\frac{1}{2}$ and is the same height as Rachel, what percentage of girls aged $15\frac{1}{2}$ are taller than Zoe? 2 marks

c What is the average height of an 18-year-old girl? 1 mark

d For adults (18 years and older), the body mass index is given by $B = \frac{m}{h^2}$, where m is the mass in kilograms and h is the height in metres.

The medically accepted healthy range for B is $21 \le B \le 25$.

What is the minimum weight for an 18-year-old girl of average height to be considered healthy? 2 marks

e The average height, C, in centimetres, of a girl between the ages of 6 years and 11 years can be represented by a line with equation $C = 6A + 79$, where A is the age in years.

i For this line, the gradient is 6. What does this indicate about the height of girls aged 6 to 11? 1 mark

ii Give ONE reason why this equation is not suitable for predicting heights of girls older than 12. 1 mark

Question 19 (5 marks) ©NESA 2013 HSC EXAM, QUESTION 28(b)

Ahmed collects data on the age (a) and height (h) of males 11 to 16 years.

He created a scatterplot of the data and constructed a line of best fit to model the relationship between the age and height of males.

a Determine the gradient of the line of best fit shown on the graph. 1 mark

b Explain the meaning of the gradient in the context of the data. 1 mark

c Determine the equation of the line of best fit shown on the graph. 1 mark

d Use the line of best fit to predict the height of a typical 17-year-old male. 1 mark

e Why would this model not be useful for predicting the height of a typical 45-year-old male? 1 mark

Question 20 (5 marks) ©NESA 2020 HSC EXAM, QUESTION 36

A cricket is an insect. The male cricket produces a chirping sound.

A scientist wants to explore the relationship between the temperature in degrees Celsius and the number of cricket chirps heard in a 15-second time interval.

Once a day for 20 days, the scientist collects data. Based on the 20 data points, the scientist provides the information below.

- A box plot of the temperature data is shown.

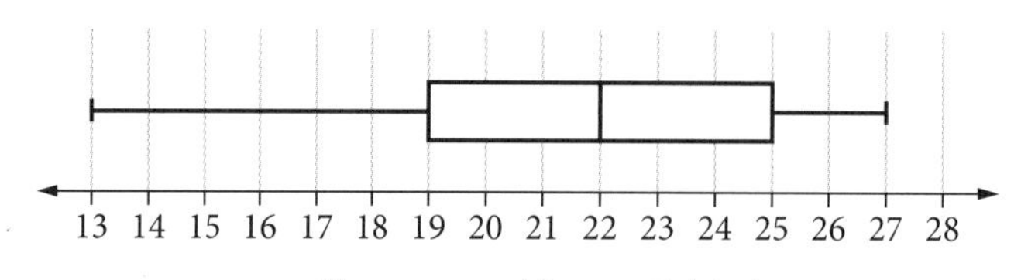

- The mean temperature in the dataset is 0.525°C below the median temperature in the dataset.
- A total of 684 chirps was counted when collecting the 20 data points.

The scientist fits a least-squares regression line using the data (x, y), where x is the temperature in degrees Celsius and y is the number of chirps heard in a 15-second time interval. The equation of the line is

$$y = -10.6063 + bx,$$

where b is the slope of the regression line.

The least-squares regression line passes through the point $(\bar{x}, \bar{y})$ where $\bar{x}$ is the sample mean of the temperature data and $\bar{y}$ is the sample mean of the chirp data.

Calculate the number of chirps expected in a 15-second interval when the temperature is 19°C. Give your answer correct to the nearest whole number. 5 marks

Practice set 1

Worked solutions

1 D

There is no pattern in the scatterplot.

2 D

The points are not linear.

3 A

The direction of the points is uphill (positive correlation).

4 B

For a perfect negative correlation, $r = -1$, all points are in a straight line in a downhill direction.

5 D

The scatterplot points are much further apart but still in a line: weak. The direction of the points is downhill (negative correlation).

6 B

The correlation coefficient fits a little more than halfway between 0 and 1 on the positive scale. r is positive, so it is a moderate, positive correlation.

7 C

The points are very close together to form a thin, straight positive line.

8 C

The more sharks at a beach, the fewer people will swim in the water.

9 A

This is symmetrical (normal distribution) because each quarter has the same range.

10 B

Raw score = mean − $2\sigma = 75 - 2 \times 10 = 55$

11 A

$$z(\text{French}) = \frac{82 - 70}{8} = 1.5$$

$$z(\text{Commerce}) = \frac{80 - 65}{5} = 3$$

$$z(\text{Music}) = \frac{74 - 50}{12} = 2$$

Commerce has the highest z-score, therefore it is his highest result. French has the lowest z-score, therefore it is his lowest result.

12 C

$$z = \frac{x - \mu}{\sigma}$$

$$0.75 = \frac{x - 65}{12}$$

$$0.75 \times 12 = x - 65$$

$$9 = x - 65$$

$$9 + 65 = x$$

$$x = 74$$

Her raw mark in Music is 74.

13 B

$$s = 1.14h - 24.86$$

$$195 = 1.14h - 24.86$$

$$195 + 24.86 = 1.14h$$

$$219.86 = 1.14h$$

$$\frac{219.86}{1.14} = h$$

$$h = 192.859$$

$$\approx 193\text{ cm}$$

14 D

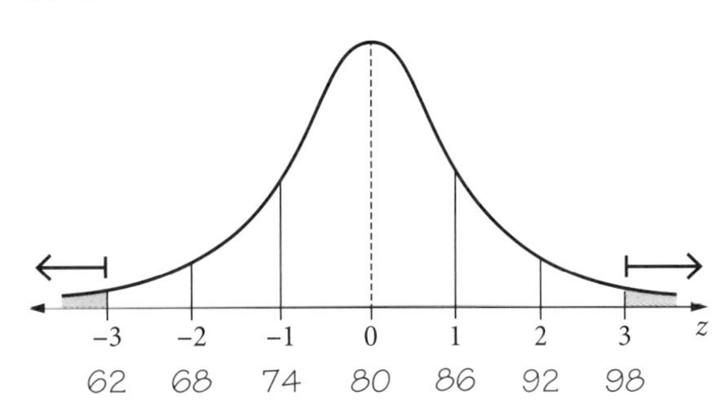

99 mm is outside this range of 62 to 98.

15 A

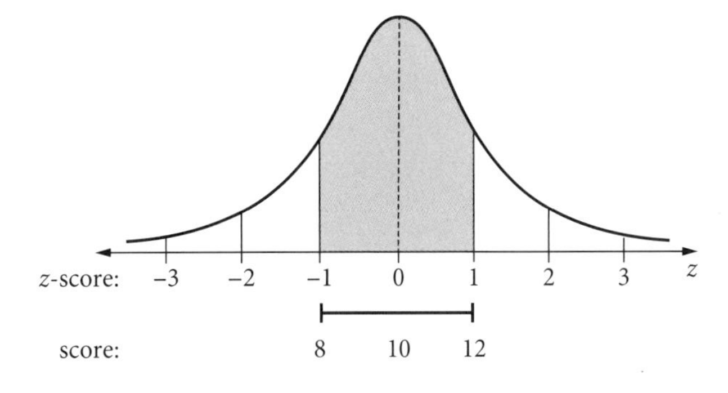

Within 1 standard deviation of the mean is approximately 68% × 125 = 85 cars.

16 B

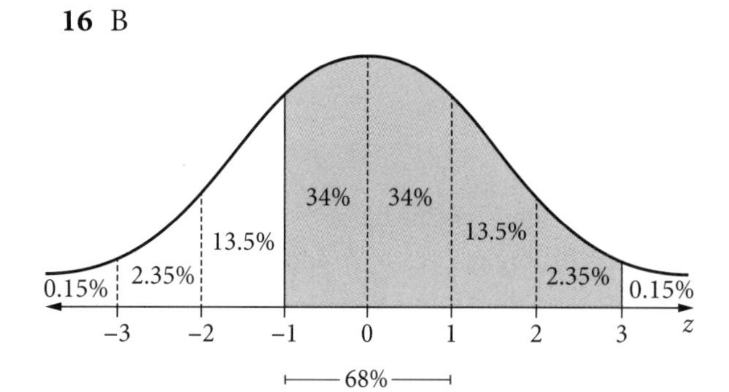

$\frac{1}{2} \times 68\% = 34\%$ of data from $z = -1$ to $z = 0$

$\frac{1}{2} \times 99.7\% = 49.85\%$ from $z = 0$ to $z = 3$

So $34\% + 49.85\% = 83.85\%$.

17 A

$$m = \frac{\text{rise}}{\text{run}}$$
$$= \frac{25 - 26.4}{10 - 8.6}$$
$$= -1$$

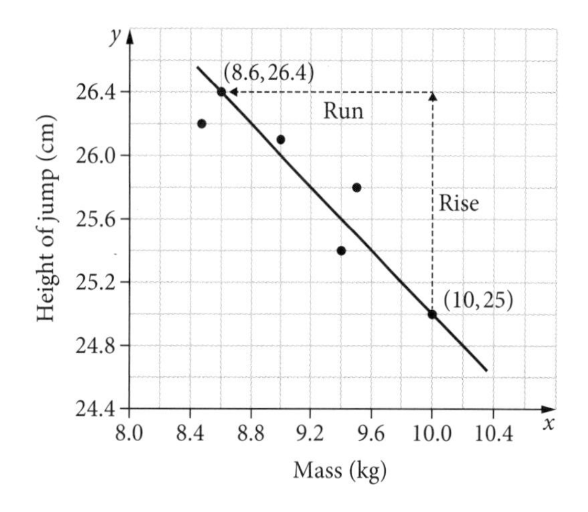

18 B

Using the STAT mode on the calculator:

y-intercept: $a \approx -150.36$

gradient: $b \approx 1.31$

The equation in the form
$y = bx + a$ is:

$y = 1.31x - 150.36$

19 A

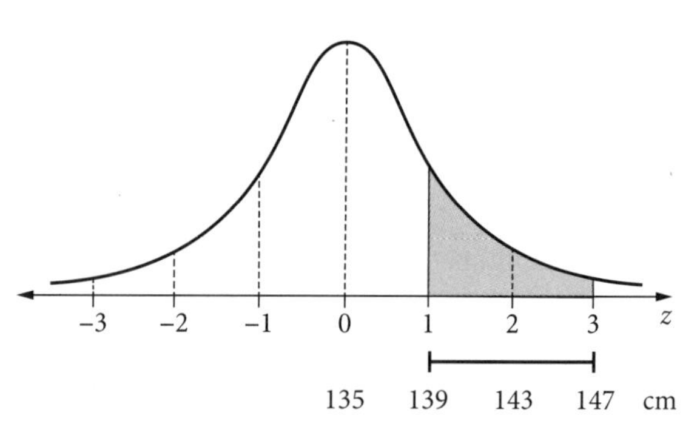

139 cm and 147 cm are between 1 and 3 standard deviations above the mean.

$\frac{1}{2} \times 99.7\% - \frac{1}{2} \times 68\% = 15.85\%$

Number of girls $= 15.85\% \times 600 = 95.1 \approx 95$

20 A

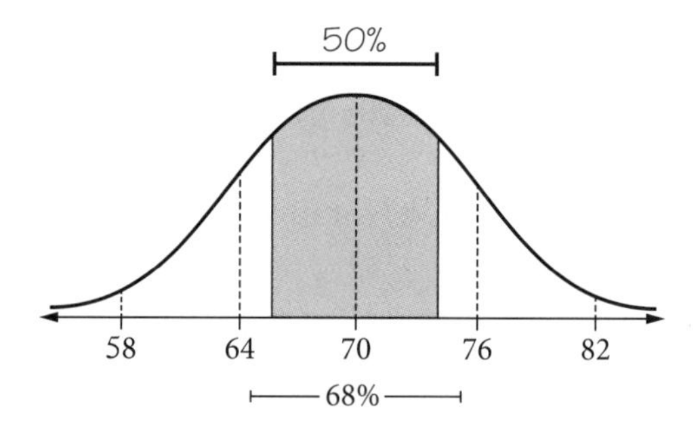

Lower quartile, Q_1, is at 25% and upper quartile, Q_3, is at 75%, which is a range of 50%. In a normal distribution approximately 68% of scores are within 1 standard deviation of the mean. Therefore, the middle 50% is within the 68%.

Practice set 2

Worked solutions

Question 1

a $r = 0.85$ **b** $r = -0.8$ **c** $r = -0.4$

Question 2

a negative. More hot soup is sold as temperature becomes lower.

b positive. With more training, more runs should be scored.

c positive. With more rainfall, more umbrellas should be sold.

d zero. The amount of water drunk has no relationship to a person's height.

Question 3

a 75 and 82 (occur twice each)

b

Stem	Leaf
5	4
6	7 9
7	0 3 4 5 5 9
8	(1 2) 2 4 7 8 9
9	3 5
10	2 7

$$\frac{81 + 82}{2} = 81.5$$

c 81.3

d It is very close to being symmetrical – a normal distribution has the same mean, mode and median.

Question 4

a $100\% - 68\% = 32\%$

b $100\% - 95\% = 5\%$

Question 5

a

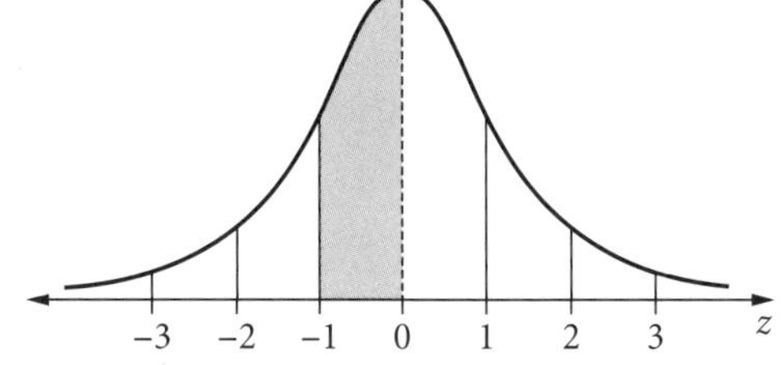

b

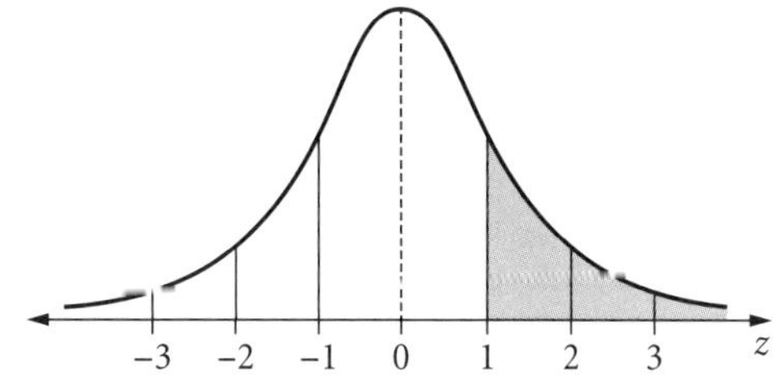

Question 6

a Within 1 standard deviation of the mean is 68%.

b Within 2 deviations of the mean is 95%.

c 3 deviations below the mean
$= 99.7\% \div 2 = 49.85\%$

Question 7

$z = \dfrac{x - \mu}{\sigma}$ so $x = \mu + z \times \sigma$

Score = mean + z-score × standard deviation

1 standard deviation above mean score
$= 18 + 4 = 22$

2 standard deviations above mean score
$= 18 + 2 \times 4 = 26$

2.75 deviations above mean score
$= 18 + 2.75 \times 4 = 29$

1.5 standard deviations below mean score
$= 18 - 1.5 \times 4 = 12$

2 standard deviations below mean score
$= 18 - 2 \times 4 = 10$

3 standard deviations below mean score
$= 18 - 3 \times 4 = 6$

***z*-score**	−3	−2	−1.5	−1	0	1	2	2.75	3
Raw score	6	10	12	14	18	22	26	29	30

Question 8

a $z = \dfrac{x - \mu}{\sigma}$

$= \dfrac{188 - 176}{10}$

$= 1.2$

b 176 cm is the mean, therefore $z = 0$.

Question 9

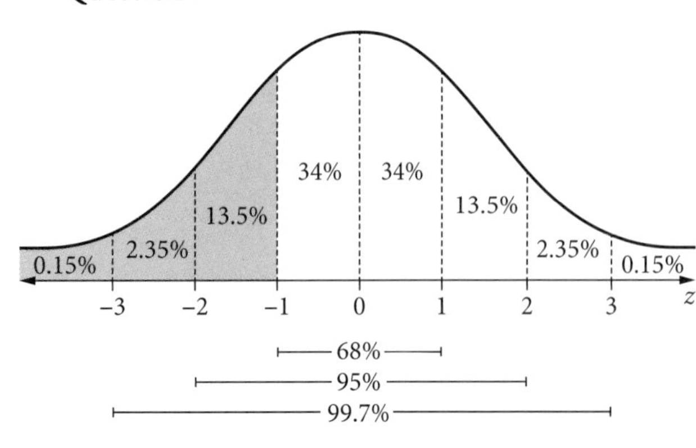

More than 1 standard deviation below mean:

$\frac{1}{2} \times (100\% - 68\%) = 16\%$

Number of students = 16% × 18 000 = 2880

Question 10

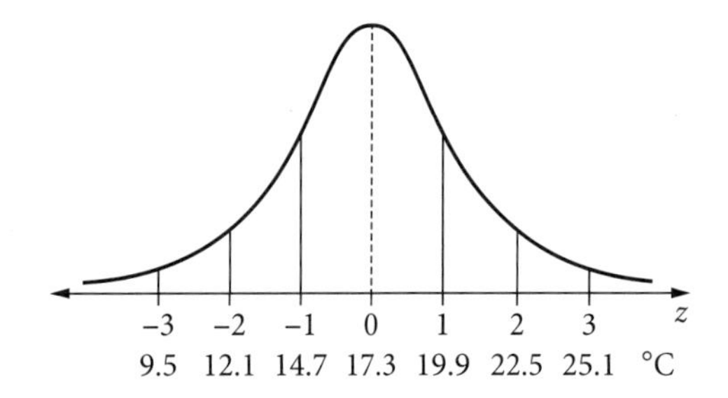

a 2 standard deviations below mean score
$= 17.3 - 2 \times 2.6 = 12.1°C$

b

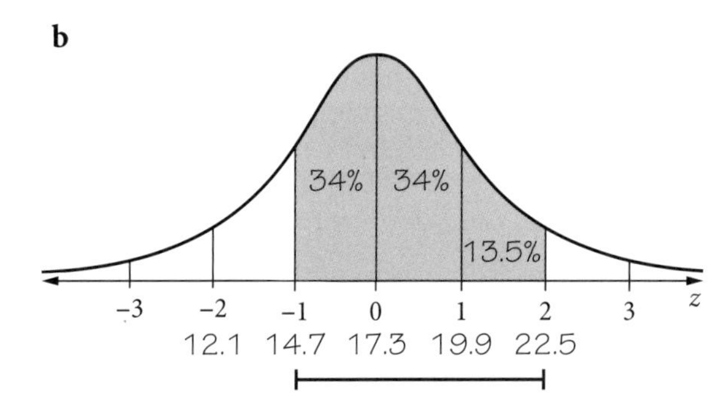

68% + 13.5% = 81.5%

c 30 days in June, 31 days in July, 31 days in August = 92 winter days

81.5% × 92 = 74.98

Approximately 75 days in winter will have a maximum temperature between 14.7°C and 22.5°C.

Question 11

$$z_m = \frac{x - \mu}{\sigma}$$
$$= \frac{72 - 60}{8}$$
$$= 1.5$$

$$z_c = \frac{x - \mu}{\sigma}$$
$$1.5 = \frac{x - 66}{10}$$
$$15 = x - 66$$
$$x = 81$$

Her mark in Chemistry is 81.

Question 12

a

−10 −10 +10

Original mark	35	45	55	65	75
z-score	−2	−1	0	1	2
Scaled mark	36	48	60	72	84

−12 −12 +12 +12

b **i** $55 + 1.8 \times 10 = 73$

ii $60 + 1.8 \times 12 = 81.6 \approx 82$

Question 13

a
$$z = \frac{x - \mu}{\sigma}$$
$$-2.75 = \frac{14 - 25}{\sigma}$$
$$\sigma = \frac{-11}{-2.75}$$
$$= 4$$

b
$$x = 25 + 2 \times 4$$
$$= 33$$

$$x = 25 + 3 \times 4$$
$$= 37$$

Sebastian could have spelled between 33 and 37 words correctly in this test.

Question 14

a
$$y = 0.36 \times 17.8 + 41.6$$
$$= 48.008$$

b
$$209.54 = 0.36x + 41.6$$
$$209.54 - 41.6 = 0.36x$$
$$167.94 = 0.36x$$
$$x = \frac{167.94}{0.36}$$
$$= 466.5$$

Question 15

95% of the bottles filled lie between 496 mL and 524 mL, which is 2 standard deviations either side of the mean, so 4 standard deviations in total.

$524 - 496 = 28$
$28 \div 4 = 7$

Standard deviation = 7 mL

Mean = $496 + 2 \times 7 = 510$ mL

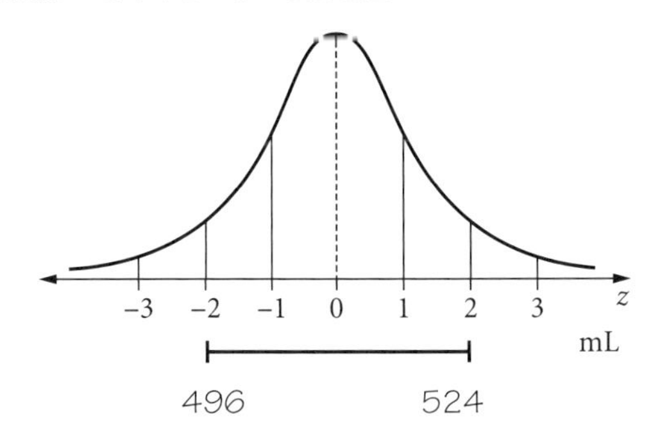

Question 16

a Insert coordinates into a calculator using STAT mode:

136	138	140	144	144	146	148
142	142	146	148	149	150	152

$r = 0.981\,11\ldots$
≈ 0.98

b strong, positive

c $H = 0.866 \times a + 23.7$

When $a = 143$:

$H = 0.866 \times 143 + 23.7$
$= 147.538$ cm

Question 17

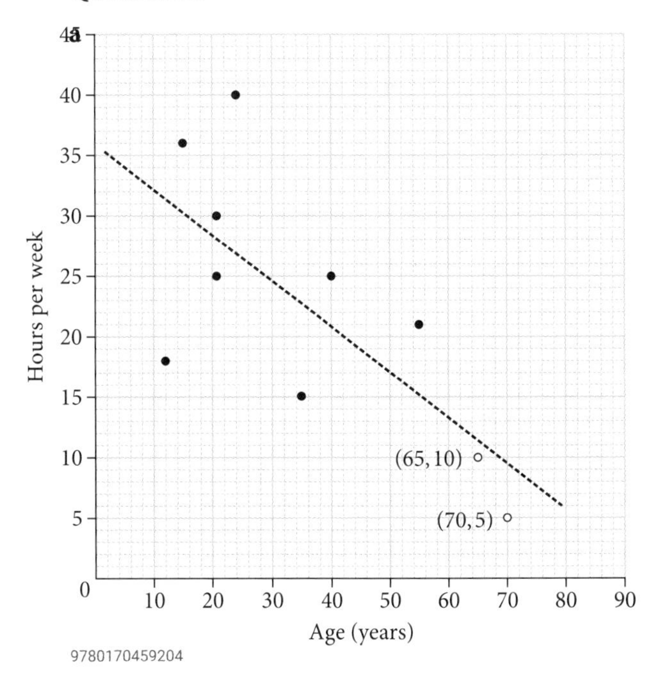

b Insert coordinates into a calculator using the STAT mode: $r \approx -0.72$.

c A moderate and negative correlation implies that there is a slight association – that the older the person, the less social media they use.

d Insert coordinates into calculator using the STAT mode:

$a \approx 36.01$
$b \approx 0.38$

$y = a + bx$
$= 36.01 - 0.38x$

e When $x = 18$:

$y = 36.01 - 0.38 \times 18$
$= 29.17$
≈ 29 hours

f This data only covers a small survey of 10 people. To better model this data, a large sample of the population should be taken. Also, social media usage depends on more factors than age, for example friendship group, monetary access and lifestyle.

Question 18

a From the graph, at 6 years and 120 cm the z-score is $z = 1$.

b i

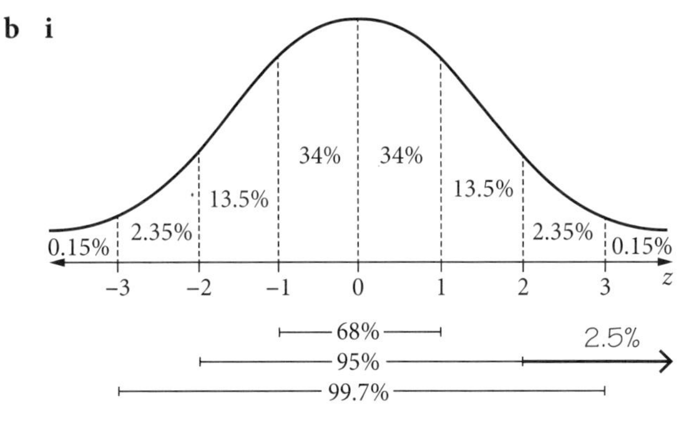

2.5% of girls taller than Rachel would indicate the z-score is 2. From the graph, the height from $10\frac{1}{2}$ years reads Rachel's height is 155 cm.

ii From the graph, at $15\frac{1}{2}$ years the height 155 cm has a z-score of -1.

From $z > -1$, 34% + 50% = 84% of the girls are taller than Zoe.

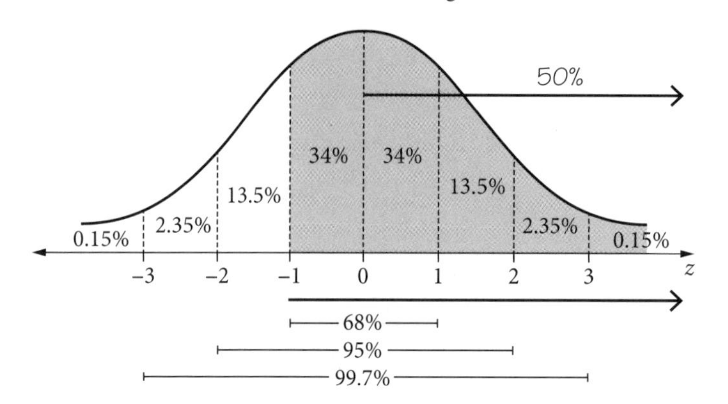

c From the graph, 18-year-old average (mean) has $z = 0$, so height is approximately 163 cm.

'Height-for-age' for girls aged 5–19 years (z-scores)

Height (cm)

Age (years)

z-scores

(i)
(ii) (1)
(ii) (2)
(iii)

Source: U.S. Census Bureau, Population Division

d $B = \dfrac{m}{h^2}$ For minimum weight, use $B = 21$.

$$21 = \frac{m}{(1.63)^2}$$

$$m = 21 \times (1.63)^2$$

$$= 55.7949$$

$$\approx 56\,\text{kg}$$

e i Girls aged 11–16 years have a height growth rate of 6 cm per year.

ii Growth rates change in teenage years, and growth stops in adulthood.

Question 19

a Use points (11, 146) and (15, 170).

Gradient $m = \dfrac{\text{rise}}{\text{run}} = \dfrac{146 - 170}{11 - 15}$

$m = \dfrac{-24}{-4}$

$= 6$

b Growth rate is 6 cm per year.

c The vertical intercept (c) is not shown on this graph because the axes do not start at 0. To find the value, substitute a point that is on the line, for example (15, 170):

$y = 6x + c$
$170 = 6(15) + c$
$c = 170 - 90$
$= 80$

Using the variables h and a, the equation of the line is $h = 6a + 80$.

d $h = 6(17) + 80$
$= 182$ cm

e The data set is only for males aged between 11 and 16. A 45-year-old male is beyond the data set and the extrapolation would be unreliable.

Question 20

Mean temperature = median − 0.525°C

Median from box plot = 22°C

Mean temperature, $\bar{x} = 22 - 0.525$
$= 21.475$°C

Mean of chirp data, $\bar{y} = 684$ chirps ÷ 20 days
$= 34.2$ chirps per day

Substitute $(\bar{x}, \bar{y})$ into $y = -10.6063 + bx$:

$34.2 = -10.6063 + b(21.475)$
$34.2 + 10.6063 = b \times 21.475$
$44.8063 = b \times 21.475$
$b = \dfrac{44.8063}{21.475}$
$= 2.08644\ldots$

When temperature, $x = 19$°C:

$y = -10.6063 + 2.08644 \times 19$
$= 29.036\ldots$
≈ 29 chirps

HSC exam topic grid (2011–2020)

This grid shows the coverage of this topic in past HSC exams by question number. The past exams can be downloaded from the NESA website (www.educationstandards.nsw.edu.au) by selecting 'Year 11 – Year 12', 'HSC exam papers'. NESA marking feedback and guidelines can also be found there.

Before 2019, 'Mathematics Standard 2' was called 'Mathematics General 2' and, before 2014, 'General Mathematics'. For these exams, select 'Year 11 – Year 12', 'Resources archive', 'HSC exam papers archive'.

	Scatterplots and correlation	**Linear regression**	**The normal distribution**	**z-scores**
2011	8			27(c)
2012	11, 29(a)	19 (replace 'median regression' with 'line of best fit')	29(b)	
2013	2	**28(b)**	20	29(b)
2014	30(b)(i)	30(b) except (v)	24	
2015	28(e)	28(e)	20	28(b)
2016	3, 29(d)(i)	29(d)(ii) (using a calculator and table of values)	13	
2017	**12**	29(d) (using a calculator and table of values for (i))	29(d)	13
2018			23	27(e)
2019 (new course)	**23(a)–(b)**	**23(c)**	**15**	38
2020	12	**36**	35	**8**, 35

Questions in **bold** can be found in this chapter.

9780170459204

CHAPTER 6 NETWORKS

MS-N2 Network concepts 167
N2.1 Networks 167
N2.2 Shortest paths 169
MS-N3 Critical path analysis 171

6

NETWORKS

Networks terminology

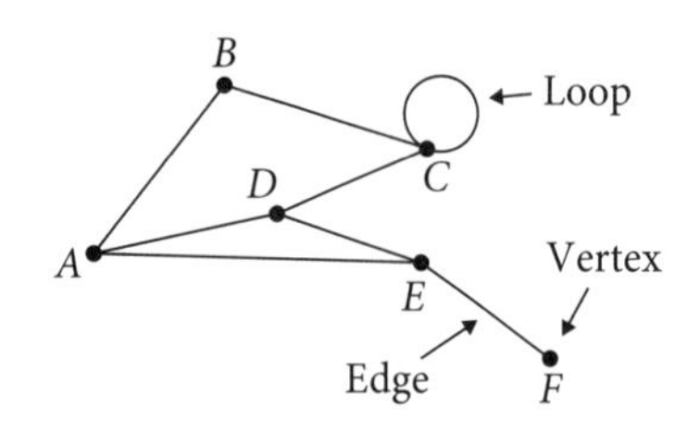

Trail

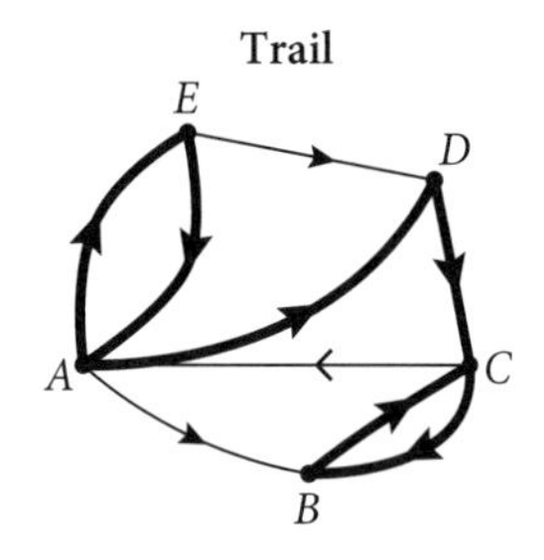

Path

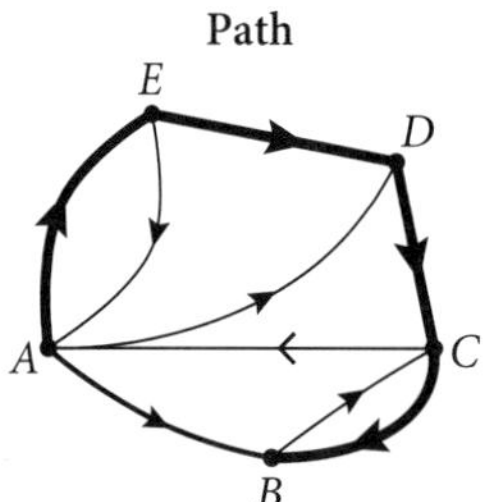

Circuit

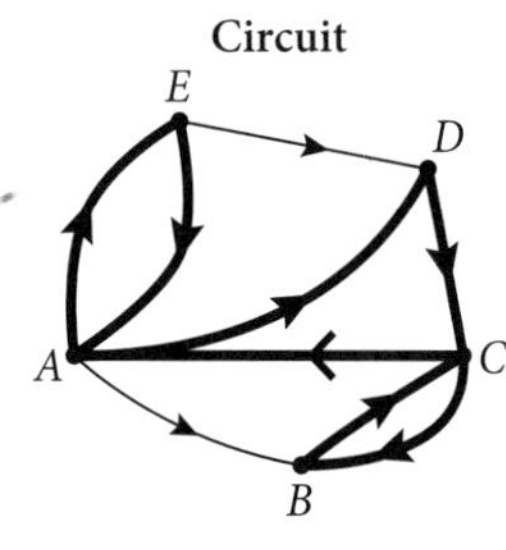

Cycle

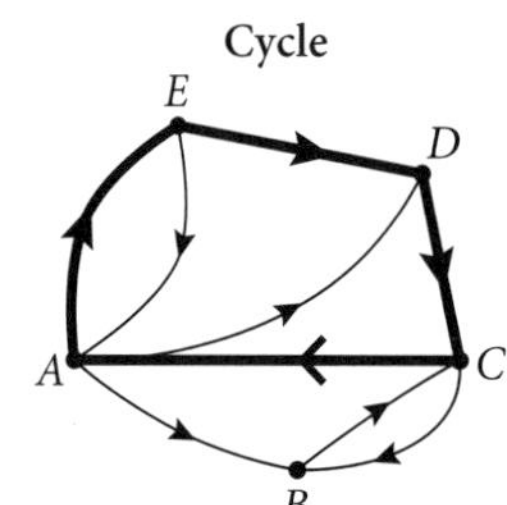

- Directed networks
- Weighted networks
- Connected networks

Minimum spanning trees

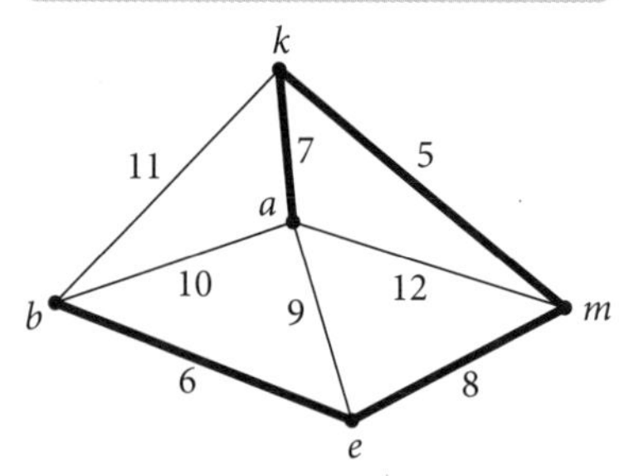

- Kruskal's algorithm
- Prim's algorithm

Shortest path problems

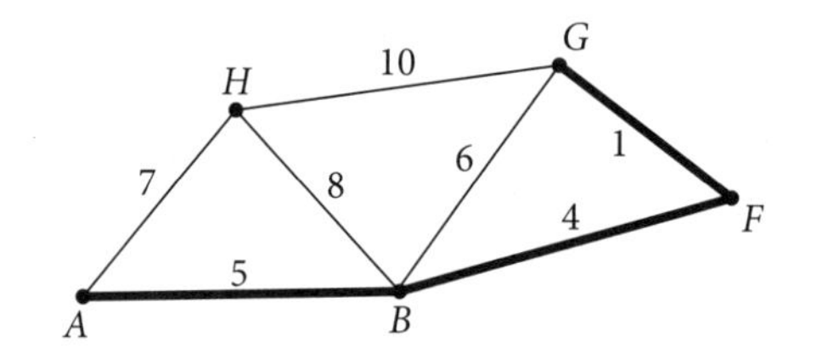

Flow networks

- 'Maximum-flow, minimum-cut' theorem
- Flow capacity

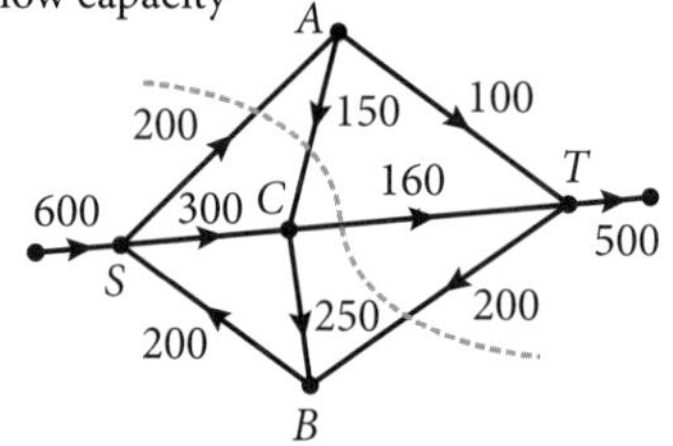

Critical path analysis

- Activity charts and network diagrams

Activity	Predecessor(s)
A	–
B	–
C	A
D	A, B
E	C

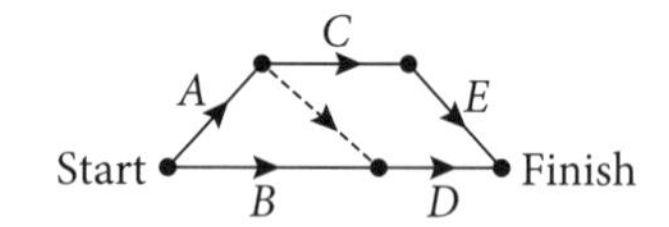

- Forward and backward scanning, float times, critical paths

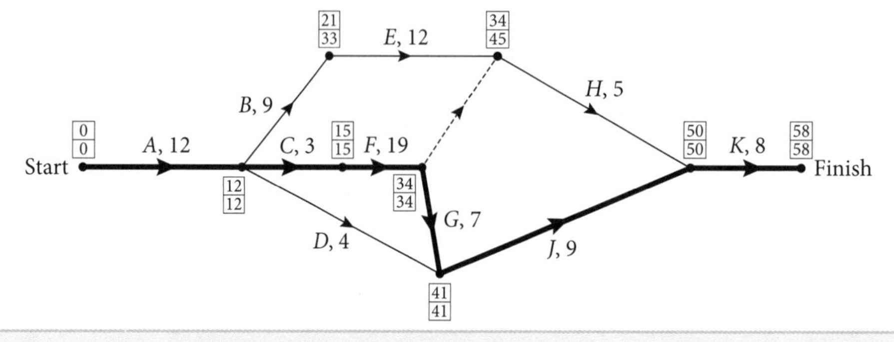

Glossary

A+ DIGITAL FLASHCARDS
Revise this topic's key terms and concepts by scanning the QR code or typing the URL into your browser.

https://get.ga/a-hsc-maths-standard-2

activity table
A table that lists a series of activities, in order, their predecessors and their completion time.

backward scanning
The method for finding the latest starting time (LST) for all activities in a directed network.

capacity
The amount of flow an edge of a flow network can hold, represented by the weight of the edge.

circuit
A closed trail that returns to the starting vertex.

connected network
A network in which every vertex can be reached. No vertices are isolated.

critical activity
An activity that lies on a critical path and must be completed on time so that the whole project is not delayed.

critical path
The longest time path in a network between the start and finish vertices; all the activities on this path must run on time.

critical path analysis
A method used to organise activities without affecting a project's completion time.

cut
A line that cuts through the edges of a flow network to block/disconnect the source from the sink.

cycle
A closed path that returns to the starting vertex.

degree
The number of edges that are connected to a single vertex.

directed network
A network in which each edge has a direction that is represented by an arrow.

disconnected network
A network that includes isolated vertices. The network is not connected.

dummy activity
An activity that has zero time used in a critical path analysis to eliminate repeated edges in a network. It is shown as an edge with zero weight or a dotted line.

earliest starting time (EST)
The earliest starting time for an activity in a directed network.

edge
A line between 2 vertices, sometimes called an arc, indicating a relationship between the vertices.

float time
The difference between the earliest starting time (EST) and latest starting time (LST) for an activity; the maximum time an activity can be delayed without affecting the time to complete the project.

flow capacity
The total flow through a flow network.

forward scanning
The method for finding the earliest starting time (EST) for every activity in a directed network.

inflow
The total capacity of all edges coming to the vertex of a flow network.

latest starting time (LST)
The latest starting time for an activity in a directed network without delaying the entire project.

loop
An edge that begins and finishes at the same vertex.

Kruskal's algorithm
A method for finding a minimum spanning tree for a weighted network by examining edges. Differs from **Prim's algorithm**.

Maximum-flow minimum-cut theorem
A method for finding the maximum flow of a flow network by cutting through edges of the network and using the minimum cut.

minimum spanning tree
The smallest total edge weight for a spanning tree of a network.

network diagram
A visual representation of a group of objects connected to one another in some way.

non-critical activity
An activity not on the critical path of a network.

outflow
The total capacity of all edges leaving the vertex of a flow network.

path
A walk with no repeated vertices or edges.

predecessor
A previous activity that needs to be completed before the current activity can begin. Also known as a prerequisite.

Prim's algorithm
A method for finding a minimum spanning tree for a weighted network by examining vertices. Differs from **Kruskal's algorithm**.

shortest path
The smallest total edge weight between 2 vertices in a network.

sink
The end of a flow network.

source
The start of a flow network.

spanning tree
A tree that connects all vertices of a network.

trail
A walk with no repeated edges.

tree
A network that is connected with no cycles. Any 2 vertices are connected by exactly 1 path.

vertex (plural 'vertices')
A point where edges meet; sometimes called a node.

walk
Any route taken through a network. Edges and vertices may be repeated.

weighted edge
An edge with a numerical quantity that represents the weight between 2 vertices, such as time, distance and cost.

weighted network
A network in which each edge has a weight (number).

Topic summary

This entire topic was new to the Mathematics Standard 2 course in 2019.

Network concepts (MS-N2)

N2.1 Networks

Networks terminology

A network is a group of objects connected to each other in some way.

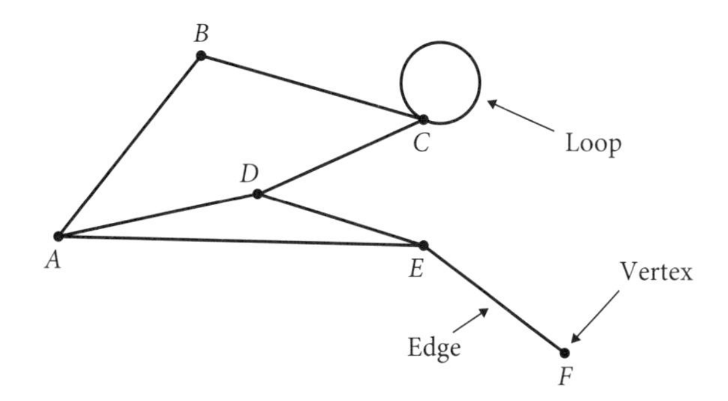

A **vertex** (or node) is a point in a network. The **network diagram** above has 6 vertices, labelled *A* to *F*.

The **edges** (or arcs) are the lines that connect vertices. The network in the diagram shown has 8 edges, including a loop at *C*.

A **loop** is an edge that starts and ends on the same vertex.

A **weighted edge** has a value assigned to it.

A **directed edge** has a direction, indicated by an arrow.

The **degree** of a vertex is the number of edges connected to it. A loop on a vertex has a degree of 2.

For the network shown above:

Vertex	*A*	*B*	*C*	*D*	*E*	*F*
Degree	3	2	4	3	3	1

The sum of the degrees of all vertices in a network is equal to twice the number of edges because each edge is part of 2 vertices.

A network can also be represented using a grid. A '1' in a grid for a network means that 2 vertices are joined by an edge. A '0' means that 2 vertices are not joined by an edge.

Diagram/graph

Table

	P	*Q*	*R*	*S*	*T*
P	0	1	0	1	1
Q	1	0	1	1	0
R	0	1	0	1	0
S	1	1	1	0	1
T	1	0	0	1	0

Routes or walks through a network have special names.

- **Walk**: Any route taken along the edges of a network. Edges and vertices may be repeated. A closed walk starts and ends at the same vertex.
- **Trail**: A walk with no repeated edges, such as *AEADCBC*.

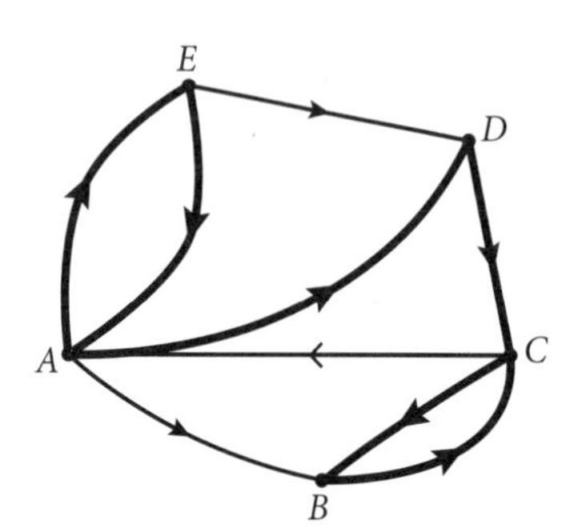

- **Circuit**: A closed trail that starts and ends at the same vertex and does not repeat any edges, such as *AEADCBCA*.

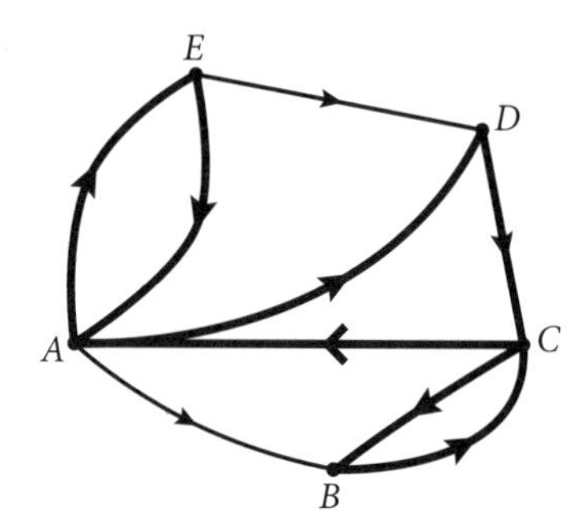

- **Path**: A walk that has no repeated vertices or edges, such as *AEDCB*.

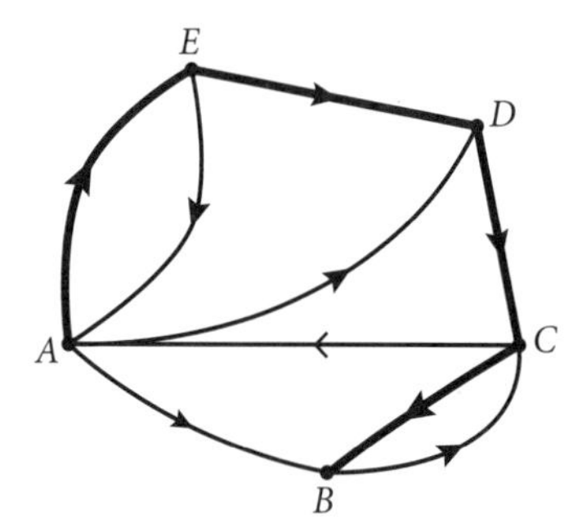

- **Cycle**: A closed path that starts and ends at the same vertex and does not repeat any vertices or edges, such as *AEDCA*.

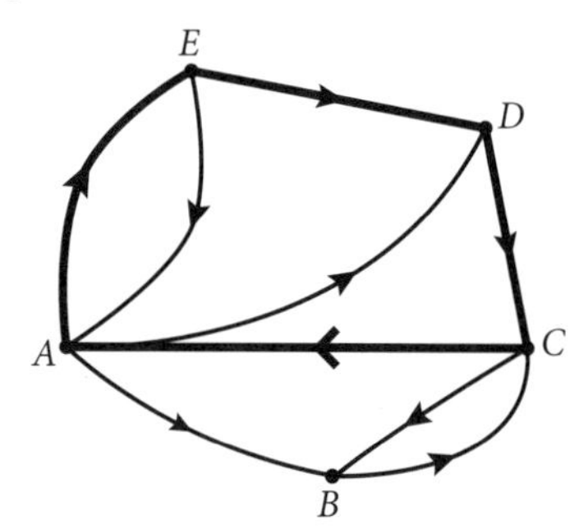

Type of walk	Repeated edges	Repeated vertices	Open / Closed
Trail	No	Yes	Open
Circuit	No	Yes	Closed
Path	No	No	Open
Cycle	No	No	Closed

Types of networks

- A **directed network** has edges with arrows to indicate direction. Edges can be travelled in one direction only.

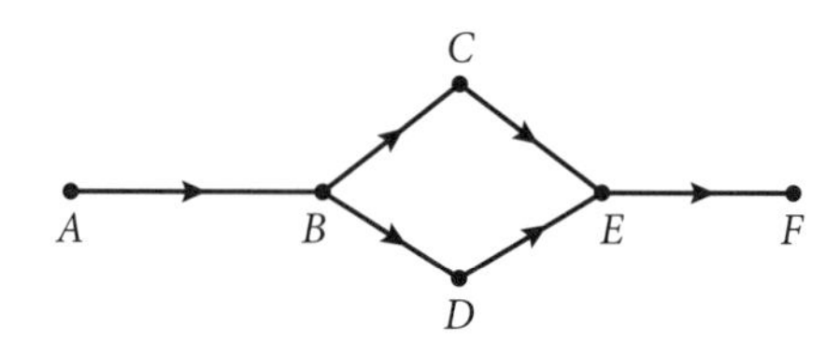

- A **weighted network** has edges with numerical values representing a quantity such as time, distance or cost.

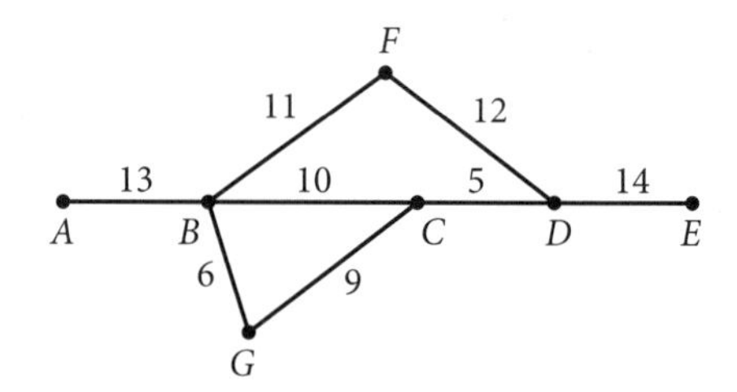

- A **connected network** has no isolated vertices. All vertices can be reached by edges.

- A **disconnected network** has vertices that are isolated from other vertices and cannot be reached by an edge.

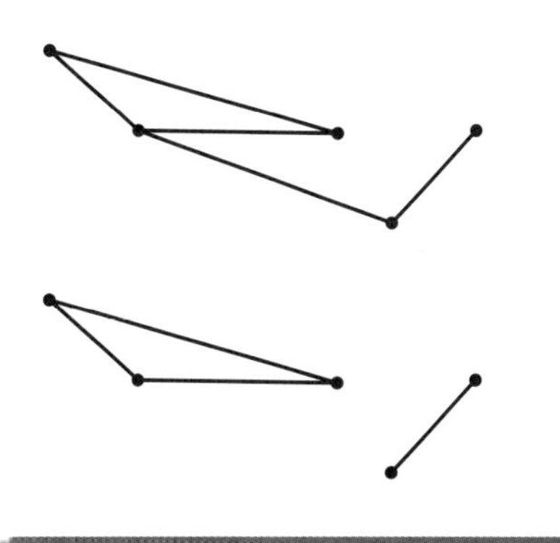

Hint
When drawing a network, use a pencil at first so that you can change it easily.

N2.2 Shortest paths

Trees

A **tree** is a network in which any 2 vertices are connected by exactly one path. There are no cycles, multiple edges, disconnected vertices or loops.

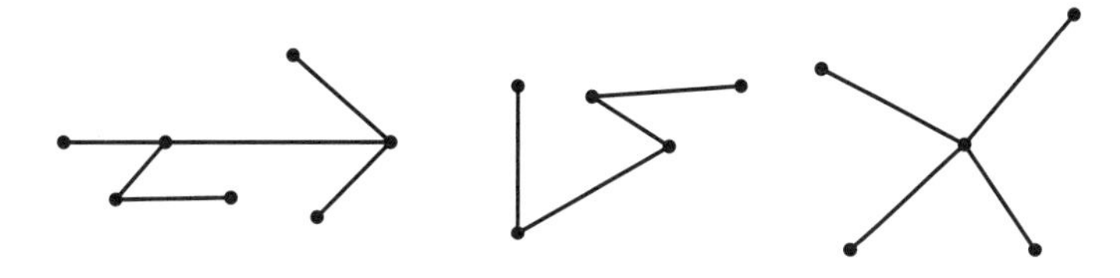

The following are not trees:

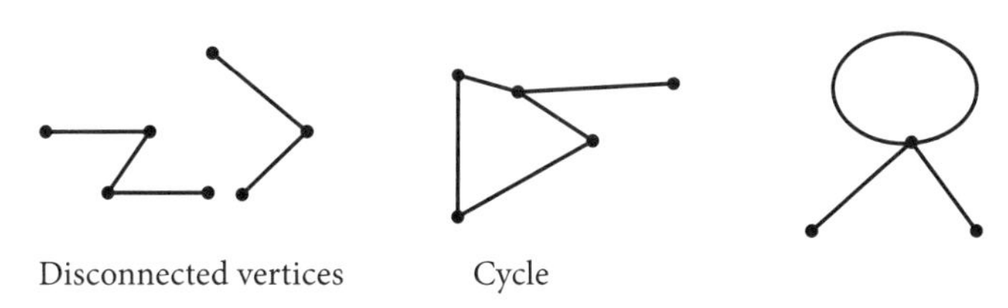

Disconnected vertices Cycle

A **spanning tree** is a subnetwork that connects all of the vertices in a network.

For example, for this network:

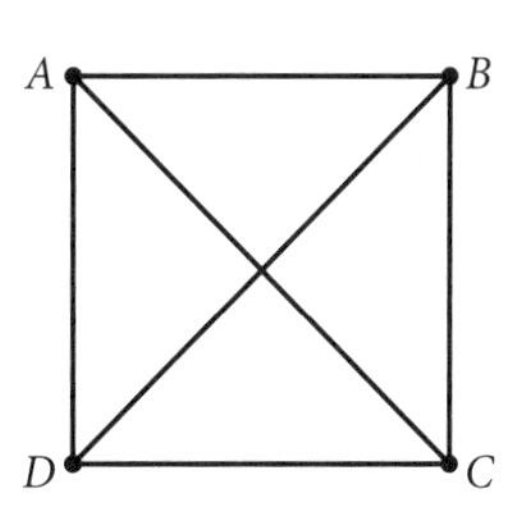

some of the spanning trees are:

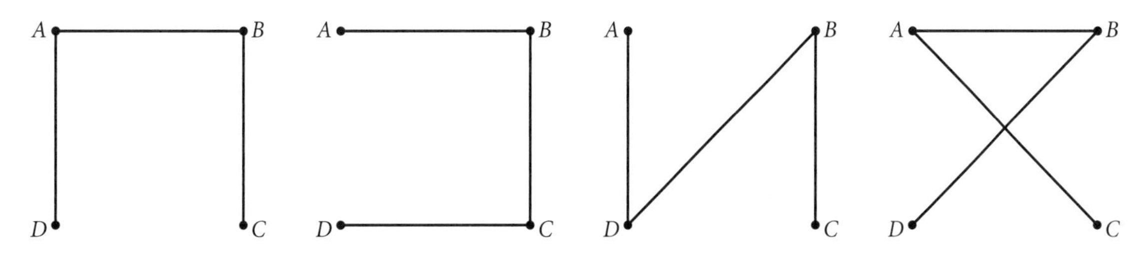

A spanning tree with n vertices has $(n - 1)$ edges. In the example above, the network has 4 vertices and each spanning tree has 3 edges.

Minimum spanning trees

A **minimum spanning tree** in a weighted network has the smallest total edge weight possible. There are 2 different methods for finding the minimum spanning tree.

Kruskal's algorithm uses edges. To use the algorithm:

1. Sort the edges in increasing order of weight.
2. Make a diagram of the network with all vertices, but no edges.
3. Choose the edge with the smallest weight.
4. Choose the edge with the next smallest weight.
5. Continue adding edges, making sure no cycles are introduced, until all vertices are connected.

Prim's algorithm uses vertices. To use the algorithm:

1. Choose any vertex within the network diagram.
2. Choose the edge with the least weight connecting this vertex to another vertex.
3. Look at all edges connecting with these 2 vertices and select the edge with the least weight to add to the tree.
4. Look at all vertices in the tree and choose the edge with least weight to add to the tree.
5. Repeat step 4 until all vertices are connected.

Example 1

Find the minimum spanning tree for this network diagram.

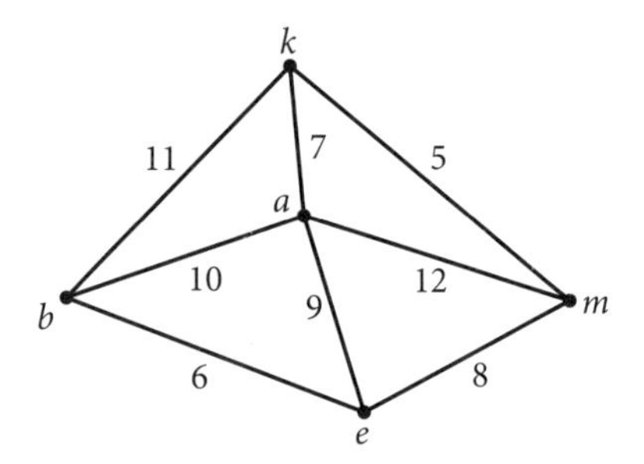

Solution

Using either Kruskal's or Prim's algorithm will give this result.

Minimum length: $5 + 6 + 7 + 8 = 26$

There are 5 vertices so the minimum spanning tree will have 4 edges.

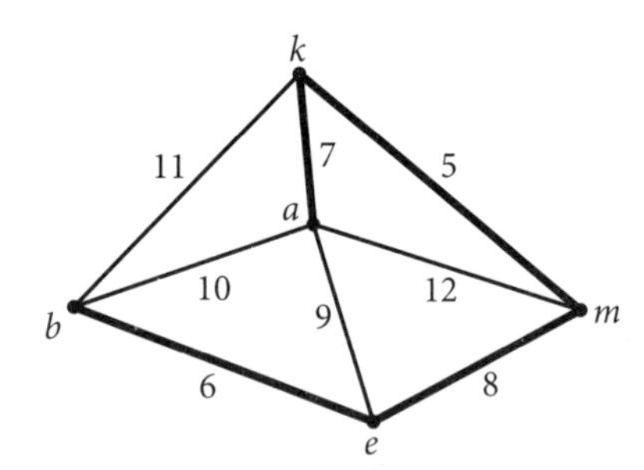

Shortest path problems

For small networks, the **shortest path** between 2 vertices can be calculated using trial and error. It involves finding the path of minimum weight between 2 specific vertices.

Note: This is not the same as a minimum spanning tree.

Example 2

Find the shortest path from A to G.

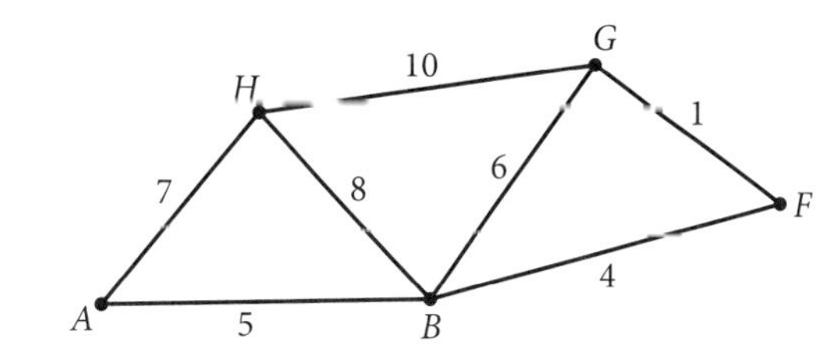

Solution

Possible paths:

- AHG: $7 + 10 = 17$
- ABG: $5 + 6 = 11$
- $AHBG$: $7 + 8 + 6 = 21$
- $ABFG$: $5 + 4 + 1 = 10$
- $AHBFG$: $7 + 8 + 4 + 1 = 20$

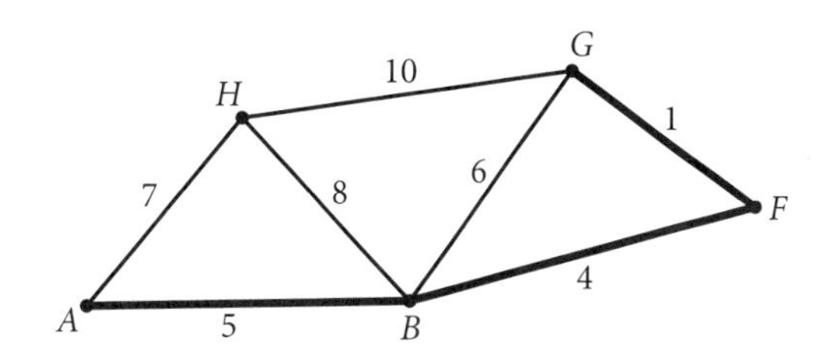

The shortest path is $ABFG$.

For larger networks, there is a method for finding the shortest path:

1. Redraw the network diagram with circles at each vertex, except at the start vertex, S.
2. For all vertices one edge away from S, write down the shortest distance inside the circle representing the closest vertex. Continue until the finish vertex, F, is reached.
3. For all vertices 2 edges away from S, write down the shortest distance inside the circle representing the closest vertex. Continue until the finish vertex, F, is reached.
4. Continue this process until the finish vertex, F, is reached.
5. Identify the shortest path by starting at F and moving backwards to the vertex from which the shortest distance at F was obtained, and continuing until S is reached.

Critical path analysis (MS-N3)

Activity tables

Networks can be used when a task or project involves a number of activities that need to be completed.

Activity tables show each activity, coded by a letter, the estimated time taken and its **predecessors** (previous activities that must be completed for the current activity to begin).

Activity	Predecessor(s)	Duration (hours)
A	–	4
B	–	1
C	A	3
D	A, B	2
E	C	3

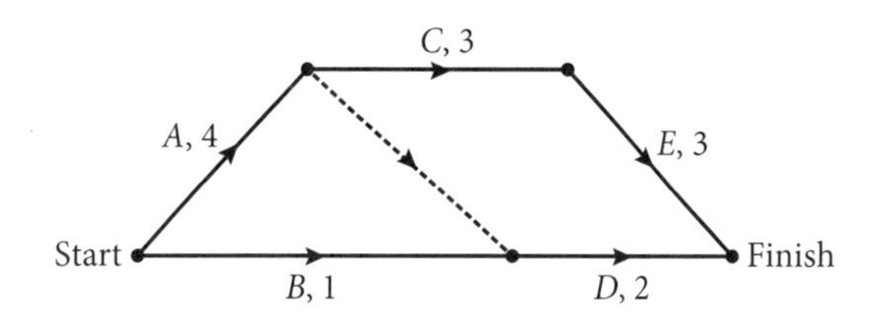

To draw a directed network diagram from an activity table:

- Label the start and finish vertices.
- Represent activities using arrows, ensuring they start and finish at a vertex.
- Connect activities that do not have predecessors to the start vertex.
- Begin activities with the same predecessor at the same vertex.
- Ensure that any 2 vertices are not connected by multiple edges.

A **dummy activity** is used to eliminate repeated edges in a network. It does not contain time and is shown as a directed edge with either a weight of zero or a dotted line, such as the one pointing to *D* in the previous example.

The **earliest starting time (EST)** is the earliest time an activity can start, depending on the completion of predecessor activities.

The **latest starting time (LST)** is the latest time an activity can start without affecting the completion time of the project.

Critical activities

The **critical path** is the longest weighted path between the start and finish vertices. It shows the shortest amount of time to complete a project.

A **critical activity** is an activity on a critical path. These activities must start and end on time or the project will be delayed. On the critical path, activities have equal EST and LST.

A **non-critical activity** is not on the critical path. These activities can be delayed without affecting the project's completion time.

Critical path analysis is a step-by-step method used to organise activities needed to complete a project in the shortest time possible.

Forward scanning calculates the EST for each activity in a network. The steps are:

1. Redraw the network diagram with two empty boxes at each vertex, one above the other.
2. Place 0 at the start vertex (in the top box).
3. Work along each path or activity from the start (left to right), writing the total time of the path at each vertex (in the top box).
4. When 2 or more paths meet, select the greatest total as the path to follow.
5. Continue until the finish vertex is reached.

Backward scanning calculates the LST for each activity in a network. The steps are:

1. Remove numbers in all vertices other than the finish vertex. At the finish, LST = EST, so write the LST in the bottom box.
2. Work backwards from right to left, subtracting the time of each activity and writing the LST at each vertex (in the bottom box).
3. If there is more than one path to the vertex, take the smallest difference as the LST.
4. Continue until you reach the start vertex. The LST at the start vertex must be 0.

Example 3

For the network in this diagram, find the EST and LST of each activity, the critical path and its length.

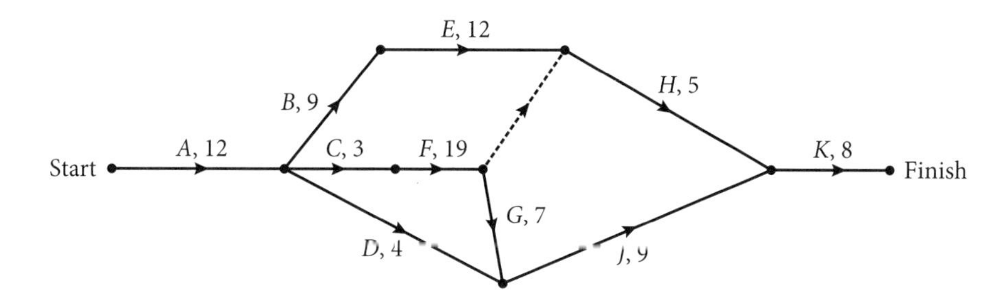

Solution

Do forward mapping to find ESTs:

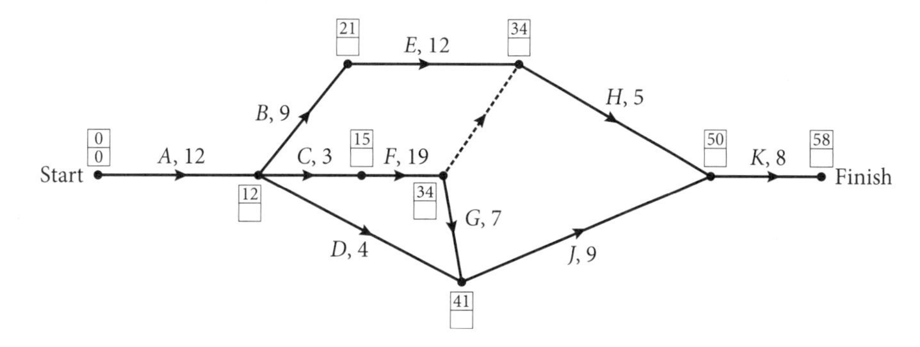

Do backward mapping to find LSTs:

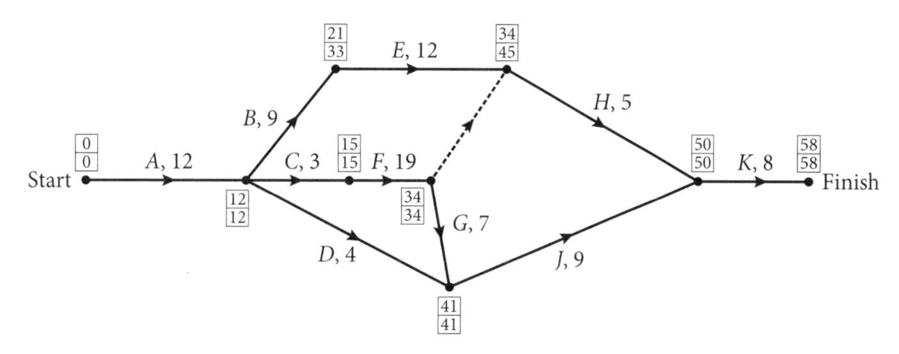

Mark the critical path for activities where EST = LST:

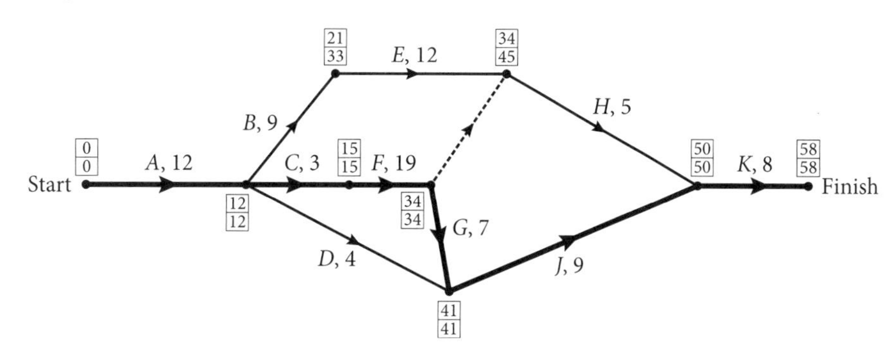

The critical path is *ACFGJK* with length 58.

Float time

Float time is the longest time a non-critical activity can be delayed before it affects the project.

$$\text{Float time} = \text{LST}_{\text{next}} - \text{EST}_{\text{next}} \text{ (if it is the only path leading to the next activity)}$$

OR

$$\text{Float time} = \text{LST}_{\text{next}} - \text{EST} - \text{activity time}$$

where EST_{next} is the EST of the next activity, which is the earliest ending time of the activity, and

LST_{next} is the LST of the next activity, which is the latest ending time of this activity.

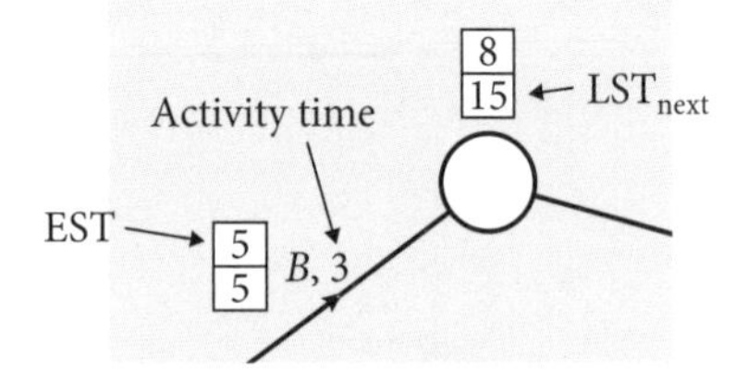

Float time of activity B
$= 15 - 5 - 3 = 7$ (or $15 - 8 = 7$)

Flow networks

Networks can be applied to flow problems such as managing deliveries, flight paths, traffic, water and passenger arrivals at airports.

Source: The start of a flow network.

Sink: The end of a flow network.

Capacity: The amount of flow an edge can hold, represented by the weight of an edge.

Flow capacity: The total flow through a network.

Inflow: The total capacity of all edges entering a vertex.

Outflow: The total capacity of all edges exiting a vertex.

Example 4

What is the maximum outflow from vertex A and from vertex C?

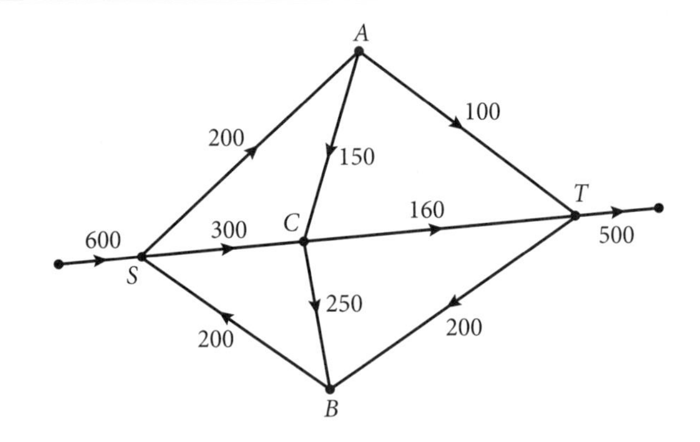

Solution

Inflow capacity to vertex A: 200

Outflow capacity from vertex A: $150 + 100 = 250$

So maximum outflow of A is 250.

Inflow capacity to vertex C: $300 + 150 = 450$

Outflow capacity from vertex C: $250 + 160 = 410$

So maximum outflow of C is 410.

Capacity of a cut

A method of finding the maximum flow of a network is to use 'cuts'.

A **cut** is a line that cuts through the edges of a flow network to disconnect the source from the sink. It acts like an arrangement of roadblocks that stops all flow from the source to the sink.

Any cut through a flow network blocks or stops the flow, so you can compare the effects of different cuts on the amount of flow being blocked.

The capacity of a cut (the amount of flow it blocks) is calculated by finding the sum of cut edges. You only add edges that are in the direction of source to sink and will block the flow of the network.

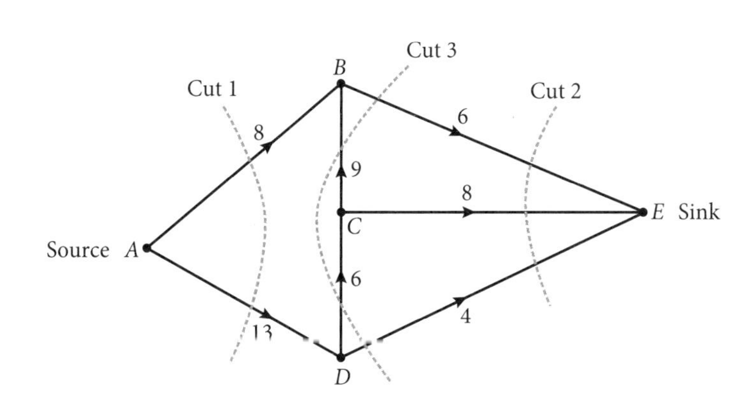

Cut 1 capacity: 8 + 13 = 21
Cut 2 capacity: 6 + 8 + 4 = 18
Cut 3 capacity: 6 + 0 + 6 + 4 = 16

In cut 3, edges leading *back* to the source are not counted.

Hint

Imagine driving a car along a cut, from bottom to top.

- Edges from left to right are counted.
- Edges from right to left are not counted.

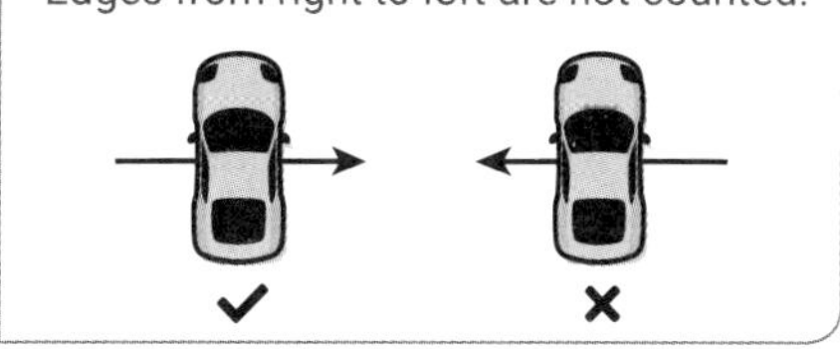

Maximum-flow minimum-cut theorem

The **maximum-flow minimum-cut theorem** is the rule that the maximum flow of a network is equal to the capacity of the minimum cut. To find the maximum flow, calculate the capacity of each cut and identify the cut that has the minimum capacity.

Example 5

Find the maximum flow for the network in this diagram from the source, *P*, to the sink, *W*.

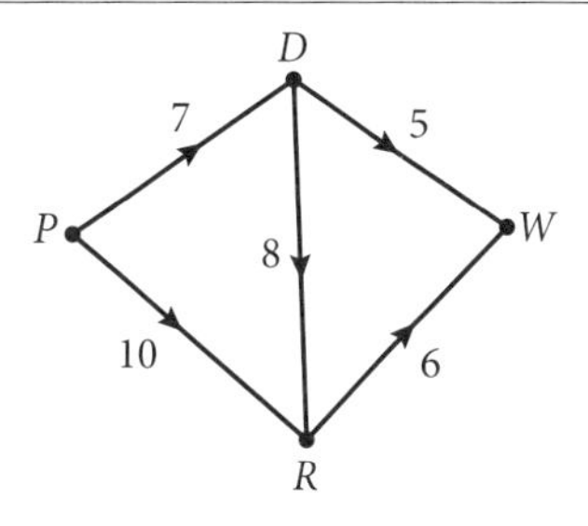

Solution

The cuts that can be made to the network are shown.

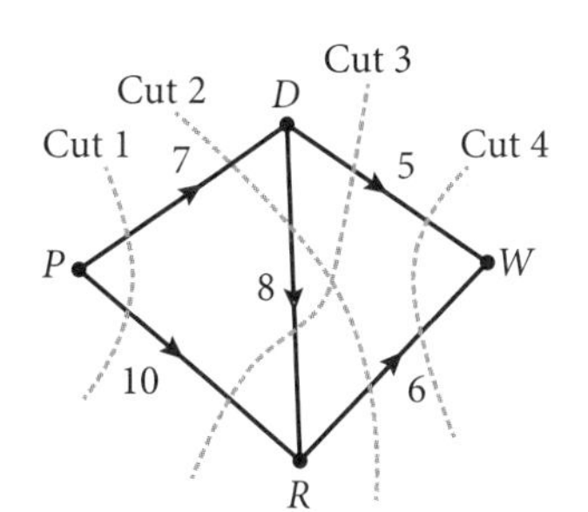

The capacities of the cuts are:

Cut 1: 7 + 10 = 17
Cut 2: 7 + 6 = 13
Cut 3: 5 + 8 + 10 = 23
Cut 4: 5 + 6 = 11

The minimum cut is 11, so the maximum flow is 11.

Hint

In Cut 2, *DR* is not included because it is in the opposite direction to *PD* and *RW*.

Alternative method

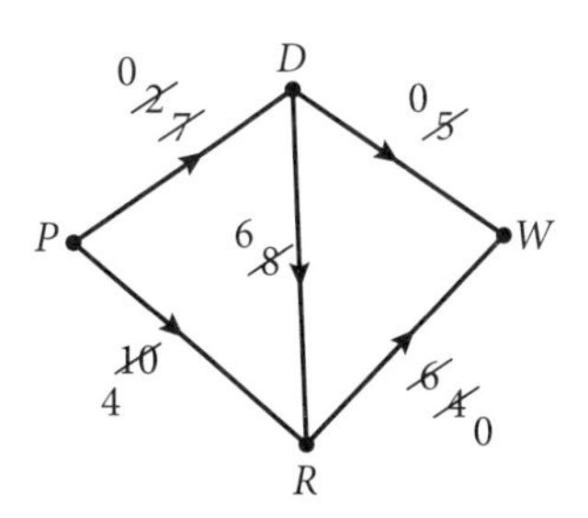

The flows that give a maximum are:

PDW	5
PDRW	2
PRW	4
Total	11

Minimum cut

The maximum flow from *P* to *W* is 11.

Hint

Look at the edges that are saturated (reduced to 0, capacity used up) to consider the minimum cut.

TOPIC SUMMARY

Practice set 1

Multiple-choice questions

Solutions start on page 190.

Question 1

Which statement about networks is correct?

A A vertex is a line connecting edges.

B A loop is an edge that connects a vertex to itself.

C The sum of degrees in a network is always an odd number.

D A cycle is an open path.

Question 2

What is the degree of vertex P in this network diagram?

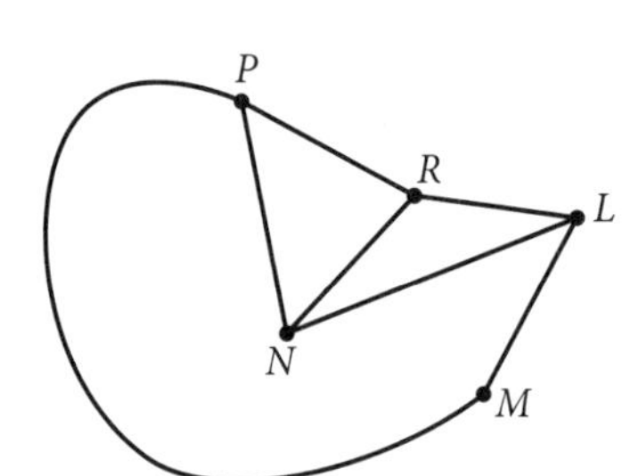

A 3

B 4

C 5

D 7

Question 3

How many edges are in this network diagram?

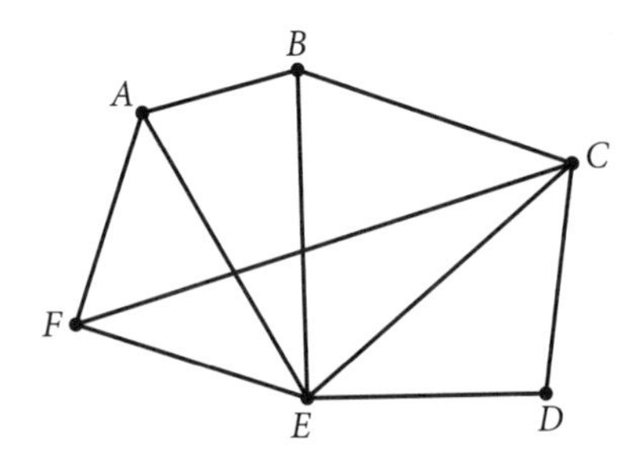

A 6

B 9

C 10

D 12

Question 4

Which of the following graphs represents a connected network?

A

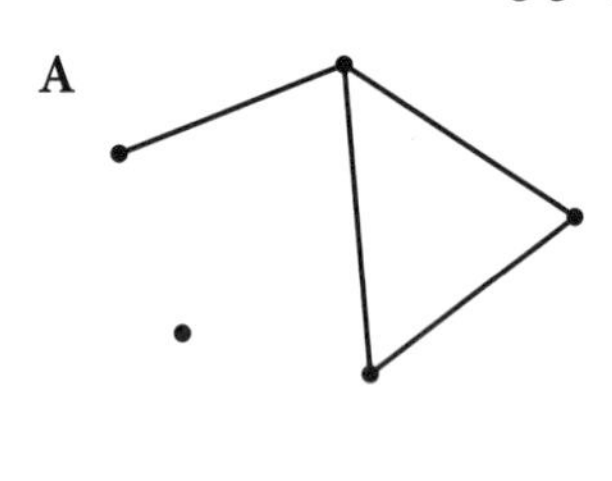

B

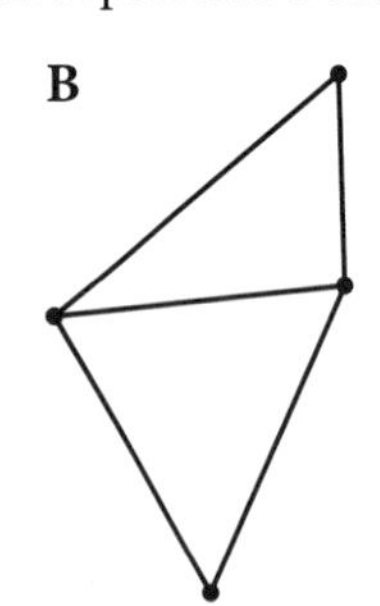

C

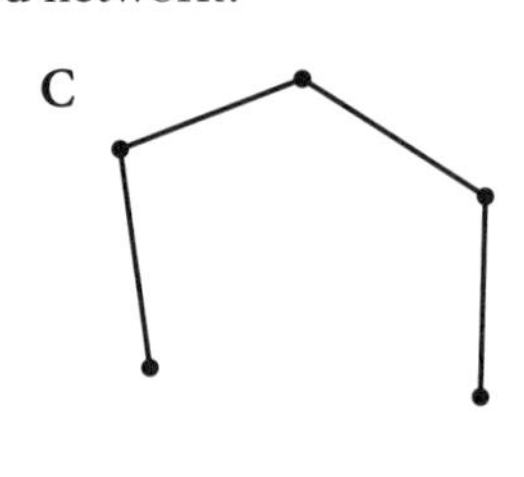

D

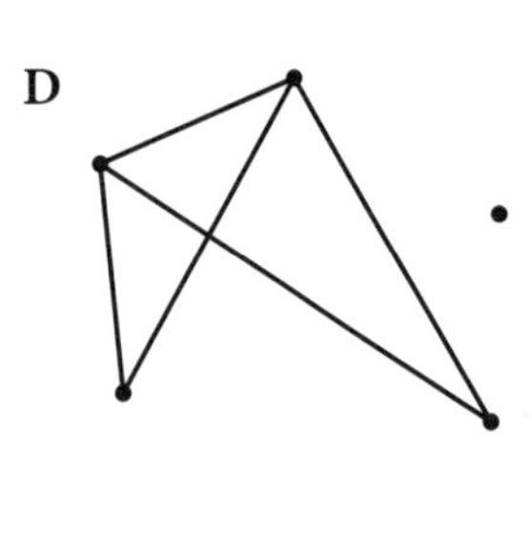

Question 5

How many edges can be removed so that this network diagram has a minimum number of edges but remains connected?

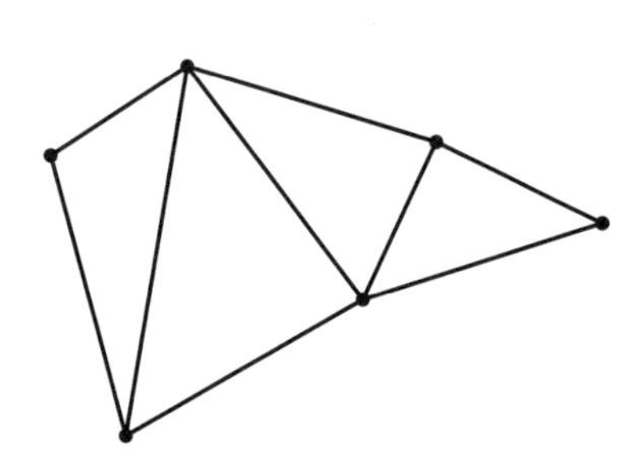

A 2

B 3

C 4

D 5

Question 6

A network diagram is shown.

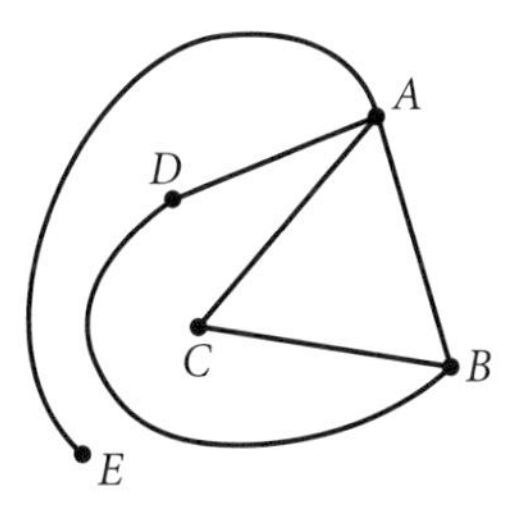

Which of the following graphs represents the same network?

A

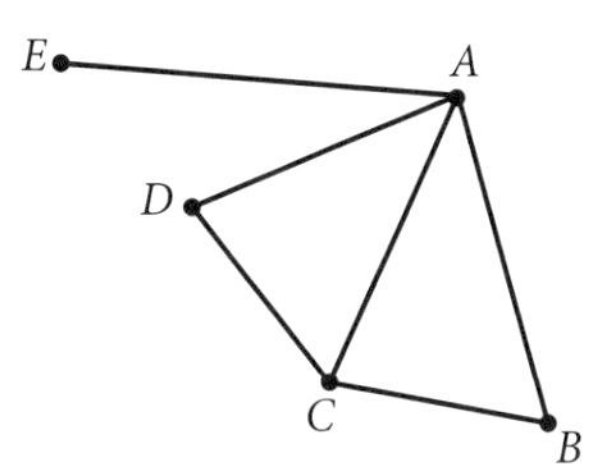

B

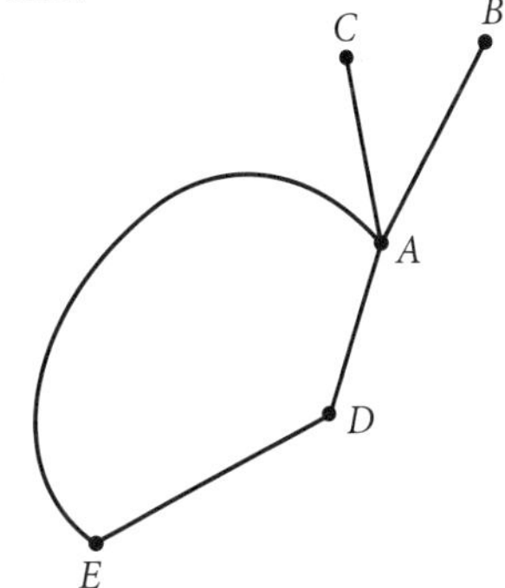

C

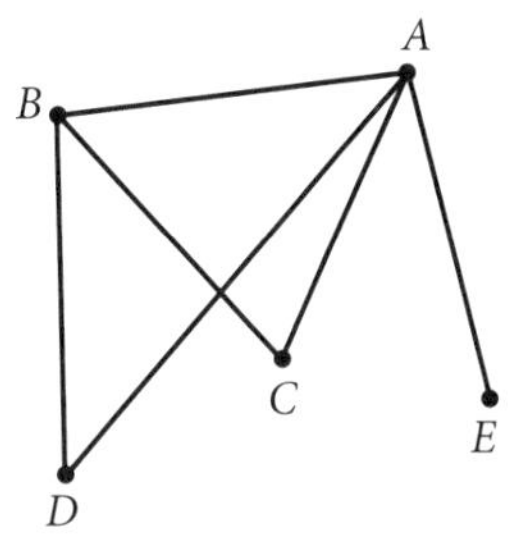

D

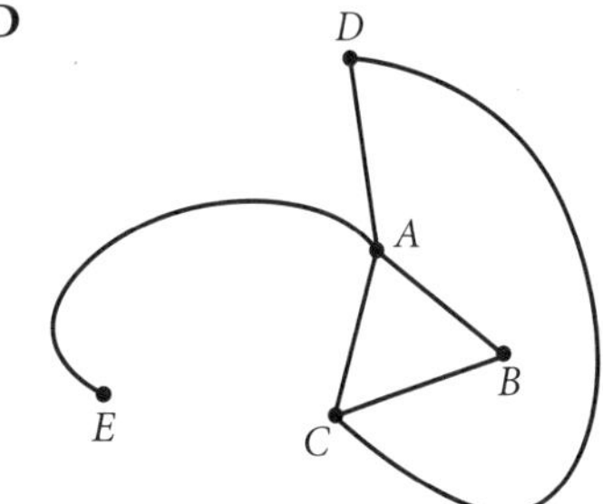

Question 7 ©NESA 2020 HSC EXAM, QUESTION 9

Team A and Team B have entered a chess competition.

Team A and Team B have three members each. Each member of Team A must play each member of Team B once.

Which of the following network diagrams could represent the chess games to be played?

A

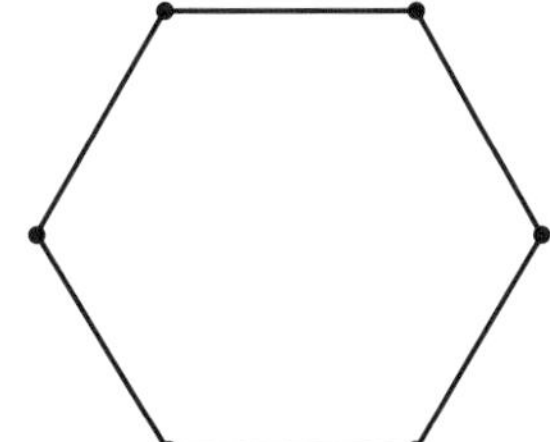

B

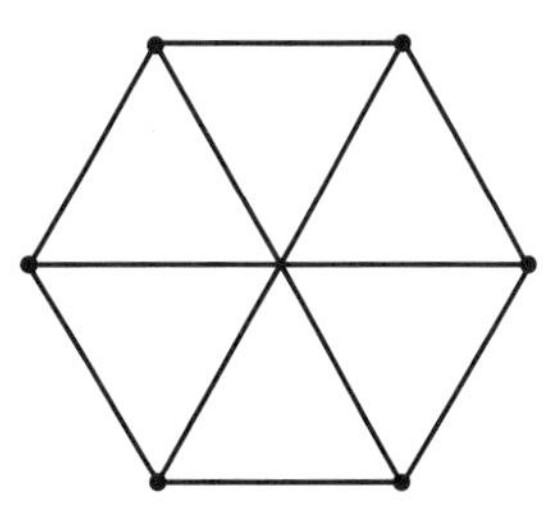

C

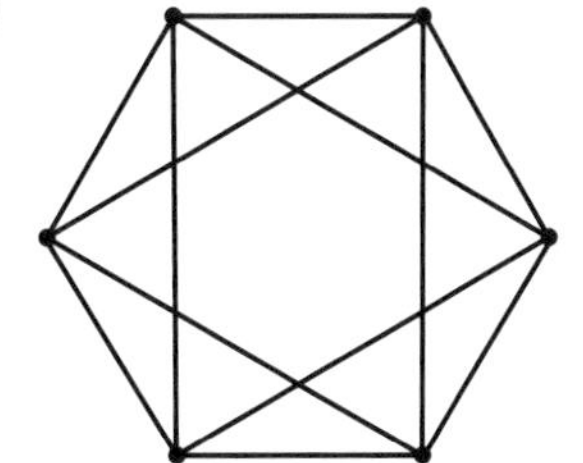

D

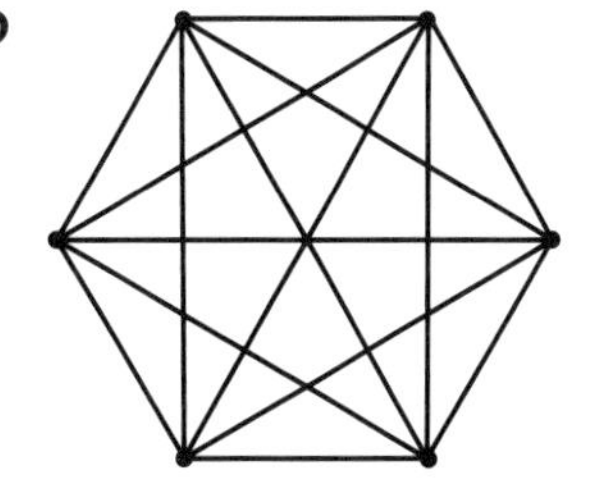

Question 8

What type of network diagram is shown?

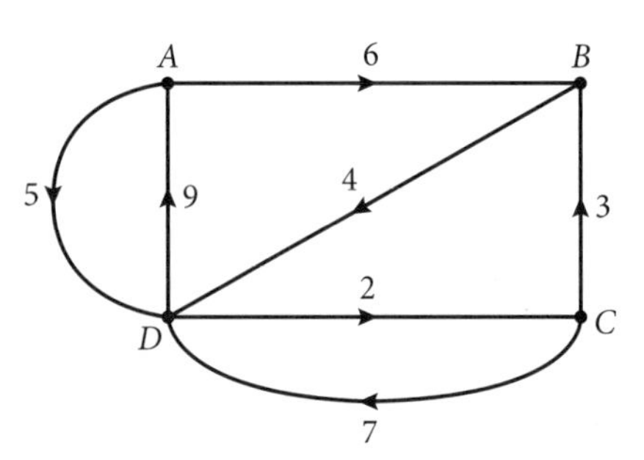

A Undirected and unweighted

B Undirected and weighted

C Directed and unweighted

D Directed and weighted

Question 9

What is the length of the shortest path from C to A in the network diagram in Question 8?

A 9 **B** 11 **C** 12 **D** 16

Question 10

Which of the following network diagrams is NOT a tree?

A

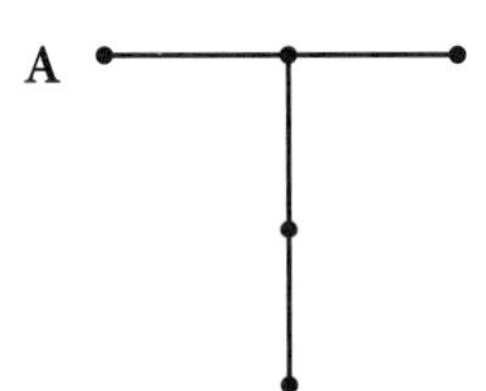

B

C

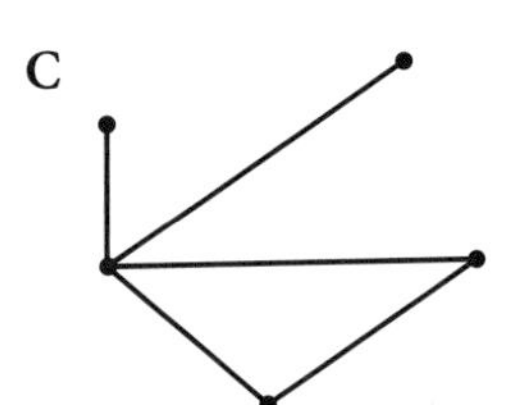

D

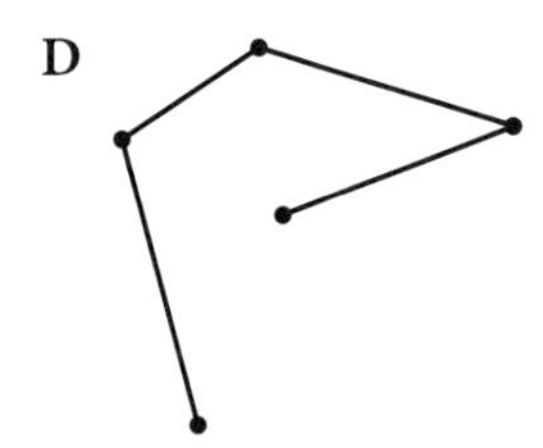

Question 11

How many edges are in a minimum spanning tree if a connected network has 8 vertices?

A 6 **B** 7 **C** 8 **D** 9

Question 12

What is the minimum spanning tree of the following network diagram?

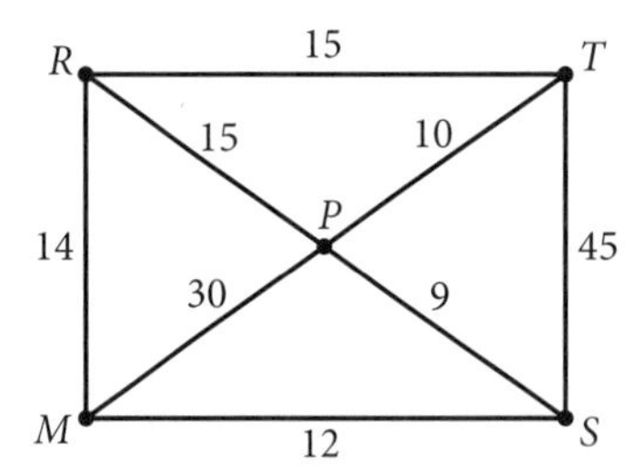

A

B

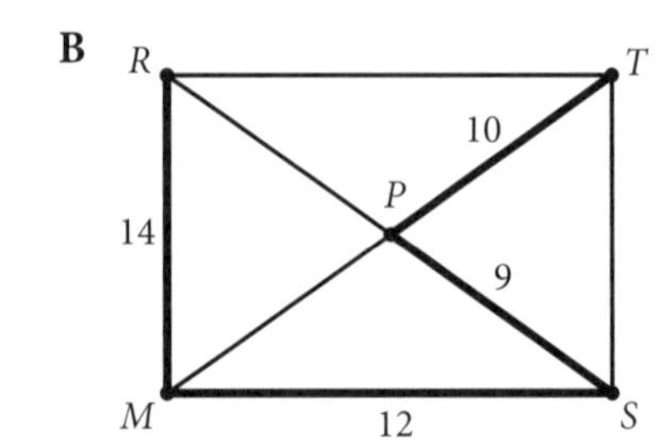

C

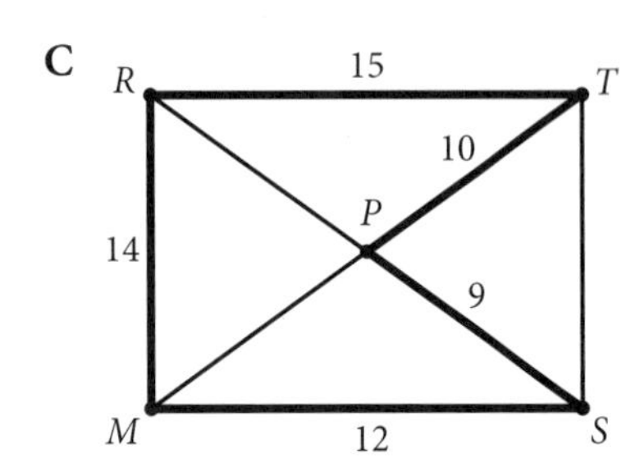

D

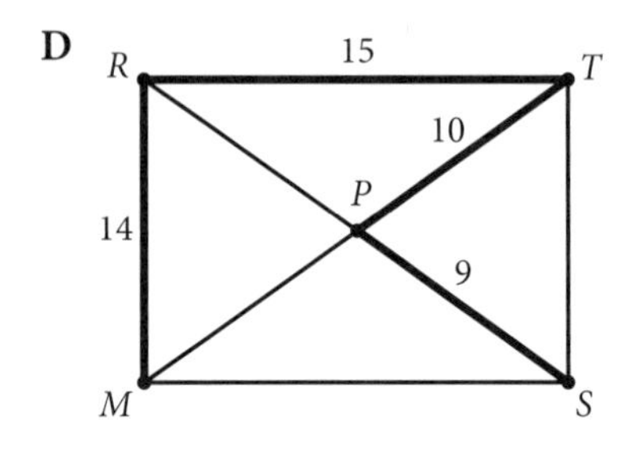

Question 13

Which statement about critical paths is true?

A The critical path is the shortest path from the start to the finish of a project.

B There is only one critical path in a project.

C Reducing the time of any activity on a critical path for a project will always reduce the minimum time to complete the project.

D A critical path includes at least two activities.

Question 14

Which activity table represents this network?

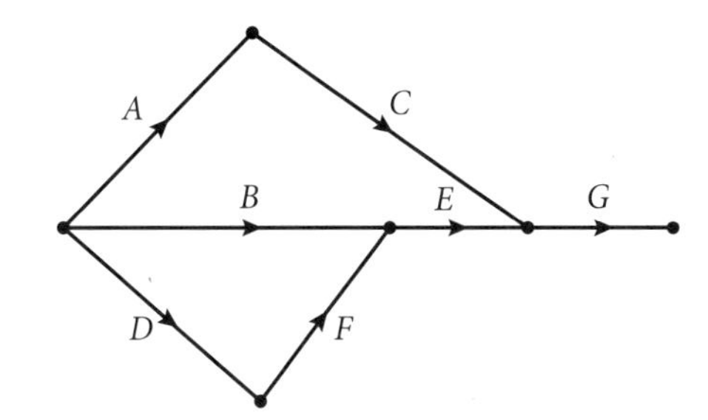

A

Activity	Immediate predecessor(s)
A	–
B	–
C	A
D	–
E	B, F
F	D
G	C, E

B

Activity	Immediate predecessor(s)
A	–
B	–
C	A
D	A
E	B, F
F	D
G	C, E

C

Activity	Immediate predecessor(s)
A	–
B	A
C	A
D	A
E	B, F
F	D
G	C, E

D

Activity	Immediate predecessor(s)
A	–
B	–
C	A
D	–
E	B
F	D
G	C, E

Question 15

What is the earliest starting time (EST) and latest starting time (LST) for activity C in the diagram?

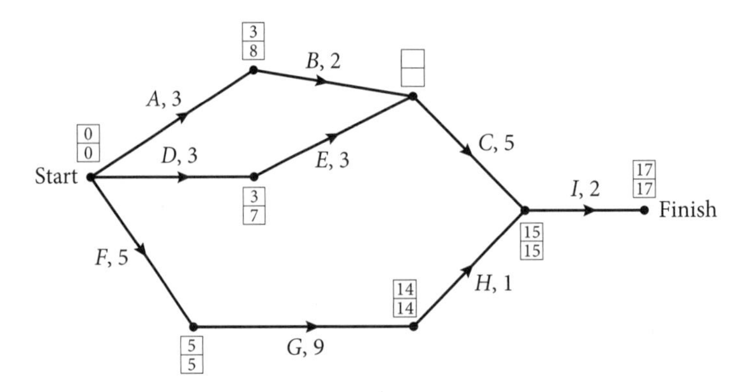

A EST = 5
LST = 10

B EST = 5
LST = 15

C EST = 6
LST = 10

D EST = 6
LST = 15

Question 16

What is the earliest starting time of activity F?

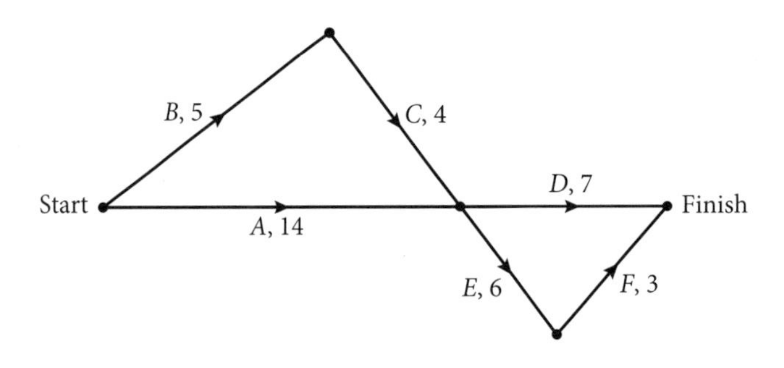

A 3

B 6

C 15

D 20

Question 17

Given the source, S, and the sink, T, which of the following is NOT a cut of the network shown?

A

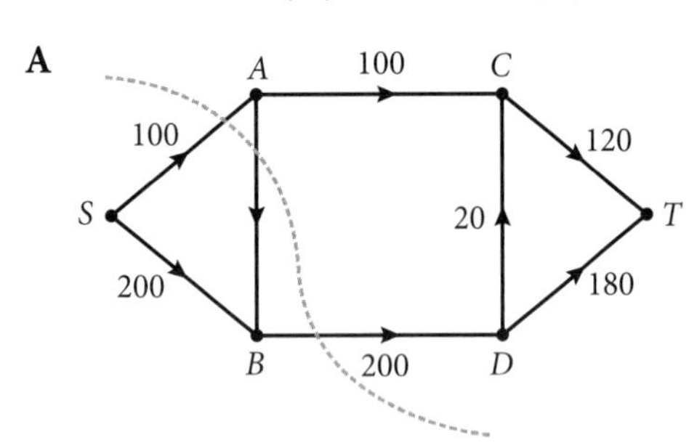

B

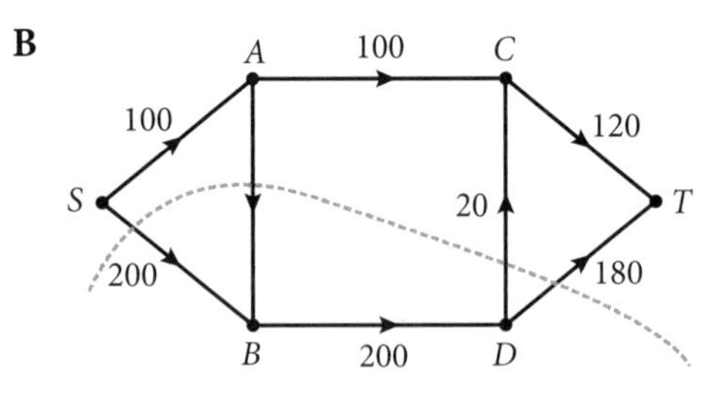

C

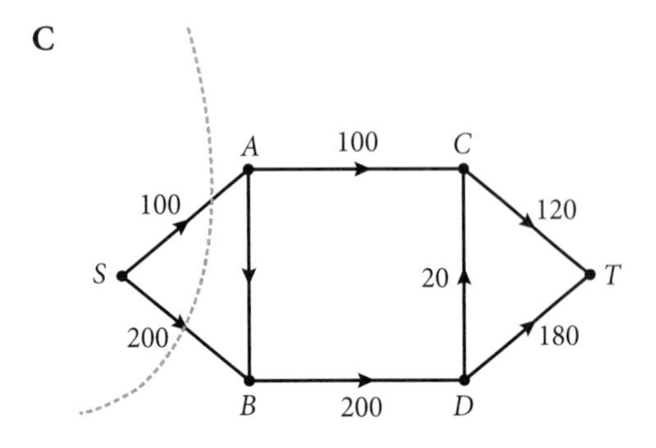

D

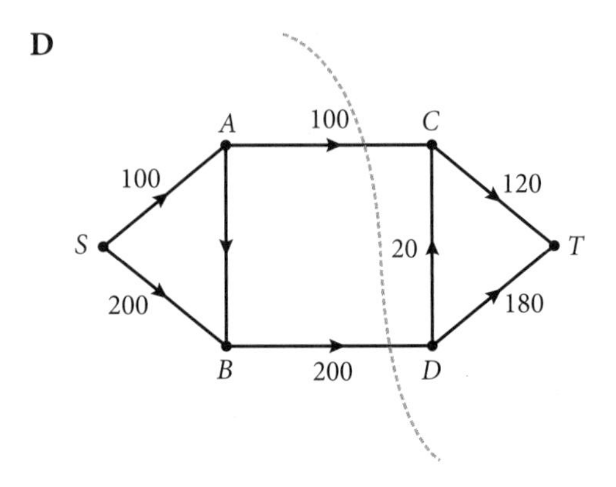

Question 18

In the network shown, what is the capacity of the cut from the source, S, to the sink, T?

A 350

B 360

C 560

D 710

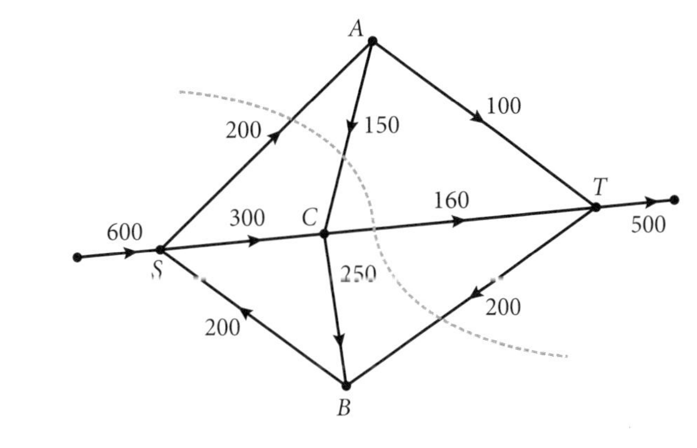

Question 19

The network diagram shows the flow of water in a garden irrigation system connecting the source to the sink. What is the maximum flow from the source to the sink?

A 12

B 13

C 21

D 27

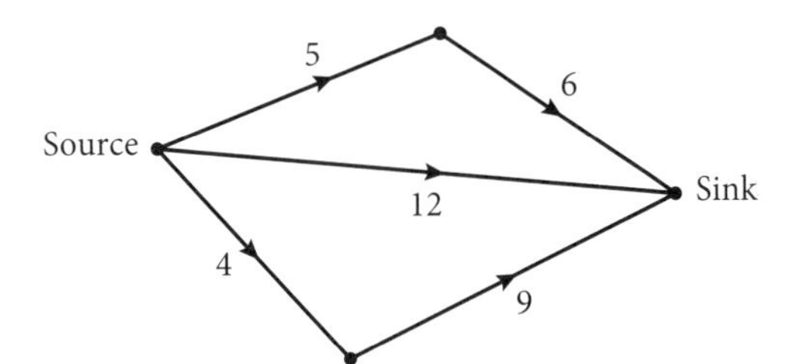

Question 20

The following network diagram displays the time, in days, to complete activities in a building project. What is the minimum completion time of the project?

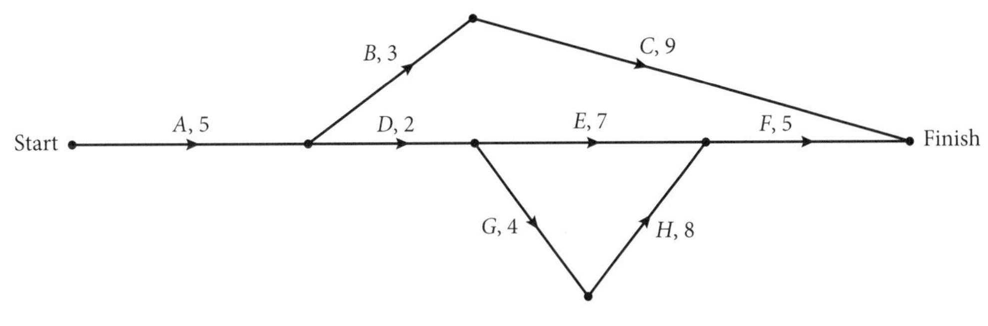

A 17 days B 19 days C 24 days D 43 days

9780170459204

Practice set 2

Short-answer questions

Solutions start on page 191.

Question 1 (2 marks)

Complete the table to indicate the degree of each vertex. 2 marks

Vertex	*A*	*B*	*C*	*D*
Degree	2			

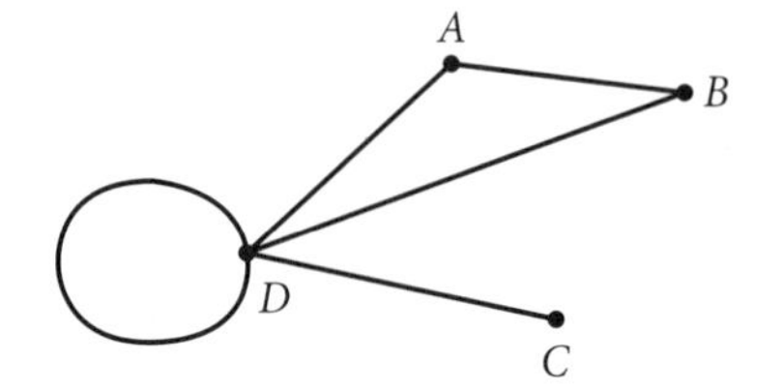

Question 2 (5 marks)

Match each description to the correct word from the list.

cycle connected path trail walk

a any route along the edges of a network, which may include repeated edges or repeated vertices 1 mark

b a walk with no repeated vertices or edges 1 mark

c a walk with no repeated edges 1 mark

d a walk with the same start and end vertex, which doesn't visit any other vertex more than once 1 mark

e describes a network where there is a path between any 2 vertices 1 mark

Question 3 (3 marks)

Using the network diagram, state whether each statement is true or false.

a *ABCDAE* is a trail. 1 mark

b *ABEDBC* is a path. 1 mark

c *ABCDA* is a cycle. 1 mark

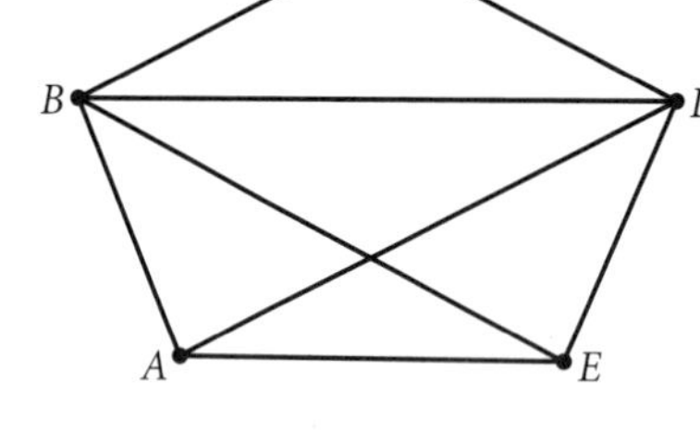

Question 4 (2 marks)

The network on the right shows the distance, in kilometres, along various roads that can connect Town *A* and Town *B*.

What is the shortest distance between the two towns? Highlight the route taken for this. 2 marks

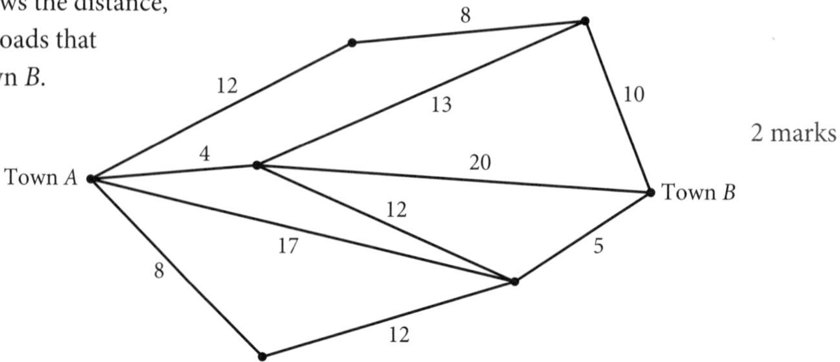

Question 5 (2 marks)

Draw 2 different spanning trees for this network. 2 marks

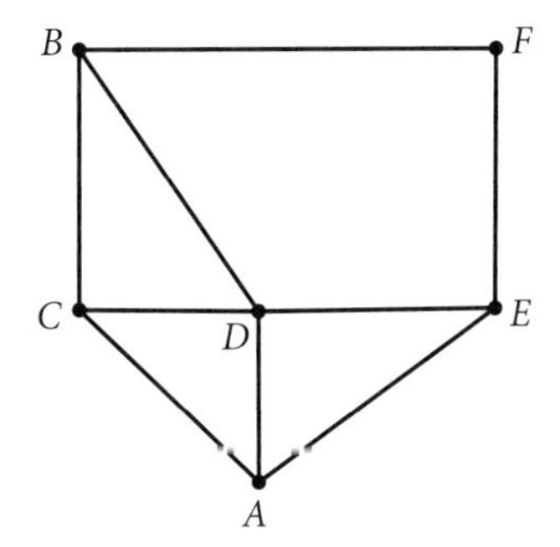

Question 6 (5 marks)

This grid shows the distances, in kilometres, of plane flights between city airports.

	Newcastle (N)	Brisbane (B)	Cairns (C)	Melbourne (M)	Sydney (S)	Adelaide (A)
Newcastle	–	770	–	1030	–	–
Brisbane	770	–	1588	1772	938	–
Cairns	–	1588	–	2810	2432	–
Melbourne	1030	1772	2810	–	850	718
Sydney	–	938	2432	850	-	1356
Adelaide	–	–	–	718	1356	–

a Complete this network diagram using the information in the grid. 2 marks

• C

N • • B

A •

• M • S

b What is the length of the shortest path from Adelaide to Cairns? 1 mark

c Draw a minimum spanning tree for the network. What is the minimum length? 2 marks

Question 7 (3 marks)

A social network includes six friends.

- Alicia is friends with Cathy and Michael.
- Cathy is friends with Alicia, Michael, Bob and Lisa.
- Lisa is friends with Cathy, Kerri and Bob.

a Draw a network diagram to represent these friendships. 2 marks

b Is the network connected or disconnected? 1 mark

Question 8 (5 marks)

a Do the bold edges on this network diagram form a spanning tree? Explain your answer.

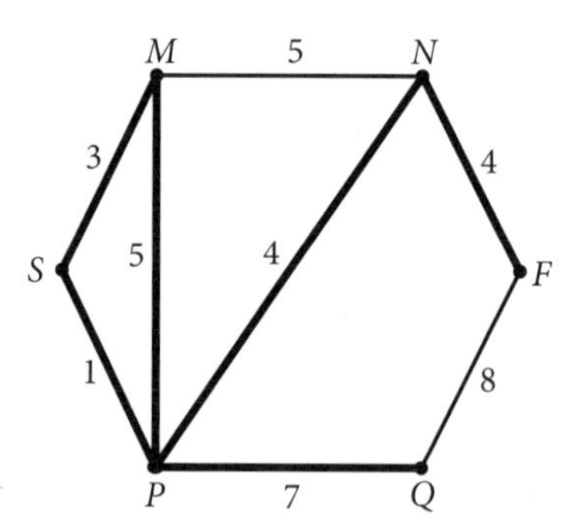

1 mark

b What is the length of the spanning tree in the network diagram on the right?

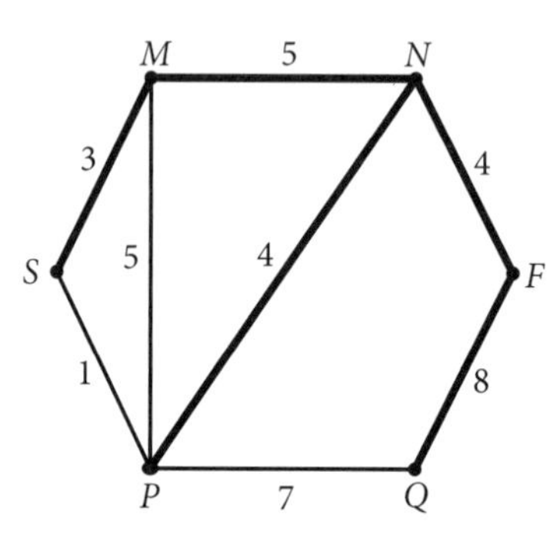

2 marks

c Draw the minimum spanning tree on the network diagram by highlighting the edges.

What is the minimum length?

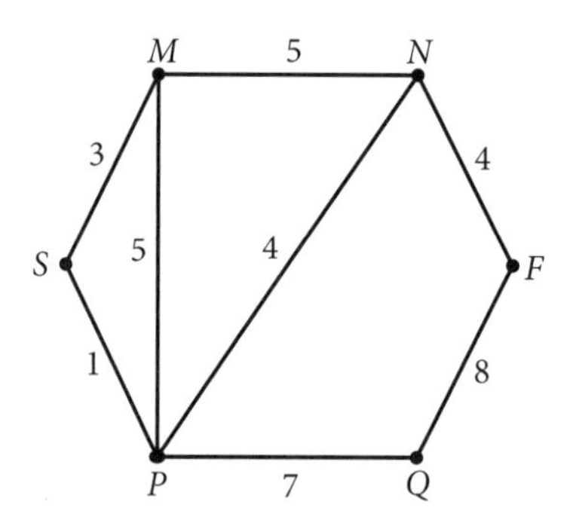

2 marks

Question 9 (4 marks) ©NESA ADDITIONAL SAMPLE EXAMINATION QUESTIONS, 2019, NW QUESTION 8

A park has five areas, A, B, C, D and E, which are connected by pathways.

The table shows the length of some of the pathways, in metres.

	A	B	C	D	E
A	–	600	–	–	?
B	600	–	?	500	–
C	–	?	–	400	–
D	–	500	400	–	300
E	?	–	–	300	–

The following network diagram is drawn to represent this information and a correct minimum spanning tree is shown by the solid lines.

Complete the network diagram including a possible value for each of the two edges AE and BC, and justify why AE and BC were not included as part of the minimum spanning tree.

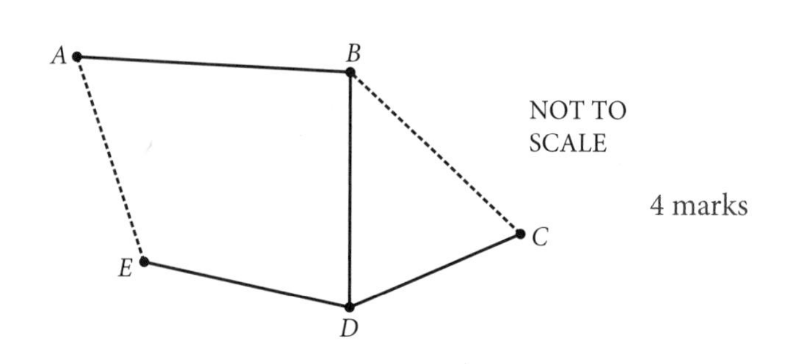

4 marks

Question 10 (4 marks)

The network shows the possible roads that a courier driver can use when delivering packages from a warehouse, W, to six stores, labelled as vertices A to F. The driver's GPS is used to find the quickest route to make deliveries. The weights on the edges shown in the network diagram are in minutes.

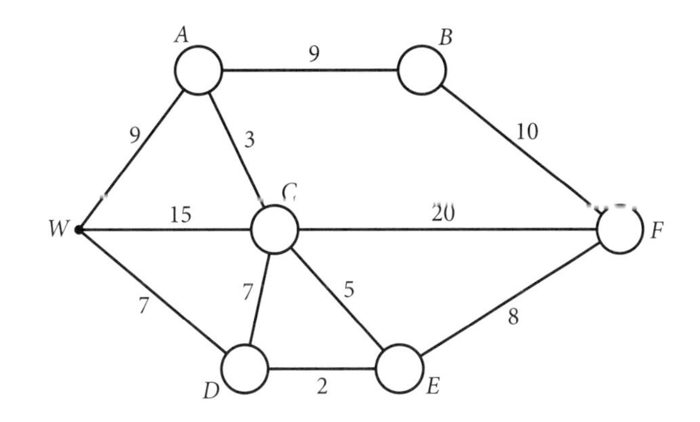

a Draw a minimum spanning tree for this network and calculate its length. 2 marks

b Does the minimum spanning tree allow the driver to deliver packages to all 6 stores in the shortest time? Justify your answer. 2 marks

Question 11 (3 marks)

The minimum spanning tree for this network diagram has a length of 43.

What lengths could be the missing values of a and b? (All lengths are whole numbers.) 3 marks

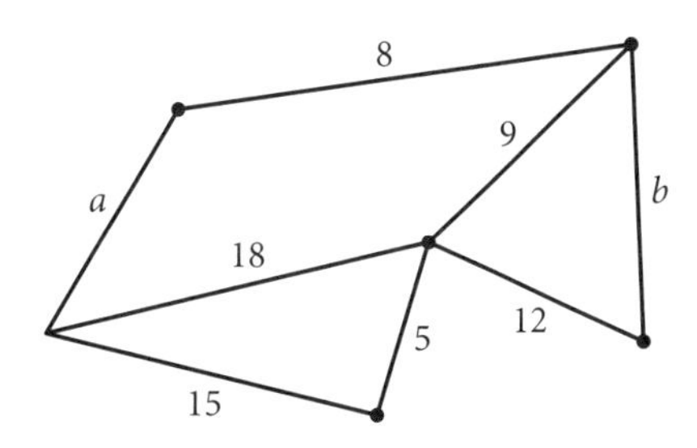

Question 12 (3 marks) ©NESA 2019 HSC EXAM, QUESTION 30

The network diagram shows the tracks connecting 8 picnic sites in a nature park. The vertices A to H represent the picnic sites. The weights on the edges represent the distances along the tracks between the picnic sites, in kilometres.

a Each picnic site needs to provide drinking water. The main water source is at site A.

By drawing a minimum spanning tree, calculate the minimum length of water pipes required to supply water to all the sites if the water pipes can only be laid along the tracks. 2 marks

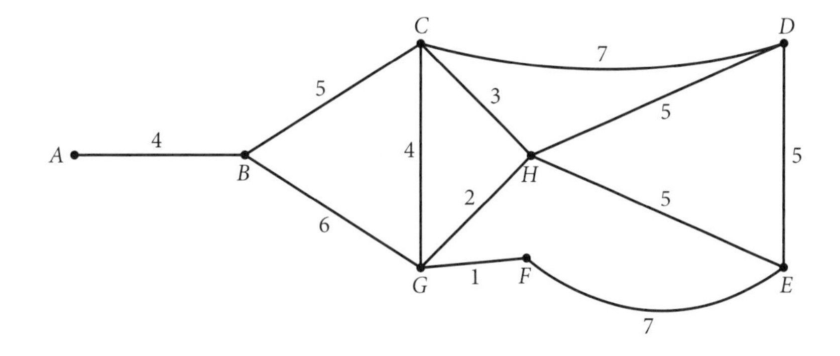

b One day, the track between C and H is closed. State the vertices that identify the shortest path from C to E that avoids the closed track. 1 mark

PRACTICE SET 2

Question 13 (2 marks)

The activity table for a project is shown. 2 marks

Activity	Immediate prerequisite(s)	Time (days)
A	–	2
B	*A*	7
C	*B*	2
D	*B*	4
E	*C*	1
F	*D*, *E*	4

Draw a network diagram for this project.

Question 14 (5 marks)

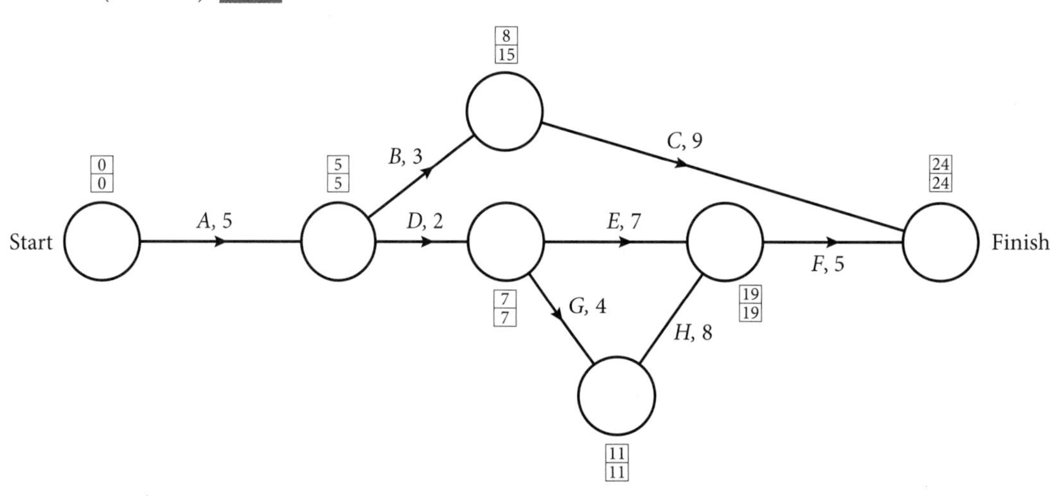

a Complete the activity table from the information in the network diagram. 4 marks

Activity	Immediate predecessor	Activity time	Earliest starting time (EST)	Latest starting time of next activity (LST_{next})	Float time
A	–	5	0	5	0
B	*A*				7
C	*B*	9	8		
D			5	7	0
E		7	7		
F			19	24	
G	*D*			11	
H	*G*		11	19	0

b Which activities are on the critical path? 1 mark

Question 15 (3 marks)

a Complete the backward scanning on the diagram by finding the latest starting time (LST) for each activity. 2 marks

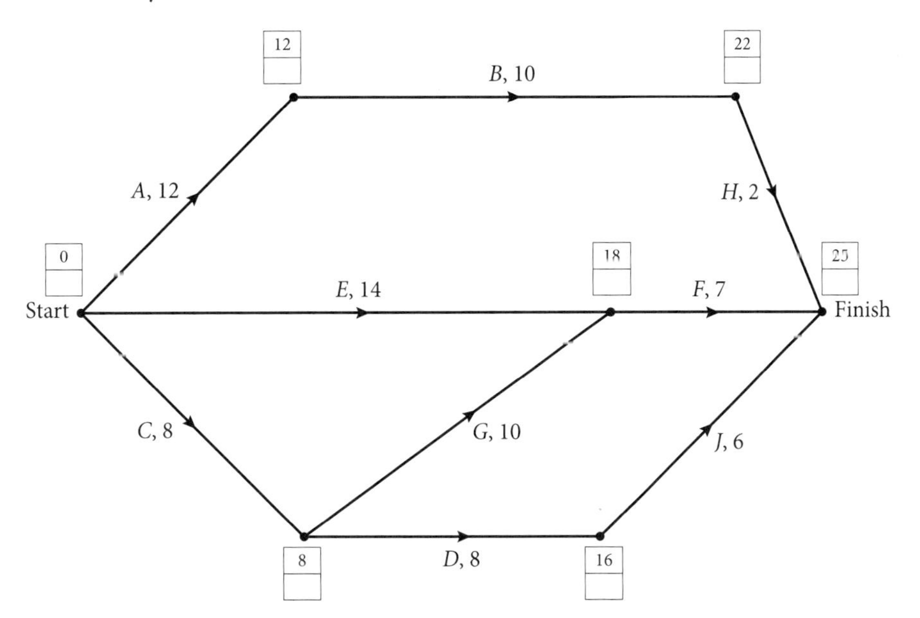

b Which activities are on the critical path? 1 mark

Question 16 (5 marks)

The following network diagram shows the activities in completing a construction project. The duration of each activity is in days.

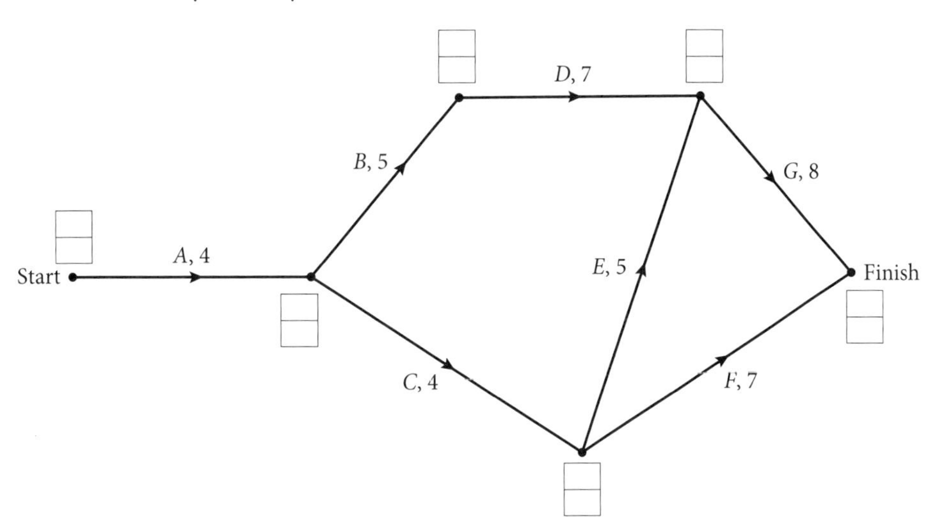

a Complete forward and backward scanning for the network diagram. 2 marks

b What is the minimum time to complete the project? 1 mark

c Name the critical path. 1 mark

d What is the float time of activity *E*? 1 mark

Question 17 (4 marks) ©NESA 2019 HSC EXAM, QUESTION 26

A project requires activities A to F to be completed. The activity chart shows the immediate prerequisite(s) and duration for each activity.

Activity	Immediate prerequisite(s)	Duration (hours)
A	–	2
B	A	6
C	A	5
D	B	2
E	C, D	4
F	E	1

a By drawing a network diagram, determine the minimum time for the project to be completed. 3 marks

b Determine the float time of the non-critical activity. 1 mark

Question 18 (5 marks) ©NESA 2020 HSC EXAM, QUESTION 26

The preparation of a meal requires the completion of all ten activities A to J. The network diagram shows the activities and their completion times in minutes.

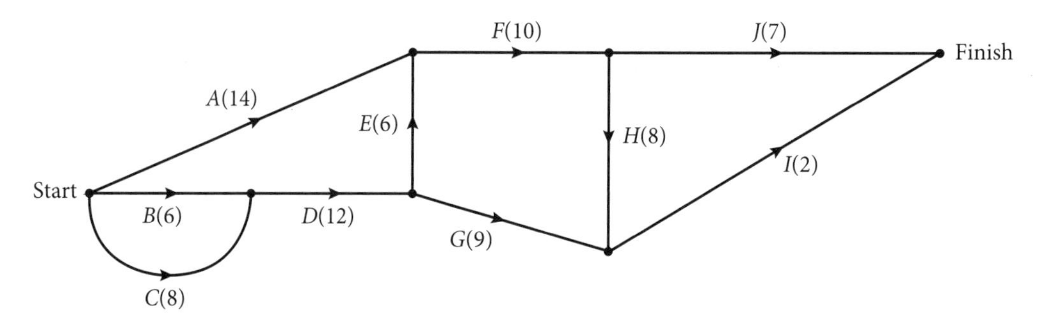

a What is the minimum time needed to prepare the meal? 1 mark

b List the activities which make up the critical path for this network. 2 marks

c Complete the table below, showing the earliest starting time and float time for activities A and G. 2 marks

Activity	Earliest starting time (minutes)	Float time (minutes)
A		
G		

Question 19 (5 marks)

The directed network diagram shows the traffic capacity, in vehicles per hour, of roads in a city suburb, where the vertices represent intersections.

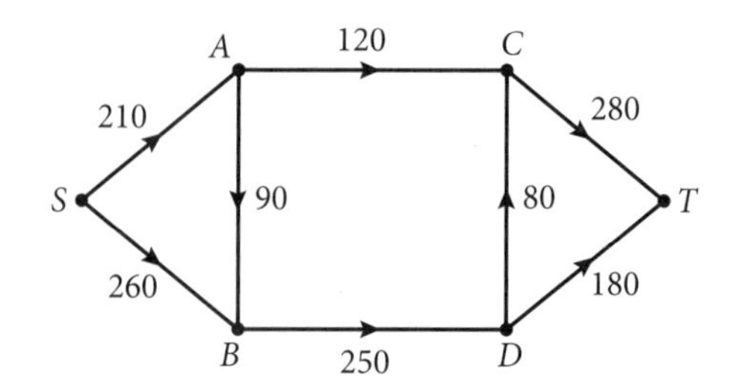

a What is the maximum outflow of cars from vertex C? 1 mark

b Determine the maximum flow from S to T for this flow network. Show the line of minimum cut. 2 marks

c To increase the maximum traffic capacity from S to T, there are plans to build a tunnel for a road directly from B to C. What should be the flow capacity of the tunnel road to maximise the traffic capacity? 2 marks

Question 20 (3 marks) ©NESA 2019 HSC EXAM, QUESTION 40

A museum is planning an exhibition using five rooms.

The museum manager draws a network to help plan the exhibition. The vertices *A*, *B*, *C*, *D* and *E* represent the five rooms. The numbers on the edges represent the maximum number of people per hour who can pass through the security checkpoints between the rooms.

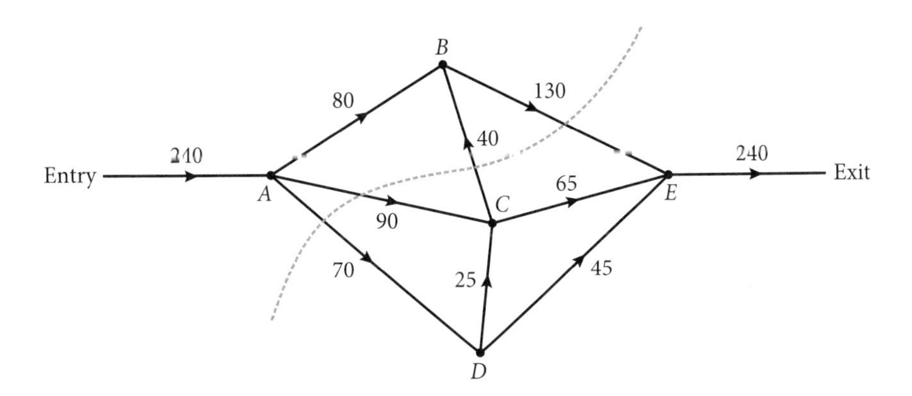

a What is the capacity of the cut shown? 1 mark

b The museum manager is planning for a maximum of 240 visitors to pass through the exhibition each hour. By using the 'minimum-cut maximum-flow' theorem, the manager determines that the plan does not provide sufficient flow capacity.

Draw the minimum cut onto the network below and recommend a change that the manager could make to one or more security checkpoints to increase the flow capacity to 240 visitors per hour. 2 marks

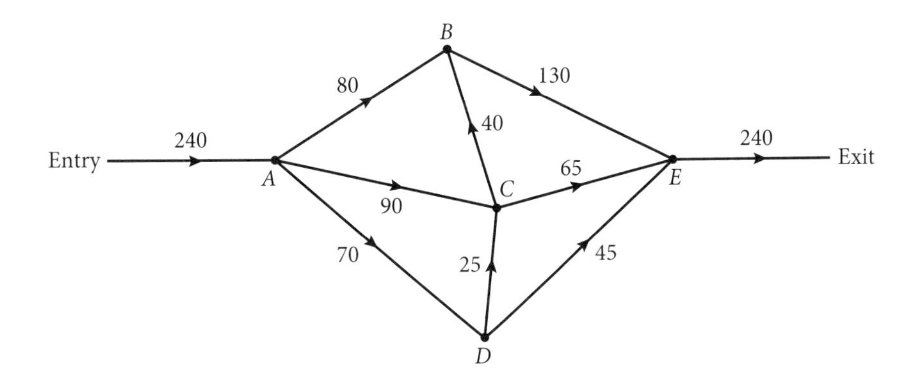

Practice set 1

Worked solutions

1 B

A is false because a vertex is a point, not a line.

C is false because the sum of degrees should always be even.

D is false because a cycle is a closed path.

2 A

3 edges connect at P.

3 C

10 lines in total

4 C

There is a path between any 2 vertices.

5 C

This network has 6 vertices and 9 edges but remove 4 edges because a network with 6 vertices only requires 5 edges to stay connected.

6 C

Each pair of vertices has the same edges joined to it.

7 B

Each team member plays 3 matches. Each vertex in the diagram has a degree of 3.

8 D

Arrows and numbers are on edges.

9 D

Following the arrows directed from C to D to A: Length = 9 + 7 = 16 (or $CBDA$ = 3 + 4 + 9 = 16)

10 C

Trees do not have a cycle.

11 B

A minimum spanning tree has 1 fewer edge than the total number of vertices.

12 B

All 5 vertices are connected with 4 edges of smallest lengths.

13 D

A path can connect 2 vertices.

14 A

A, B and D have no predecessor. E has B and F as predecessors.

15 C

EST of activity C is 3 + 3 = 6 because the predecessor activity E has an EST of 3 as well as an activity length of 3 (higher than activity B of 2) and must take the longest time path.

LST of activity C is 15 − 5 = 5 because the length of activity C is 5 and LST of activity I is 15.

16 D

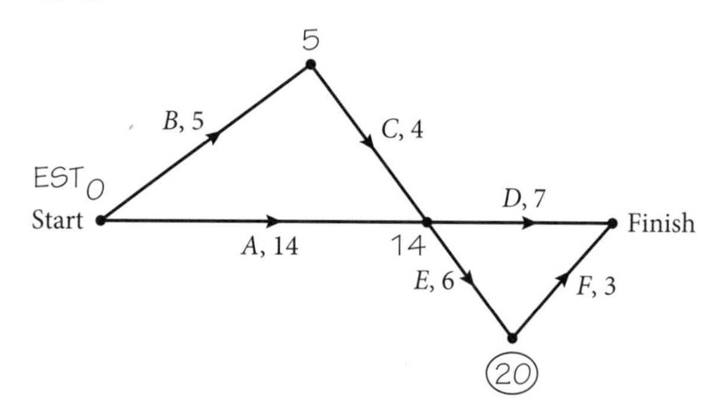

17 B

The cut does not separate the source from the sink.

18 B

Cut capacity = $SA + CT$ = 200 + 160 because AC and BT both flow in the opposite direction.

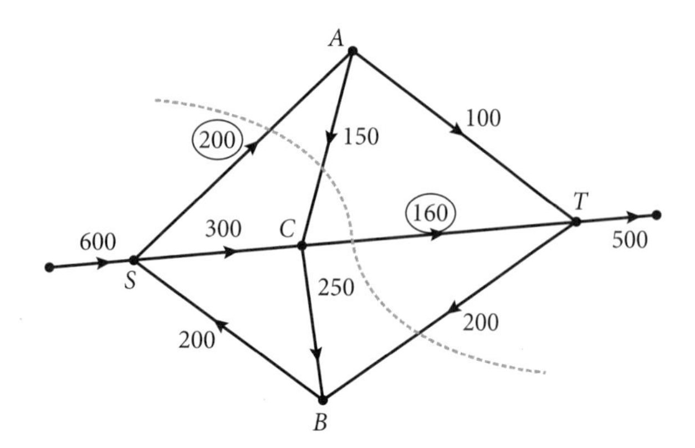

19 C

Max flow:

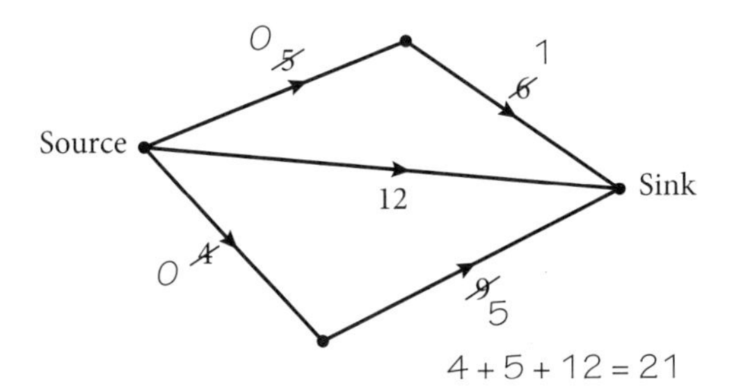

or min cut:

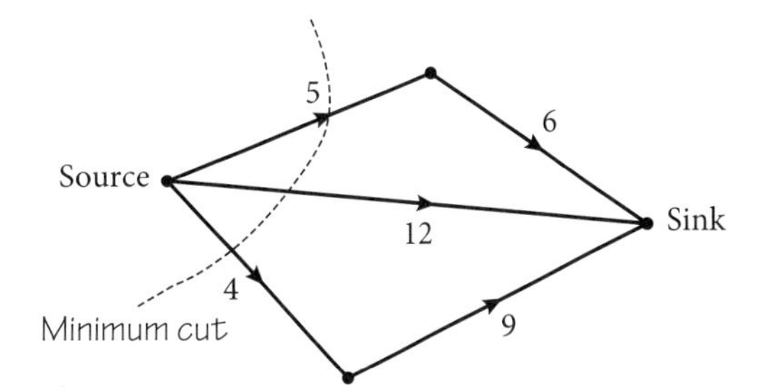

20 C

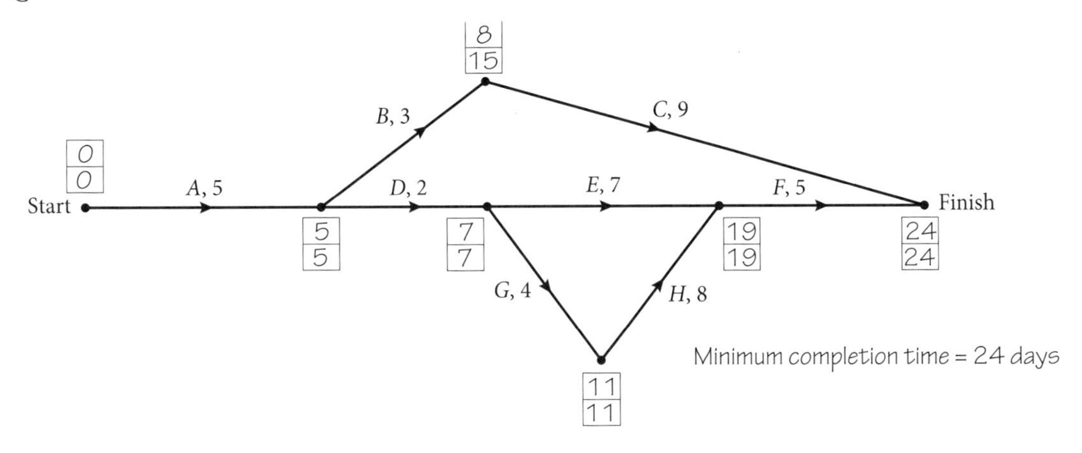

Practice set 2

Worked solutions

Question 1

Vertex	A	B	C	D
Degree	2	2	1	5

Question 2

a walk **b** path **c** trail

d cycle **e** connected

Question 3

a true – no repeated edges

b false – vertex B is repeated

c true – a closed path (no repeated edges or vertices)

Question 4

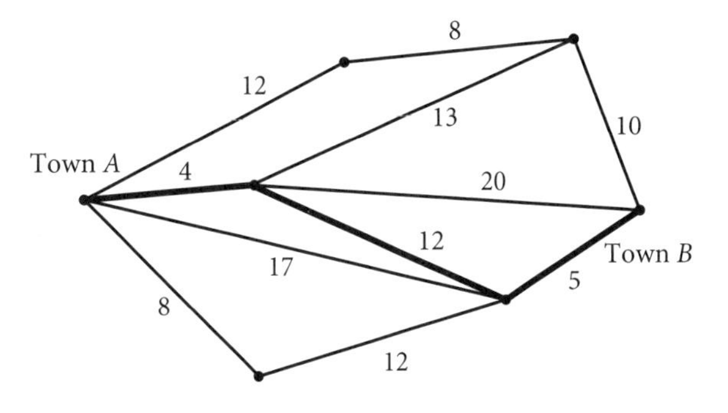

The shortest distance from Town A to Town B is $4 + 12 + 5 = 21$ km.

Question 5

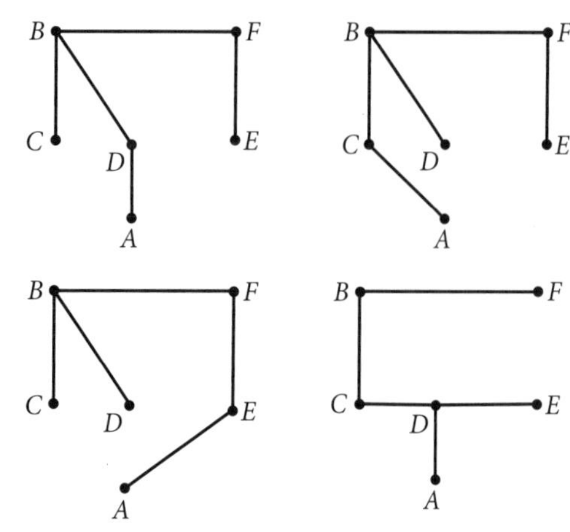

A variety of answers are possible. The spanning tree must connect all 6 vertices with 5 edges only, with no cycle.

Question 6

a

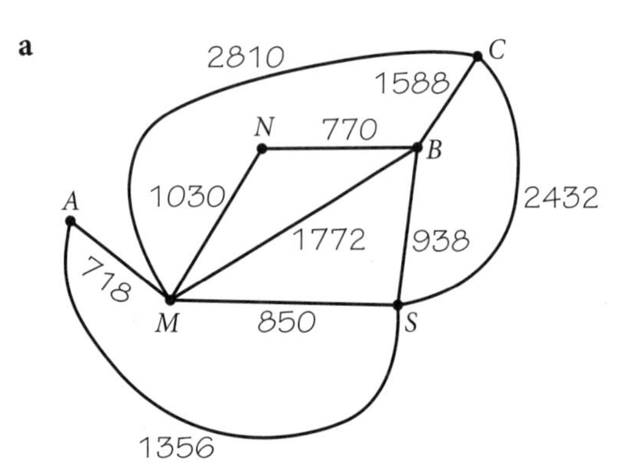

b Adelaide to Melbourne to Cairns:
$718 + 2810 = 3528$ km

c Minimum spanning tree length:
$718 + 770 + 850 + 938 + 1588 = 4864$ km

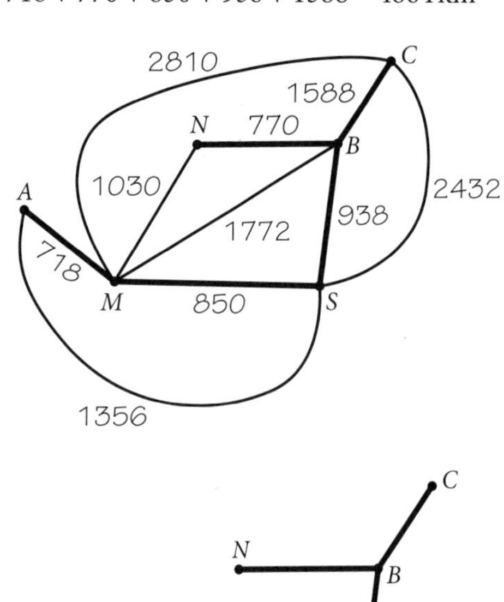

Question 7

a

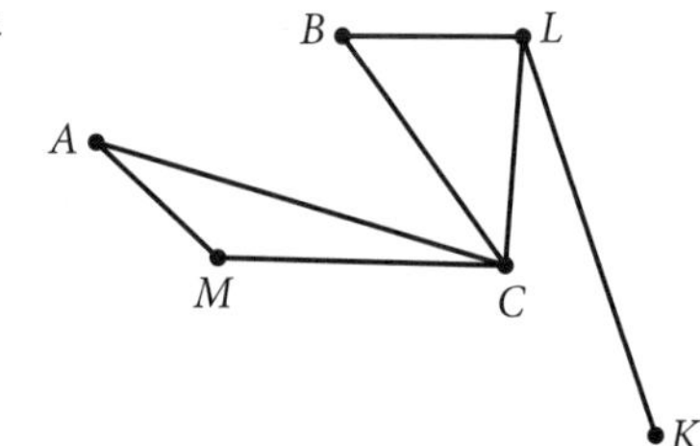

b It is connected – no vertices are unattached.

Question 8

a It is not a spanning tree: only 5 edges should be highlighted for the 6 vertices, and a spanning tree should not have a cycle.

b Length $= 3 + 4 + 4 + 5 + 8 = 24$

c

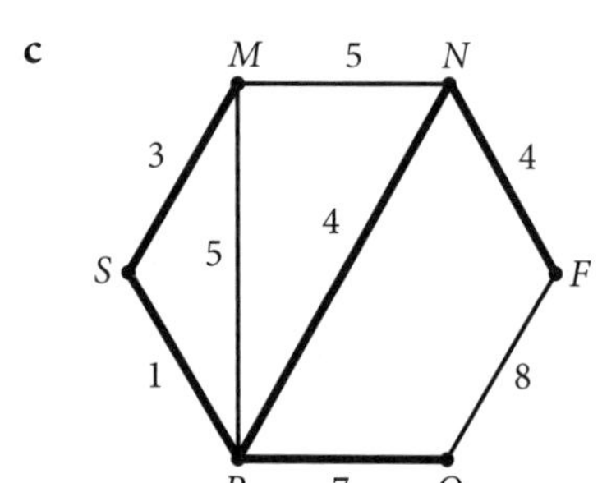

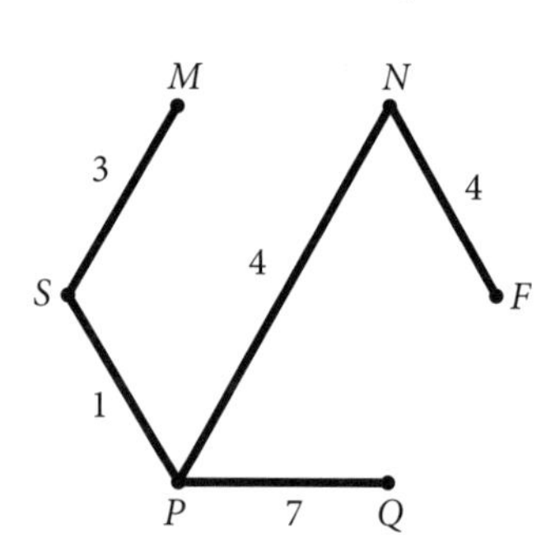

Minimum length $= 1 + 3 + 4 + 4 + 7$
$= 19$

Question 9

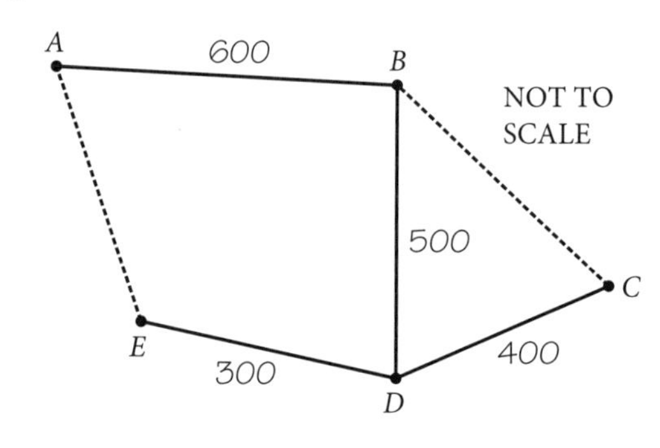

AE is any value greater than or equal to 600 and *BC* is any value greater than or equal to 500. If they were less, those edges would have to be on the minimum spanning tree or form a cycle.

Question 10

a Minimum spanning tree length:
$2 + 3 + 5 + 7 + 8 + 9 = 34$ minutes

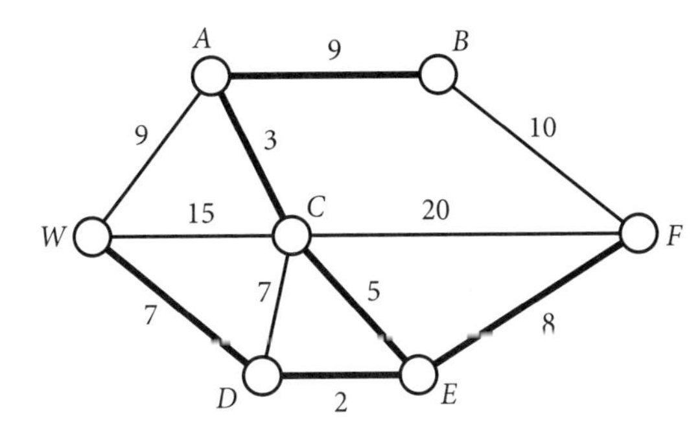

b No, the minimum spanning tree cannot be used as the quickest route because the courier would repeat edges to cover all vertices. The shortest route would be *WDECABF*, which takes 36 minutes ($7 + 2 + 5 + 3 + 9 + 10$).

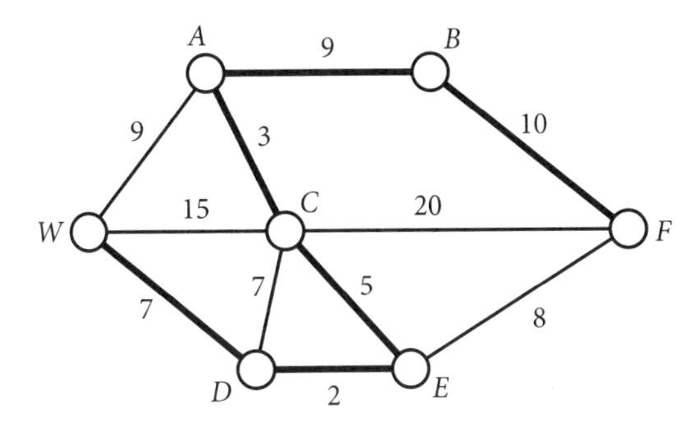

Question 11

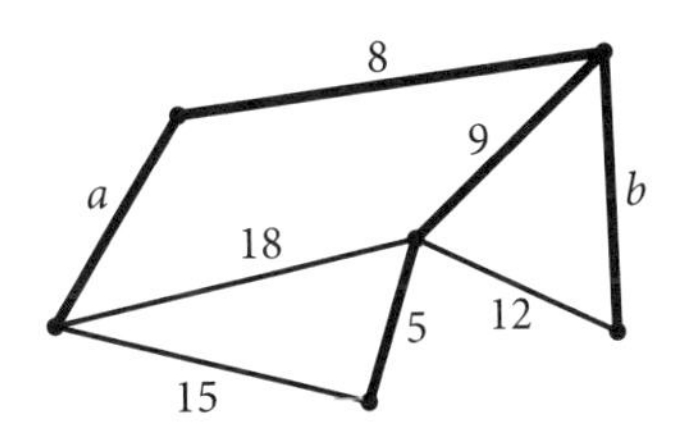

If the minimum spanning tree includes the edges 5, 8, 9 and a and b:

Length: $5 + 8 + 9 + a + b = 43$
$22 + a + b = 43$
$a + b = 21$

There is an edge of length 12 not on the tree, therefore a and b must be less 12 so that the edges are chosen on the minimum spanning tree.

Hence, $a = 10$ and $b = 11$ or $a = 11$ and $b = 10$.

If the minimum spanning tree includes shortest edges 5, 8, 9, 12 and a, but not b as it would form a cycle, then:

Length: $5 + 8 + 9 + 12 + a = 43$
$34 + a = 43$
So, $a = 9$
and $b > 12$

Question 12

a Given 8 vertices, 7 edges are needed for a spanning tree.

There are 3 possible minimum spanning trees for this network, all with minimum length:

$1 + 2 + 3 + 4 + 5 + 5 + 5 = 25$ km

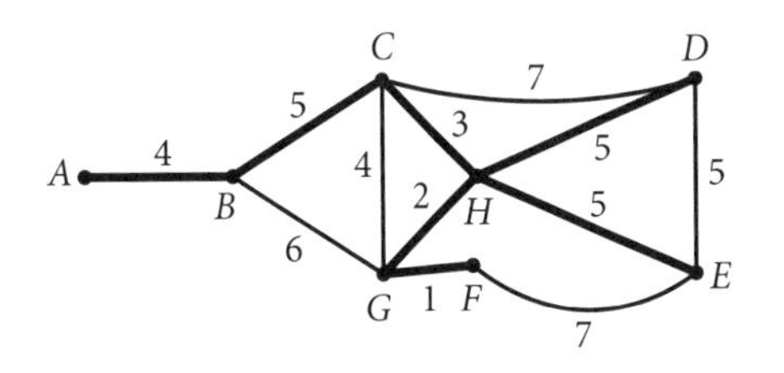

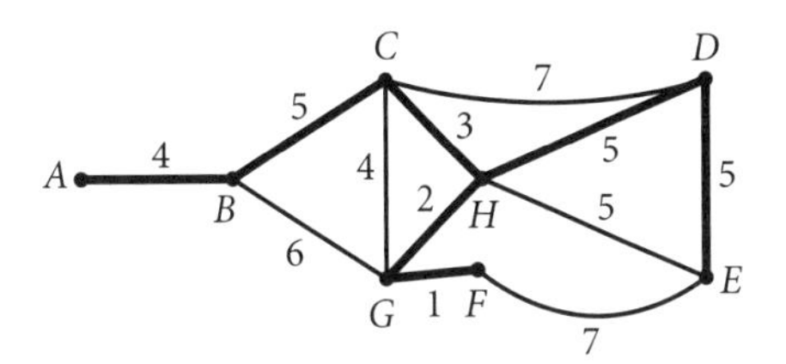

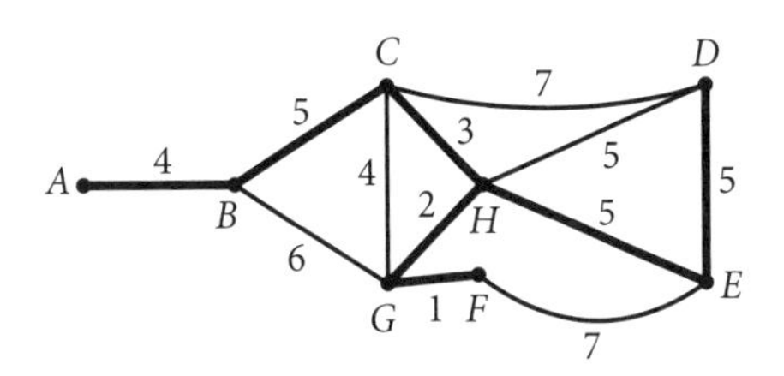

b *CGHE* is the shortest path, at 11 km.

Question 13

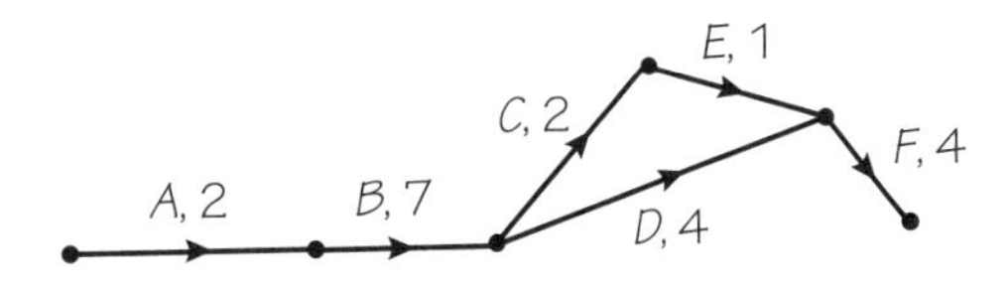

Question 14

a

Activity	Immediate predecessor(s)	Activity time	Earliest starting time (EST)	Latest starting time of next activity (LST_{next})	Float time
A	–	5	0	5	0
B	*A*	3	5	15	7
C	*B*	9	8	24	7
D	A	2	5	7	0
E	D	7	7	19	5
F	E, H	5	19	24	0
G	*D*	4	7	11	0
H	*G*	8	11	19	0

b The critical path is *ADGHF*.

Question 15

a

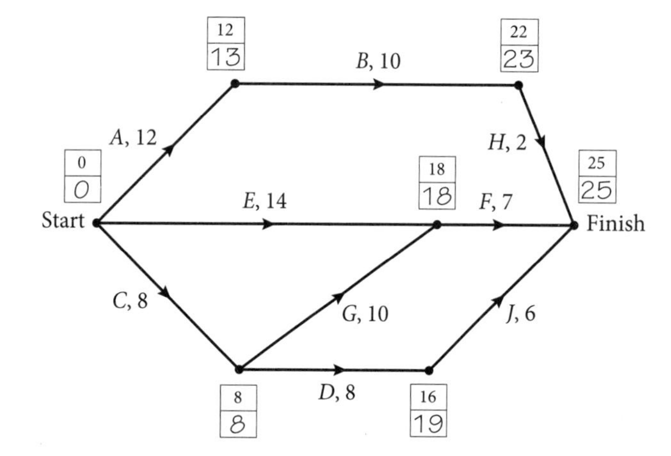

b The critical path is *CGF*.

Question 16

a

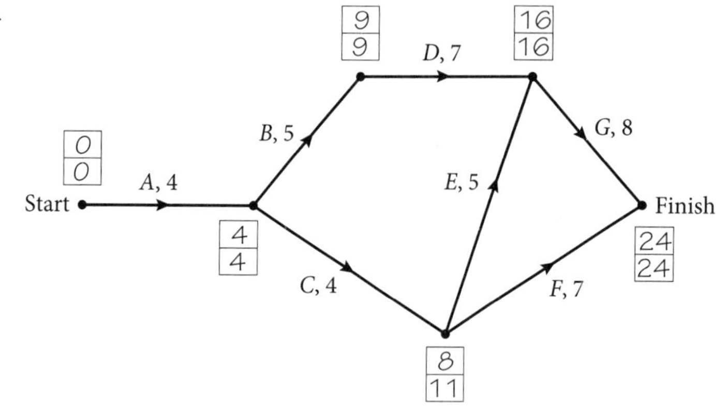

b 24 days

c *ABDG*

d LST – EST – activity time
= 16 – 8 – 5
= 3 days

Question 17

a

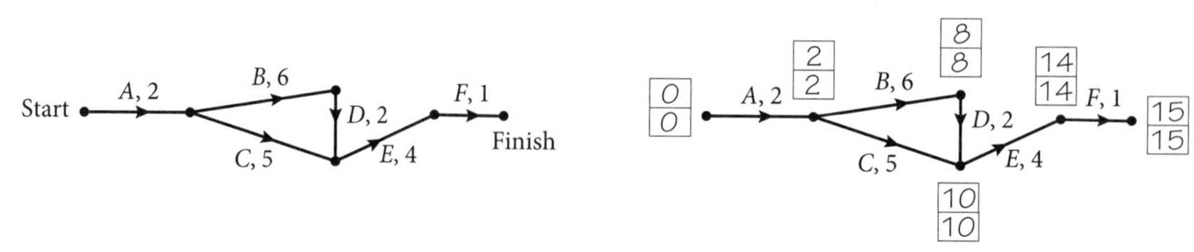

The minimum time is 15 hours.

b The critical path is *ABDEF*. So the non-critical activity is *C*, with a float time of 10 – 2 – 5 = 3 hours.

Question 18

a

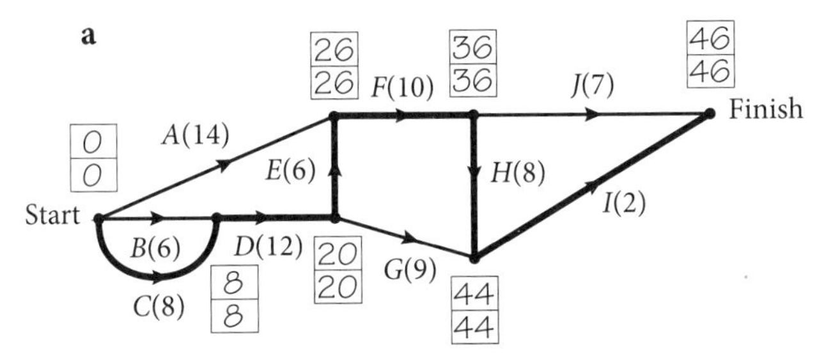

The minimum time is 46 minutes.

b The critical path is *CDEFHI*.

c

Activity	Earliest starting time (minutes)	Float time (minutes)
A	0	26 – 0 – 14 = 12
G	20	44 – 20 – 9 = 15

Question 19

a The inflow capacity of vertex *C* is 120 + 80 = 200.

The outflow capacity of vertex *C* is 280.

Hence, vertex *C* has a maximum outflow of 280 vehicles per hour.

b Maximum-flow minimum-cut is 370.

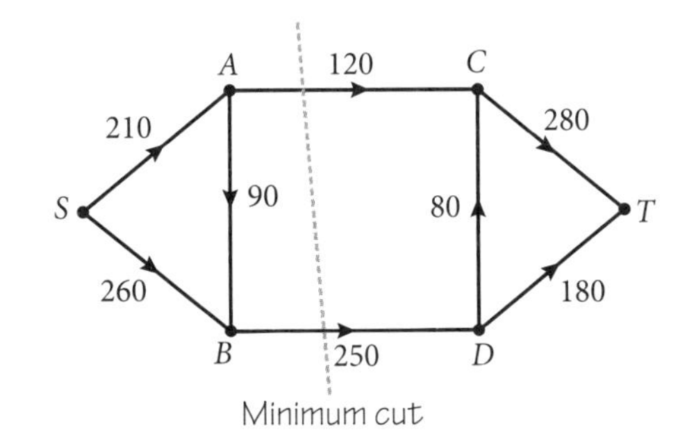

c

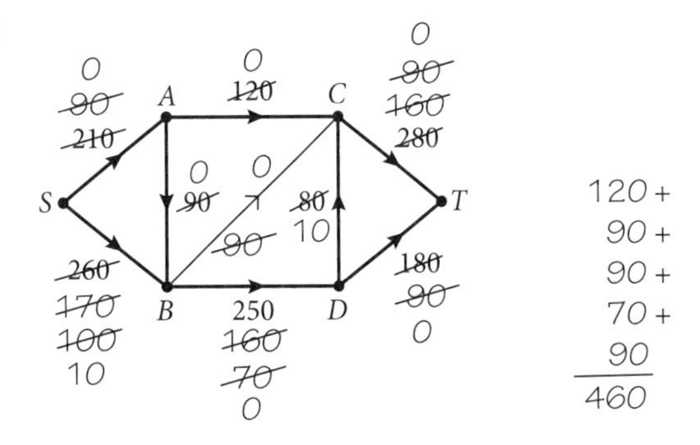

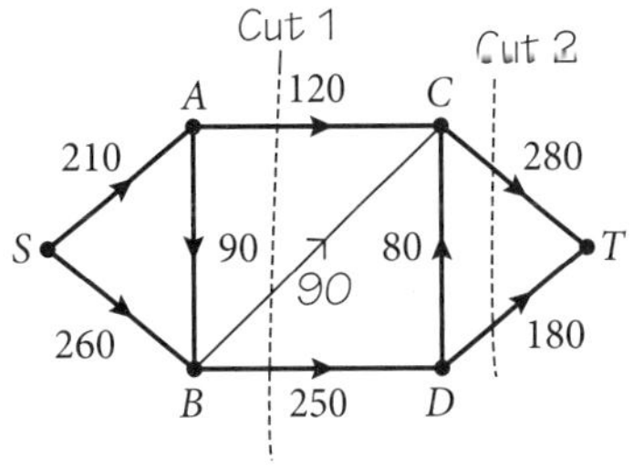

Cut 1: $120 + 90 + 250 = 460$
Cut 2: $280 + 180 = 460$

The new road *BC* will need a flow capacity of 90 vehicles per hour to increase the maximum flow capacity from *S* to *T* to be 460 vehicles per hour.

Question 20

a Cut capacity = 70 + 90 + 130 = 290

b

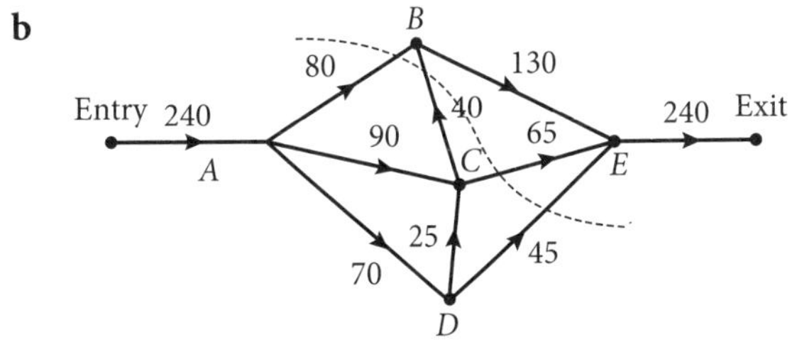

Minimum cut = 80 + 40 + 65 + 45 = 230

To achieve a maximum flow of 240, could increase edge *BC* to 50 or *DE* to 70, for example.

HSC exam topic grid (2019–2020)

This grid shows the coverage of this topic in past HSC exams by question number. The past exams can be downloaded from the NESA website (www.educationstandards.nsw.edu.au) by selecting 'Year 11 – Year 12', 'HSC exam papers'. NESA marking feedback and guidelines can also be found there.

The Networks topic was introduced to the Mathematics Standard 2 course in 2019.

	Terminology and diagrams	Minimum spanning trees	Shortest paths	Critical path analysis	Flow networks
2019 new course		**30(a)**	**30(b)**	**26**	**40**
2020	**9**	18		**26**	30

Questions in **bold** can be found in this chapter.

HSC exam reference sheet

Mathematics Standard 1 and 2

Measurement

Limits of accuracy

Absolute error $= \frac{1}{2} \times$ precision

Upper bound = measurement + absolute error

Lower bound = measurement – absolute error

Length

$$l = \frac{\theta}{360} \times 2\pi r$$

Area

$$A = \frac{\theta}{360} \times \pi r^2$$

$$A = \frac{h}{2}(a + b)$$

$$A \approx \frac{h}{2}(d_f + d_l)$$

Surface area

$$A = 2\pi r^2 + 2\pi rh$$

$$A = 4\pi r^2$$

Volume

$$V = \frac{1}{3}Ah$$

$$V = \frac{4}{3}\pi r^3$$

Trigonometry

$$\sin A = \frac{\text{opp}}{\text{hyp}}, \quad \cos A = \frac{\text{adj}}{\text{hyp}}, \quad \tan A = \frac{\text{opp}}{\text{adj}}$$

$$A = \frac{1}{2}ab\sin C$$

$$\frac{a}{\sin A} = \frac{b}{\sin B} = \frac{c}{\sin C}$$

$$c^2 = a^2 + b^2 - 2ab\cos C$$

$$\cos C = \frac{a^2 + b^2 - c^2}{2ab}$$

Financial Mathematics

$$FV = PV(1 + r)^n$$

Straight-line method of depreciation

$$S = V_0 - Dn$$

Declining-balance method of depreciation

$$S = V_0(1 - r)^n$$

Statistical Analysis

An outlier is a score

less than $Q_1 - 1.5 \times \text{IQR}$

or

more than $Q_3 + 1.5 \times \text{IQR}$

$$z = \frac{x - \mu}{\sigma}$$

Normal distribution

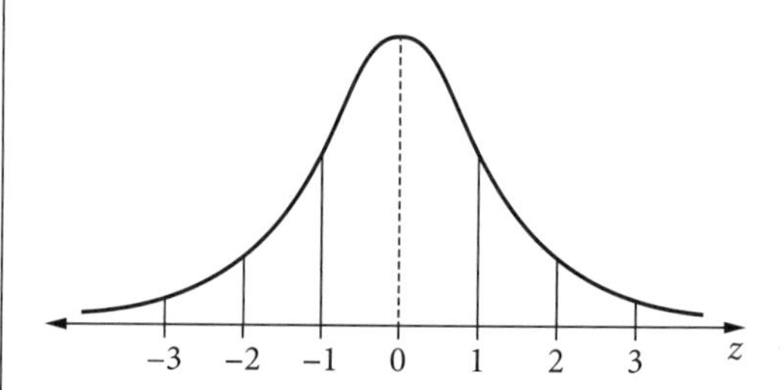

- approximately 68% of scores have z-scores between –1 and 1
- approximately 95% of scores have z-scores between –2 and 2
- approximately 99.7% of scores have z-scores between –3 and 3

Index

A

activity table 165, 171–172
acute angle 41
adjacent side 41
algebraic modelling 3, 9
angle of depression 41, 43–44
angle of elevation 41, 43–44
annuity 103, 110–114, 130
appreciated value 103
appreciation 103, 106
area
  unit conversions 74
area of a triangle 70
  sine formula 46
asset 103, 107
asymptote 3, 14, 17
axis of symmetry 12

B

backward scanning 165, 172
balance owing 108–109
bearings 41, 48–49, 70
  compass 41, 48
  true 41, 48–49
best buys 73, 81
bivariate data 133, 134, 136
bivariate data analysis 134–138
blood alcohol content 5, 22
box plot 146
break-even point 3, 10
brokerage 103, 123

C

capacity 75, 174
  unit conversions 75
capacity of a cut 174
capture–recapture method 73, 83
cash advance 109
circuit 165, 168
closed walk 168
coefficient 3, 133
  See also Pearson's correlation coefficient
compass radial survey 41, 50
compounded 104, 105, 110
compounding period 104, 105
compound interest 15, 16, 103–105, 108–110
concave down 3, 4, 12
concave up 3, 4, 12
cone 6
connected network 165
constant 16
constant term 3
correlation 133, 134
  See also Pearson's correlation coefficient
cosine rule 41, 45, 70
credit card 103, 109
critical activity 165, 172
critical path 165, 172
critical path analysis 165, 171–175, 195
critical time 165
cut 165
cycle 165, 168

D

declining-balance depreciation 16, 103–107
degree (of a network) 165, 167
dependent variable 133, 137
depreciation 15, 16, 107, 130
  See also declining-balance depreciation
directed edge 167
directed network 165, 168, 172
direct variation 3, 18–19, 38
disconnected network 165
distribution
  See normal distribution; shape of a distribution
dividend 103, 106
dividend yield 103, 106
dummy activity 165, 172

E

earliest starting time (EST) 165, 172
edge (of a network) 165, 167
elevation view 73, 84
empirical rule 133, 139
energy 79
energy usage 73
equation 3, 4, 38
  algebraic 6
  exponential 14–18
  linear 5–11, 18
  quadratic 12–13, 30
  See also simultaneous equations
exponential curve 3, 14
exponential decay 3, 16
exponential function
  See function, exponential
exponential growth 3, 15, 16
exponential model 27
expression 5
extrapolation 133, 137

F

float time 165, 174
flow capacity 165, 174
flow networks 174–175, 195
formula 4
forward scanning 165, 172
fuel consumption 73, 80
function 3
  exponential 3, 12, 14–16, 38
  linear 4, 6, 8, 9, 38
  non-linear 4, 12, 38
  quadratic 3, 4, 12, 13
  reciprocal 4, 12, 16–17
future value 103, 104–105

G

gradient 3, 6–7, 9, 11
gradient–intercept form 6

H

heart rate 73, 79
horizontal intercept
  See x-intercept
hyperbola 3, 16, 18
hypotenuse 41, 42

I

independent variable 133
inflation 103, 106
inflow 165, 174
initial value 3, 9, 16
interest-free loan 17
interest-free period 103, 109
interpolation 133, 137–138
intersect 11
inversely proportional 18
inverse variation 3, 16, 18–19, 38
investments 104–106, 130

K

kilowatt-hour 73
Kruskal's algorithm 165, 170

L

latest starting time (LST) 165, 172
least-squares regression line 133, 138
length
  unit conversions 74
linear equation 5
linear model 4, 9
linear modelling 9, 11
linear regression 162
  See also line of best fit
linear relationship 26
line of best fit 133, 136–137
loop 165, 167

M

mass
  unit conversions 76
maximum-flow minimum-cut theorem 165, 175
maximum value 4, 13
  See also turning point
mean 133, 139–143
mean score 157
median 133, 139
metric units
  conversion tables 74–76
minimum spanning tree 165, 170, 195
minimum value 4, 13
  See also turning point
minute (measure of an angle) 41
mode 133, 139, 143

N

network diagram 165, 167–169
networks 167–169
non-critical activity 166, 172, 174
non-linear relationships 12–18
non-right-angled trigonometry 44–47
normal curve
  See normal distribution
normal distribution 133, 139–141, 150, 162
number plane 4, 8

O

obtuse angle 47
origin 17
outflow 166, 174
outlier 133

P

parabola 4, 12, 13
path 168
Pearson's correlation coefficient 133, 135–136, 138
perimeter 30, 50
point of intersection 4, 10–11
  See also break-even point
point symmetry 17
power 79
present value 103, 105, 112
Prim's algorithm 166, 170
principal 103, 104
probability 141
profit 10
pronumeral 4
proportion
  See direct variation
Pythagoras' theorem 42

Q

quadrant 16–17
quadratic function
  See function, quadratic

R

radius 6
rate 73, 76–81
rate of change 6, 9
ratio 73, 81–84
raw score 133, 140
reciprocal function
  See function, reciprocal; *See also* hyperbola; inverse variation
recurrence relation 103, 110–111, 124
reducing balance loan 103, 108–109, 130
repayment 17
revenue 4, 10
right-angled triangle 42
right-angled trigonometry 42–44, 70
rise 4
rotational symmetry
  See point symmetry
run 4

S

salvage value 16, 103, 107
scale drawing 73
scale ratio 73
scatterplots 133–135, 162
shape of a distribution 143
  negatively skewed 143
  positively skewed 143
  symmetrical 143
shares 103, 106, 130
shortest path 166, 169–171, 195
simple interest 103, 104–105
simultaneous equations 4, 38
  linear 10–11
sine rule 41, 44–45, 70
sink 166, 174
sketch 13, 14
skewed
  See shape of a distribution
source 166, 174
spanning trees 166, 169
  See also minimum spanning trees
speed 73, 77–79
standard deviation 133, 138–142
subject (of a formula) 4, 6
symmetry 133, 139

T

table of values 8, 9, 11, 13
tables for present and future values 112–114
time
  unit conversions 76
trail 168
trapezoidal rule 73, 87, 100
trigonometric ratio 42
  cosine 41
  sine 41
  tangent 41
turning point 4
  maximum 4, 12–13
  minimum 4, 12–13

U

unit price 73, 81

V

variable 4, 5, 11
variation equation 19
vertex (of a network) 165–167
vertex (of a parabola) 4, 12, 13
  See also turning point
vertical axis
  See y-axis
vertical intercept 4, 11
  See also y-intercept
volume 6
  unit conversions 75

W

walk 168
weighted edge 166, 167
weighted network 166, 168

X

x-axis 4
x-intercept 4, 13

Y

y-axis 4
y-intercept 4, 6, 9
  See also vertical intercept

Z

z-score 133, 140–141, 162